EVOLUTIONARY MODELS OF CONVERGENT MARGINS - ORIGIN OF THEIR DIVERSITY

Edited by **Yasuto Itoh**

Evolutionary Models of Convergent Margins - Origin of Their Diversity
http://dx.doi.org/10.5772/63702
Edited by Yasuto Itoh

Contributors

Atsushi Noda, Ayumu Miyakawa, Shigekazu Kusumoto, Tetsuya Tokiwa, Makoto Takeuchi, Yusuke Shimura, Kazuho Shobu, Akari Ota, Koshi Yamamoto, Hiroshi Mori, Yumi Kaneko, Toru Takeshita, Yuto Watanabe, Norio Shigematsu, Ko-Ichiro Fujimoto, Yasuto Itoh, Paul Green, Keiji Takemura, Tomotaka Iwata, Naoto Ishikawa, Takashi Suzuki, Shiro Ishida, Hiroyuki Arato, Daisuke Miura, Kiyoshi Toshida, Ken'Ichi Arai, Yutaka Wada, Takeshi Wachi

Published by InTech
Janeza Trdine 9, 51000 Rijeka, Croatia

Publishing Process Manager Ana Pantar
Technical Editor SPi Global
Cover InTech Design team

First published June, 2017
Printed in Croatia
Legal deposit, Croatia: National and University Library in Zagreb

Additional hard copies can be obtained from orders@intechopen.com

Evolutionary Models of Convergent Margins - Origin of Their Diversity, Edited by Yasuto Itoh
p. cm.
Print ISBN 978-953-51-3287-5
Online ISBN 978-953-51-3288-2

For EU product safety concerns:
IN TECH d.o.o., Prolaz Marije Krucifikse Kozulić 3, 51000 Rijeka, Croatia,
info@intechopen.com or visit our website at intechopen.com.

Meet the editor

Yasuto Itoh was born in Osaka. He received his PhD degree in Kyoto University. Now he is the professor of the Graduate School of Science of Osaka Prefecture University. He conducts research into tectonics, stratigraphy, structural geology, and paleo-/rock magnetism and has published more than 100 papers in the fields of deformation process of active plate margins, seismic hazard assessment of active faults, paleoenvironment of East Asia, and mechanism of sedimentary basin formation/development. His professional memberships include the American Geophysical Union, American Association of Petroleum Geologists, Japan Geoscience Union, and Japanese Association for Petroleum Technology. He is also a member of the Comprehensive Research and Study on Active Fault Systems organized by the Research and Development Bureau of the Ministry of Education, Culture, Sports, Science and Technology (MEXT), Japan.

Contents

Preface

This book deals with recent developments in evolutionary models for convergent margins. Reflecting transient modes for oceanic plate convergence, such boundaries are sites of varied tectonic processes, which provoke vigorous material recycling and frequent natural disasters such as massive earthquakes and catastrophic volcanism. Therefore, the origin of their diversity has long been one of the most significant themes in Earth science.

Section 1 of this book focuses on the formative processes of forearc basins at convergent margins, which are significant buffers for material recycling in arc-trench systems. In Chapter 1, Noda and Miyakawa present a comprehensive review of well-known modern forearc basins and suggest a causal relationship between basin formation and accretionary wedge growth. Their results indicate that forearc basins can be classified into three types: a single-wedge model characterized by landward tilting of the basin strata, a two-wedge model marked by contractional deformation linked to symmetric doubly vergent wedge uplift, and an additional strike-slip model represented by transpressional/transtensional deformation due to oblique subduction. In Chapter 2, Arato describes the development process for a tectonically erosive margin based on an interpretation of extensive seismic data covering an entire basin range. He emphasizes the significance of massive rearrangement of forearc components related to transcurrent faulting, which is generally applicable to oblique subduction margins around the globe.

Section 2 is a special section dealing with the evolution of arc-bisecting faults, which are commonly found on obliquely converging plates. The objective of the researchers involved is to elucidate the formation timeline for the Median Tectonic Line (MTL) dividing the southwestern Japan arc since the late Cretaceous. This ancient lateral fault has become intermittently active in response to temporal shifts in the rate of convergence of the oceanic plates in the northwestern Pacific basin. Following a short prologue that provides the readers with a spatiotemporal perspective of the regional fault, in Chapter 3, Itoh and others describe the 3D architecture of the MTL based on a dataset of reflection on seismic surveys and borehole stratigraphy. They identify a Quaternary half-graben overlain by a large wedge of Cretaceous rocks and interpret this feature as a retroactive crustal failure juxtaposed on an active dextral trace of the MTL. The discrepancy between the geophysical low-angle view and the high-angle view expected as a result of predominantly strike-slip motion on the fault is highlighted by Kusumoto and others in Chapter 4, based on the latest analysis of gravity anomalies. They performed numerical modeling of the active geomorphology and concluded that the thrust is not a mechanical boundary in the upper crust associated with a recent shear stress but is instead a dormant material boundary that developed during an older contraction phase. A study of the incipient evolution of the MTL is presented by Itoh and others in Chapter 5, based on a fission track (FT) thermochronological analysis of sandy clastics burying a gigantic pull-apart basin formed during Cretaceous sinistral activity at the fault. Considering the bimodal distribution of FT ages determined for apatite and zircon grains extracted from sandstone samples, they argue for two distinct provenances in southwest Japan. Thermodynamic modeling of annealing of apatite FTs led them to distinguish three cooling (exhumation) events, the youngest of which is linked to the most recent Miocene inversion within the arc, together with intensive reverse faulting at the MTL. A noteworthy study on structural buildup and material circulation at the incipient MTL is presented by Kaneko and others in Chapter 6. Based on a detailed structural and geochemical analysis of the eastern segment of the regional fault, they identified a normal faulting phase during the early Paleogene period and material migration resulting from vigorous fluid circulation within the fault-brecciated zone. Finally, in Chapter 7, Itoh and Iwata describe complex structural features of the MTL that were formed during recurrent tectonic episodes beginning in the late Cretaceous era. Their phase stripping method was designed to identify a single deformation phase and can successfully extract the neotectonic structural features of the fault.

Section 3 is a collection of studies on the southwestern Japan arc, but the researchers' interests are not restricted to local geology. In Chapter 8, Ishikawa and others present a chronicle of arc rotation processes related to Miocene back-arc rifting. The opening process of the Japan Sea is known to have been characterized by an extraordinarily rapid spreading rate with a fan-shaped mode. Their important paleomagnetic data place constraints on the timing of each step of the back-arc opening sequence. In Chapter 9, Miura and others discuss volcanic activity associated with rifting events and present significant tectonic results based on a dike swarm analysis. As a result of extensive field observations, they advocate a clear relation between the physicochemical properties of magmas and the morphology of dikes, which should be taken into account when estimating ancient stress fields. In Chapter 10, Tokiwa and others report on an investigation of frontal tectonic processes to determine the tectono-stratigraphic order of Cretaceous accretionary complexes along the Nankai Trough. They succeeded in determining the settlement ages of accreted bodies based on a quantitative cluster analysis of high-resolution U-Pb dating results for clastic rocks. The striking coincidence of their radiometric age estimates with previous biostratigraphic ages endorses the versatility of their analytical method.

The important scientific results obtained by these prominent researchers pave the way for further in-depth studies on mobile belt frontiers, where harsh conditions hinder efforts to understand the Earth's spatiotemporal changes.

Yasuto Itoh
Graduate School of Science,
Osaka Prefecture University,
Osaka, Japan

Section 1

Pursuit of Formative Processes of Forearc Basins on Convergent Margins

Chapter 1

Deposition and Deformation of Modern Accretionary-Type Forearc Basins: Linking Basin Formation and Accretionary Wedge Growth

Atsushi Noda and Ayumu Miyakawa

Additional information is available at the end of the chapter

http://dx.doi.org/10.5772/67559

Abstract

Since a comprehensive review of forearc basins was published by Dickinson more than 20 years ago, a significant amount of new data about them have been published. These recent studies revealed details of depositional and deformation styles in the forearc basins, suggesting the formation processes were not unique. In this chapter, we reviewed modern forearc basins to understand how is the basin stratigraphy related with growth of accretionary wedges. The results indicate forearc basin can be classified into two (single- and two-wedge models) plus one (strike-slip model): (1) the single-wedge model which is characterized by landward tilting of the basin strata ascribed to asymmetrical doubly vergent (single-vergent) uplift of the outer arc high with forethrusts (seaward-vergent thrusts in the pro-wedge); (2) the two-wedge model which is marked by contractional deformation caused by symmetrical doubly vergent uplift of the wedge with forethrusts in the prowedge and back-thrusts (landward-vergent thrusts) in the retro-wedge; and (3) the strike-slip model which is an additional one being represented by transpressional and/or transtensional deformations due to oblique subduction. We speculate that these models spatially and temporally depend on material fluxes at the plate interfaces that affect geometry and mechanical strength of backstops.

Keywords: forearc basin, accretionary wedge, deposition, deformation, plate convergent margin

1. Introduction

Forearc basin is one of the major elements of plate convergent margins (e.g., [1, 2]). The formation is considered to be closely associated with accretion or erosion at frontal and basal parts of accretionary wedge (e.g., [3–5]). Especially, the seaward margin of the basin is directly affected by uplift and subsidence of the outer arc high of accretionary wedge, which acts as topographic barrier of sediments deposited in the basin. Forearc basin generally develops at which interplate coupling is strong [6, 7], meaning the location is underlain by an area where strain release during earthquakes is large. Therefore, styles of deposition and deformation of forearc basins may record long-term average stresses on the plate interfaces over millions of years.

However, the formation and deformation processes of forearc basins are poorly known. Such basin is incorporated into a dynamic system of subduction zone composed of backstop, basin sediments, accretionary wedge, trench fill deposits, subducting plate with conduit for the subducting materials. This system is spatially and temporally influenced by changes of plate configuration (slap dip, obliquity, or convergent rate), activity of volcanic arc (sediment production), topography of subducting slab, strength and geometry of backstop, and material flux at the plate boundary.

The purpose of our study is to develop a tool for reconstruction of paleostress fields by lightening how forearc basin formations respond to subduction zone dynamics that include frontal/basal accretion/erosion and mechanical relationships among the basin, backstop, accretionary wedge, and subducting plate. For the first step of this purpose, we collected modern examples of deposition and deformation of forearc basins in the world. The studied areas are Sunda (Sumatra-Java), Japan, Aleutian-Alaskan, Lesser Antilles, South American (Columbia-Ecuador-Peru-Chile), and Tonga-Kermadec-Hikurangi margins (**Figure 1**). Most of them are classified into the compressional accretionary-type forearc basin [2], which is characterized by growth of accretionary wedge by frontal/basal accretion and compressional deformation in

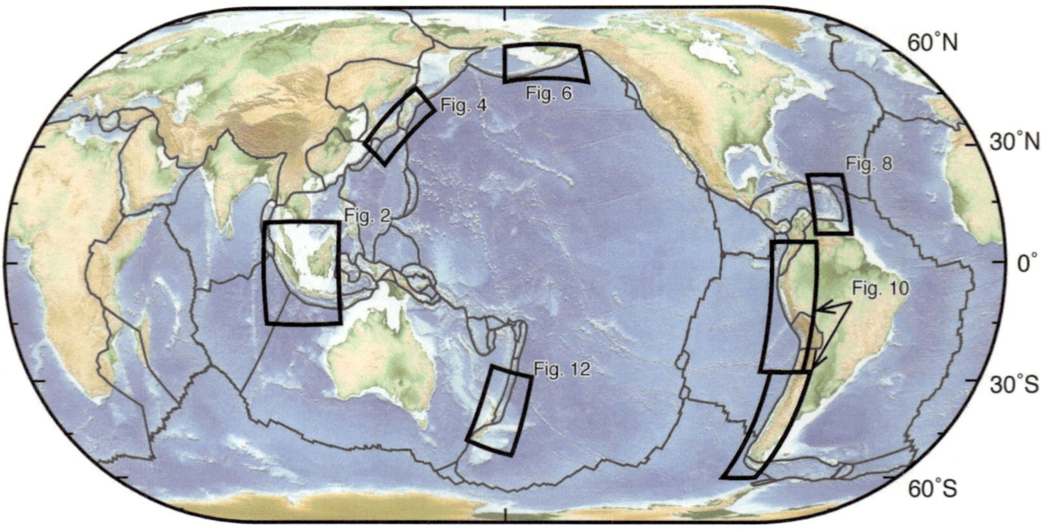

Figure 1. Index map of the study areas. Plate boundaries are based on Ref. [8].

the basins. This paper can contribute a basic understanding about the formation processes of forearc basins along plate convergent margins.

2. Sunda arc (Sumatra-Java)

2.1. Plate configuration

Along the Sunda (Sumatra-Java) arc, the India-Capricorn-Australian plate subducts beneath the Sunda microplate (Eurasian continental plate) with 5–7 cm/yr in the southeastern Asia (**Figure 2**). Under the present configuration, the angle of subduction direction gradually varies from nearly orthogonal off Java Island to almost parallel off Andaman Islands farther north, which leads to development of arc-parallel strike-slip fault systems [9]. Water depth of the trench floor also gradually decreases from east (6000 m water depth off eastern Java) to northwest (4000 m water depth off Sumatra), which is ascribed to an enormous sediment input from the Bay of Bengal [10]. Age of the subducting plate is younging from ca. 130 Ma of east to ca. 50 m.y. of northwest [11].

In the northern part of the forearc (off Sumatra), multiple subbasins separated by structural highs have been formed, indicating transtension and transpression regimes are repeated due to strike-slip deformation along this margin [12]. The central Sunda margin (off western Java to southern Sumatra) is marked by a transition zone from orthogonal convergence in the east to oblique subduction to the northwest.

The subduction along this arc is active since Eocene [13]. The oceanic plate initially subducted beneath continental-type basement rock off Sumatra and oceanic-type one off Java [14, 15]. The convergence rate increased from 5 cm/yr (Eocene to Miocene) to 7 cm/yr during the last 10 My [16].

2.2. Forearc basins

2.2.1. Simuelue basin

The Simuelue basin (**Figure 3**) is a moderate-sized (100 × 260 km) basin with sediments of up to 5 s two-way travel time off Sumatra Island (e.g., [15, 17]). The basement of this basin is considered as the continental backstop of the Sumatran continental crust with a regional horst and graben structures which might be generated from the Late Eocene to Early Oligocene [21] (**Figure 3A**). The northern part of this basin (**Figure 3A**) shows depositional and deformational styles changed between Units 3 and 4. A contractional strain ascribed to landward-vergent thrusts in the retro-wedge was recorded before the deposition of unit 4. On the other hand, the strata of Units 4 and 5 are characterized by seaward tilting and seaward migration of the depocenter, suggesting the seaward side of the basin subsided relative to the landward side during that time.

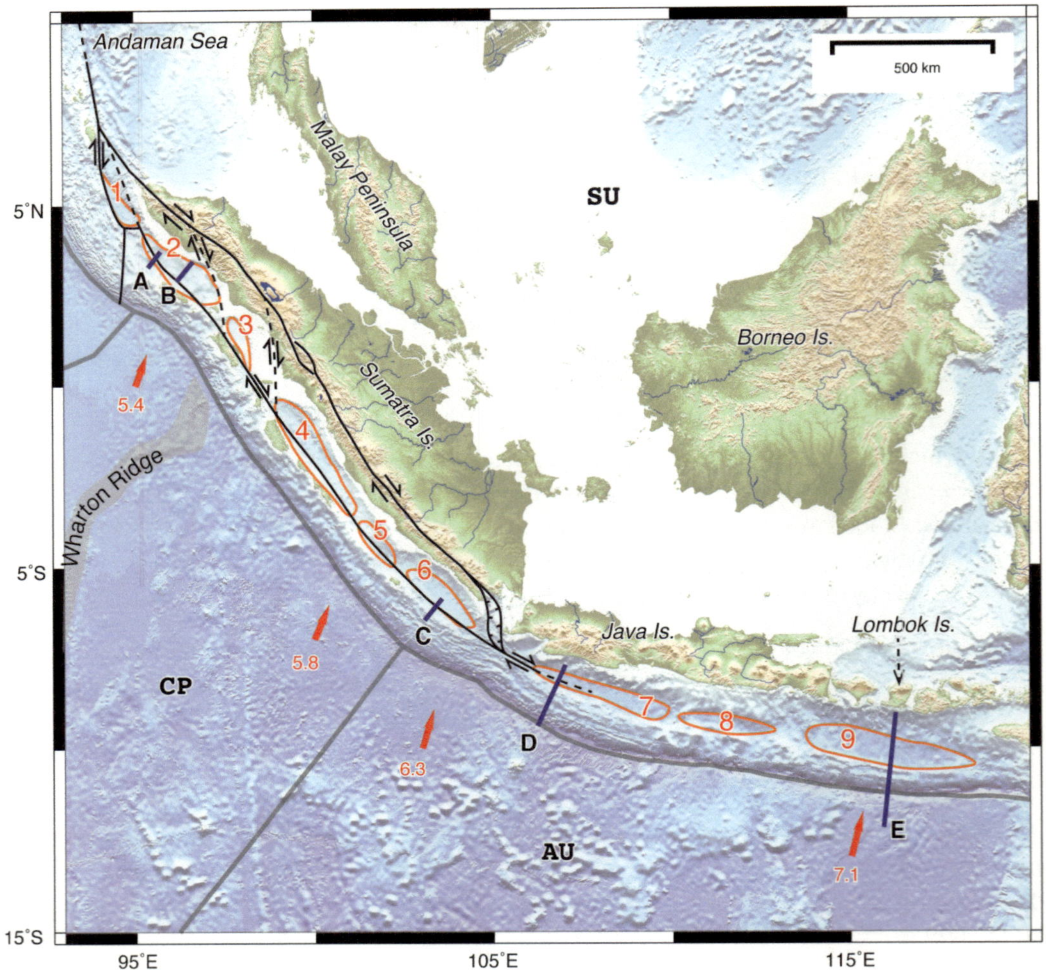

Figure 2. Index of the Sunda margin. Blue lines with labels A–E indicate the locations of survey lines shown in **Figure 3**. Forearc basins: 1—Aceh, 2—Simuelue, 3—Siberut, 4—Bengkulu, 5—Enggano, 6—Mentawai, 7—Western Java, 8—Eastern Java, 9—Lombok basins. Plate names: SU—Sunda, CP—Capricorn, AU—Australia plates. Red arrows indicate the direction and velocity [cm/yr] of plate motion relative to the Sunda plate based on Ref. [8].

The depositional history of the central part (**Figure 3B**) was divided into four stages [17]. At the end of Paleogene time (ca. 40 Ma), a regional erosional surface was developed after slope sediments covered the topographic lows (Unit 1 in **Figure 3B**). An Early/Mid Miocene stage is marked by a rapid subsidence in the trenchward and thick deposition occurred along the seaward margin of the basin (Unit 2). During the Late Miocene/Pliocene, subsidence continued in the northern and the trenchward region of the Simeulue basin (Unit 3). The basin from Pleistocene to recent was under an influence of strike-slip faults due to the oblique subduction evidenced by a wrench faulting, and subsidence expanded significantly landward (Unit 4).

http://dx.doi.org/10.5772/67559

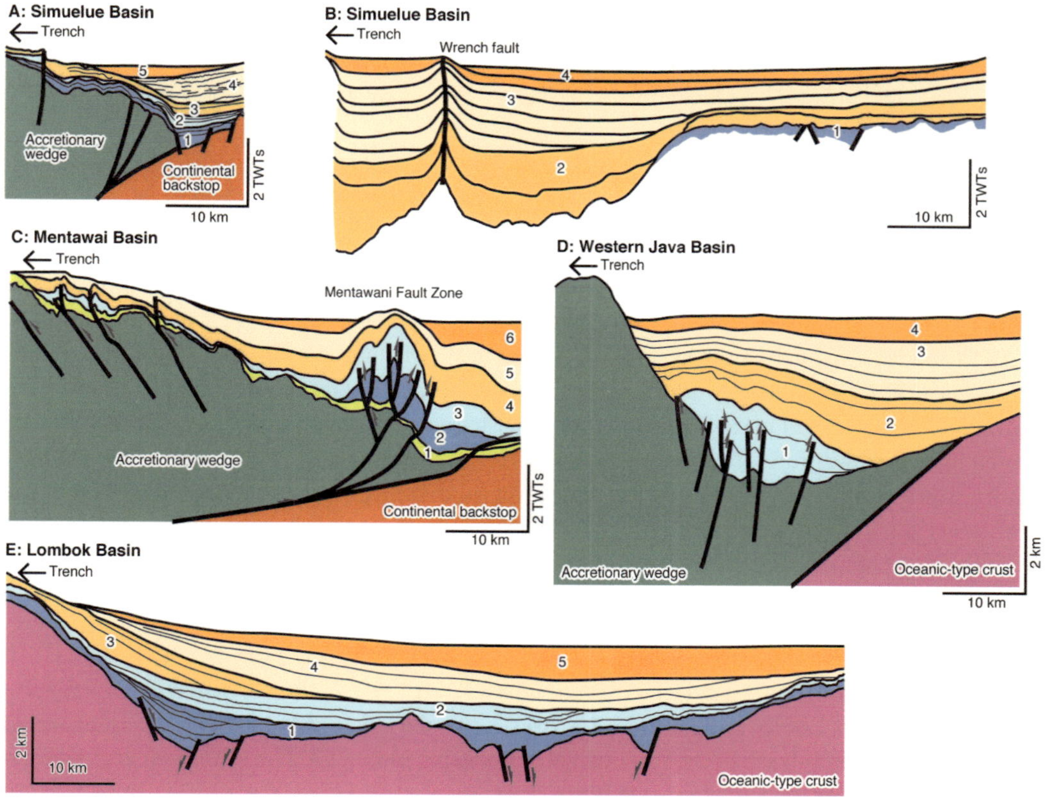

Figure 3. (A) Northern part of the Simuelue basin [15]. (B) Central part of the Simuelue basin [17]. (C) Mentawai basin [18]. (D) Western Java basin [19]. (E) Lombok basin along the Java Trench [20]. Locations of the survey lines are shown in **Figure 2**.

2.2.2. Mentawai basin

Deposition and deformation of the Mentawai basin have been strongly influenced by imbricated seaward-vergent fore-thrusts in the accretionary wedge and landward-vergent back-thrusts in the Mentawai fault zone [18, 22, 23] (**Figure 3C**). This fault zone has both characters of strike-slip fault (e.g., [23]) and back-thrust [18]. The fault zone may have initiated during Early-Middle Miocene and indicates the arc-parallel transpressional features along the boundary between the accretionary wedge and the continental backstop [18]. A continuous contraction developed the fold-thrust belt during Pliocene time.

The oldest deposits in this basin are the Middle Oligocene-Early Miocene, which thinly cover the forearc slope (Unit 1 in **Figure 3C**). Since the latest Early Miocene, uplift of the outer arc high began with imbricated fore-thrusts in the accretionary wedge, which reduced the thickness of forearc basin sediments trenchward to the outer arc high (Units 2–4). At the same time, landward-vergent back-thrusts also developed near the boundary between the accretionary wedge and the continental backstop. Although the activity of these thrusts waned in the early

Late Miocene (Unit 3), it reactivated in the late Late Miocene (Units 4 and 5). This intense deformation might be induced by accretion of thick Bengal Fan sediments and increase of plate convergence rate since the Late Miocene.

2.2.3. *Western Java basin*

The basement of the forearc basin is considered as an oceanic-type crust underlain by the shallow mantle deduced from large seismic velocities [14, 24] (**Figure 3D**). The seaward margin of the basin is bounded by the outer arc high with a sharp and steep scarp. Up to 4 km, sediments are accumulated along the outer arc high [19]. Depositional ages and sedimentary faces in the forearc basin are unknown, because of no drilling core in and around the basin. Growth of the accretionary wedge might cause a compressional deformation in association with landward-vergent back-thrusts in the early stages of the basin formation (Units 1 and 2), which folded and uplifted the basin strata. The deformation and thickness variations reduced during the following stages (Units 3 and 4). A total displacement is smaller than that in the Mentawai Basin. The steep landward slope of the outer arc high implies that the vertical growth might be contributed by basal accretion beyond the frontal wedge [19].

2.2.4. *Lombok basin*

The Lombok forearc basin is a large-sized (600 × 120 km) forearc basin with 3.5–4.5 km sediment thickness off Lombok Island [13, 16, 20]. The basement of this basin is also interpreted as an altered, heavily fractured piece of an older oceanic plate [25]. The basement is characterized by horst and graben structures (**Figure 3E**).

A progressive landward retreat of the depocenter in this basin is an evidence of a strong influence by growth of accretionary wedge (**Figure 3E**). Before the formation of the basin (Late Eocene), sediments filled topographic lows of horst and graben structures under an extensional tectonic regime (Unit 1 in **Figure 3E**). After the beginning of the uplift of the outer arc high (Unit 2 of the Early Oligocene), the basin subsided by a pull of the subducting oceanic plate (Unit 3 of the Late Oligocene), resulting in down lapping strata on the unconformity. From Late Oligocene to present (Units 4 and 5), the depocenter of the forearc basin continuously shifted landward without any compressional deformation or thrusting in the seaward margin of the basin.

3. Japan

3.1. Plate configuration

Two contrasting oceanic plates of the Pacific plate (old, cold, and steeply dipping) along the northern Japan and the Philippe Sea plate (young, hot, and gently dipping) along the southern Japan are subducting beneath the Okhotsk (North American) and Eurasian plates, respectively (**Figure 4**). Subduction of the Pacific plate is at a rate of ~9 cm/yr with almost orthogonal angle to the trench. The Japan Trench is one of the well-known examples of tectonically erosive margin (e.g., [26]).

http://dx.doi.org/10.5772/67559

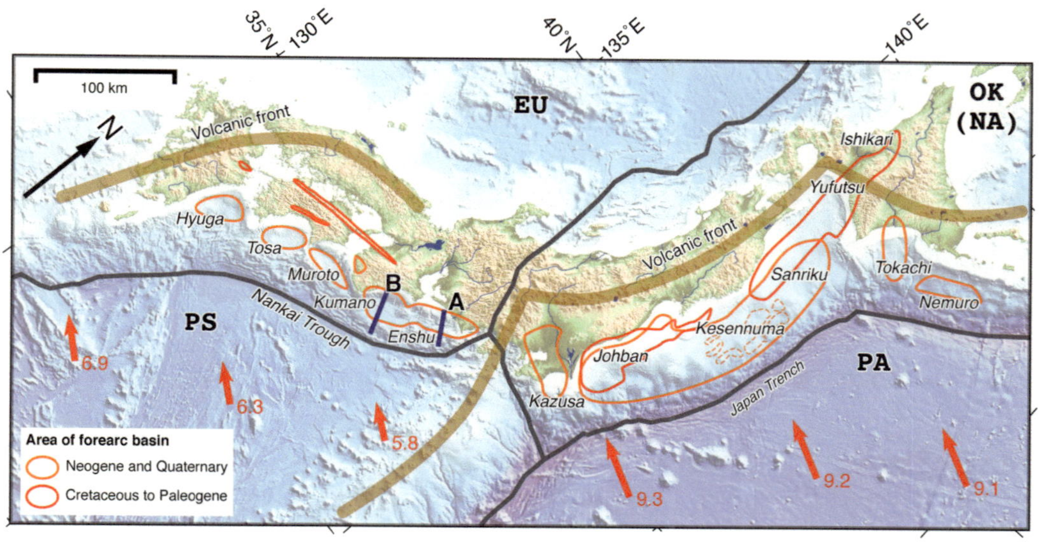

Figure 4. Index map of the Japan margin. Series of forearc basins are developed between volcanic fronts and trenches. Blue lines with labels A and B indicate locations of survey lines shown in **Figure 5**. Plate names: PA—Pacific, PS—Philippine Sea, OK—Okhotsk, NA—North American, EU—Eurasian plates. Red arrows indicate the direction and velocity [cm/yr] of plate motion relative to the Okhotsk or Eurasian plate based on Ref. [8].

In the southwestern Japan margin, the Philippine Sea plate subducts at a rate of 5–7 cm/yr with moderate obliquity at present. In this margin, the subducting plate changed from the Pacific to the Philippine Sea plate as a migration of a triple junction [27]. Thick accretionary wedge (the Shimanto Complex) and segmented forearc basins have been built from Cretaceous to present between the volcanic front and the Nankai Trough (e.g., [28]).

3.2. Enshu and Kumano basins

Enshu and Kumano forearc basins have been developed along the landward side of prominent outer arc highs associated with regional megasplay faults along the Nankai Trough (**Figure 5**) [29, 30]. Huge amounts of studies about the basins in terms of seismic interpretations (e.g., [31]) and sediment cores obtained from IODP expeditions (e.g., [32, 33] and references therein).

The Kumano forearc basin of the ~3 km sediments has been evolved by combined influences of frontal accretion and megasplay faults (out-of-sequence thrusts). A brief history about the basin can be summarized: (1) small trench-slope basins developed during 6–2 Ma on older consolidated inner accretionary wedge of the Miocene-Pliocene (Units 1 and 2 in **Figure 5**). (2) Splay fault development caused by a shallowing of the decollement (2–1.3 Ma) uplifted the slope basins and merged them into one large forearc basins (Unit 3). The basin floor was draped by submarine-fan turbidities derived from gravity flows through submarine canyons. (3) Diachronous depositional surfaces were formed by spatial and temporal variations of the fault system around the basin during 1.3–0.9 Ma (Unit 4). Hemipelagic sediments were derived from the outer arc high in the seaward side of the basin, while terrigenous sediments were

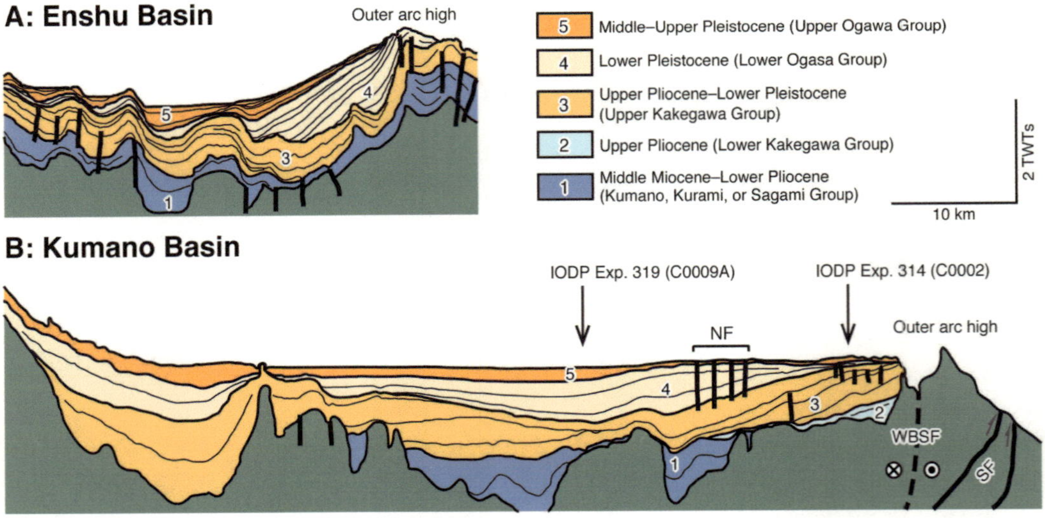

Figure 5. Interpretations of seismic profiles across (A) the Enshu and (B) the Kumano basins (modified from Ref. [37]). Seaward margins of the basins are characterized by landward tilting strata resulted from uplift of the outer arc highs. *Abbreviations*: NF—normal faults, SF—splay faults, WBSF—wedge boundary strike-slip fault. Locations of the profiles are shown in **Figure 4**.

deposited in the landward side. (4) Continuous uplift of the outer arc high by the splay faults resulted in a stable environment as the depocenter migrated landward (Unit 5 of <0.9 Ma). Normal faults and strike-slip faults overprinted the deformation in the seaward margins of the basins under the present condition [34–36]. No contractional deformation was recorded in the seaward side of the Kumano basin.

4. Aleutian-Alaskan forearc

4.1. Plate configuration

The Aleutian-Alaskan margin is a long (~3800 km) and arcuate subduction zone. The present configuration is marked by subduction of the Pacific plate beneath the North American plate at a rate of 6–7 cm/yr with increasing the obliquity from east (Alaskan margin) to west (central Aleutian margin). This margin might be established as a transition from erosional to accretionary margin occurred at ca. 6–3 Ma, triggered by an increase in sediment supply from glacial erosion in Alaska [38, 39].

4.2. Tugidak basin

The eastern Aleutian-western Alaskan margin is an example of double forearc basins where seaward (Sanak, Shumagin, Tugidak, Albatross, Stevenson) and landward (Cook Inlet and Shelikof Strait) basins are separated by the uplifted Kodiak Island and Kenai Peninsula

(**Figure 6**). This separation of forearc basins into two zones were explained by vertical growth of the accretionary wedge resulted from underplating [40] or flexural subsidence of the inland induced by a flat slab subduction [41].

The Tugidak basin is situated between Kodiak and Semidi Islands and at the landward side of the Tugidak anticline (**Figures 6** and **7**). The basin has a broad width and contains as much as 5 km of sediments from the late Miocene to the recent. The strata show that the depocenter shifted landward as the basin was shallowed from slope to shelf [42, 43]. The basin strata are broadly synclinal, but are contracted and thrusted near the Tugidak anticline that uplifted rapidly in Pliocene. On the other hand, normal faults are extensively developed in the landward side of the basin.

Although the depositional age is poorly constrained, a history of the basin development is as follows. The first unit (Unit 1 in **Figure 7B**) is slope-cover deposits unconformably covered on the lower Miocene or older rocks, before the Tugidak anticline uplifted. Deposition of Unit 2 corresponds to growth of the anticline in the Pliocene. The depocenter gradually shifted landward with onlap on both landward and seaward sides of the basin as the anticline uplifted. Nearly constant thickness of the Unit 3 indicates a seaward shift of the depocenter, suggesting that the sediment supply exceeded the rate of uplift or relative uplift of the forearc basement (inner wedge) to the anticline (outer wedge). Units 4 and 5 represented by a broad basin with

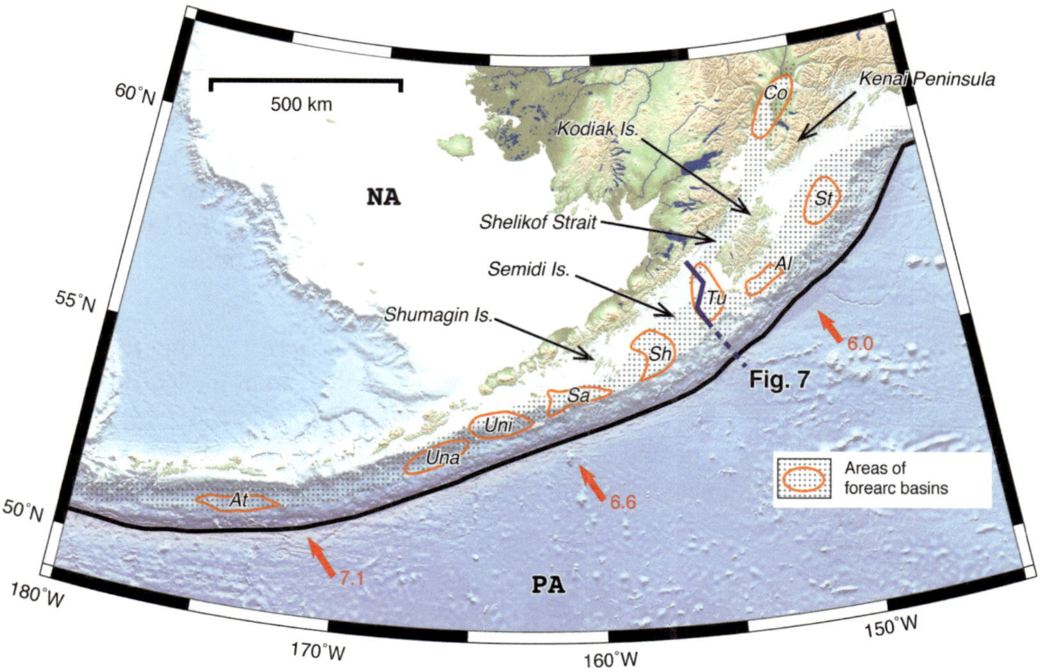

Figure 6. Index map of the Aleutian-Alaskan margin. Forearc basins: At—Atka, Una—Unalaska, Uni—Unimak, Sa—Sanak, Sh—Shumagin, Tu—Tugidak, Al—Albatross, St—Stevenson, Co—Cook Inlet basins. Plate names; PA—Pacific and NA is North American. Red arrows indicate the direction and velocity [cm/yr] of plate motion relative to the North American plate based on Ref. [8].

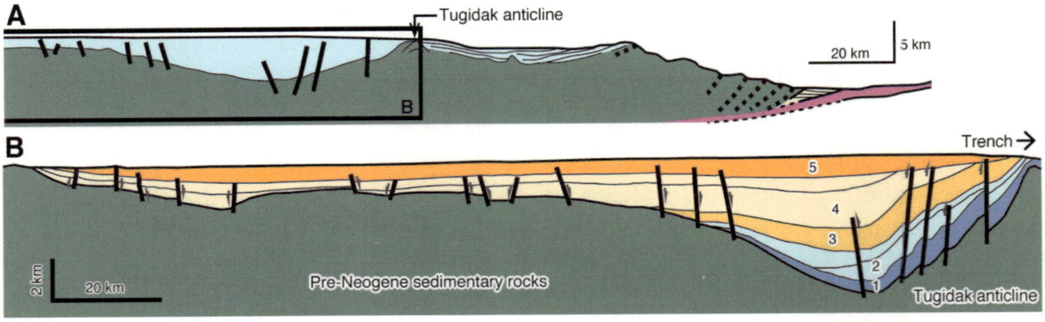

Figure 7. Interpretations of seismic profiles across the Tugidak basin (modified from Ref. [42]). The seaward margin is cut and tilted by several reverse faults related with growth of the Tugidak anticline. On the other hand, central and landward sides of the basin are characterized by numerous normal faults. Locations of the profiles are shown in **Figure 6**.

a landward shift of the depocenter, implying renewed growth of the anticline and regional subsidence of the forearc basement with numerous normal faults.

5. Lesser Antilles margin

5.1. Plate configuration

Caribbean plate was generated in the eastern Pacific as a Late Cretaceous oceanic plateau and advanced eastward between the North American and South American plates since Eocene. Subduction magmatism has constructed the Lesser Antilles island arc on the Caribbean plate during the past 40 m.y. A chain of volcanic islands stretches 800 km, forming a magmatic-arc platform (**Figure 8**). The present configuration in this region is characterized by the subduction of the North American plate (the Late Cretaceous Atlantic oceanic crust) beneath the Caribbean plate at a rate of ~2 cm/yr.

The Barbados ridge accretionary complex (**Figure 8**), which is the largest accretionary complex on Earth, has formed by frontal and basal accretion of sediments from the North American plate to the Caribbean plate since Eocene (e.g., [44]). It is <40 km wide and 7 km maximum thickness in the northern margin and is 300 km wide and up to 20 km thickness in the southern margin. Most of the sediments accreted are sourced from Orinoco River along the trench.

5.2. Forearc basins

The Tobago forearc basin (Tobago Trough) is one of the largest forearc basins in the world [44–47], which has 100 km wide, 200 km long, and more than 10 km-thick sediments (**Figure 9A** and **B**). The basin has formed between the crystalline platform of the Lesser Antilles arc and the Barbados Ridge accretionary complex. Barbados Island is a part of the outer arc high bounding the seaward margin of the basin. The strata near the outer arc high are deformed by landward-vergent thrusts in the retro-wedge (Unit 1 in **Figure 9B**). Development of these thrusts probably began in early Miocene or earlier, which imbricated forearc basin strata as a duplex between the

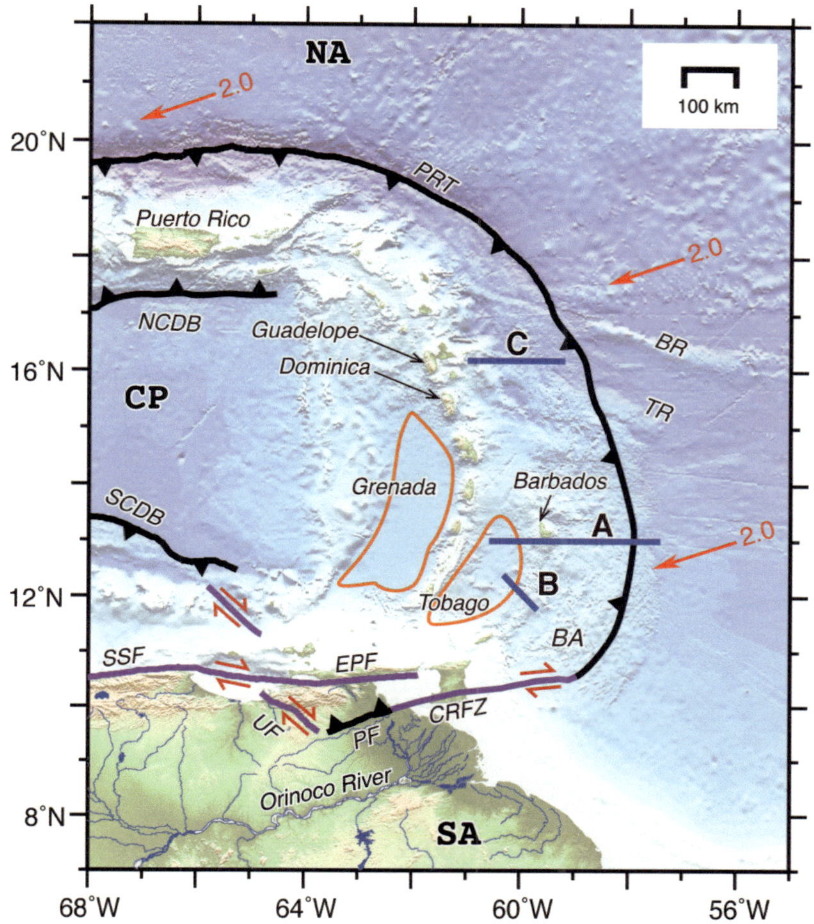

Figure 8. Index map of the Lesser Antilles margin. Blue lines with labels A–C indicate locations of survey lines shown in **Figure 9**. Plate names: NA—North American, CP—Caribbean, SA—South American plates. Other abbreviations: PRT—Puerto Rico Trench, BR—Barracuda Ridge, TR—Tiburon Ridge, BA—Barbados Ridge accretionary complex, NCDB—Northern Caribbean deformed belt, SCDB—Southern Caribbean deformed belt, SSF—San Sebastian Fault, EPF—El Pilar Fault, CRFZ—Central Range fault zone, UF—Urica Fault, PF—Pirital Fault. Red arrows indicate the direction and velocity [cm/yr] of plate motion relative to the Caribbean plate based on Ref. [8].

accretionary wedge and the early detachment in late Miocene or Pliocene time. The horizontal contraction of the basin was estimated to be between 4 and 45%, increasing southward [45]. Thickened outer arc high due to this contraction might increase tectonic loading, resulting in a large subsidence and trenchward shift of the depocenter in Pliocene and Pleistocene (Unit 3 in **Figure 9B**).

Forearc basins off Guadeloupe and Dominica have been developed in the northern part of this margin where a seismic Barracuda and Tiburon ridges are obliquely subducting (e.g., [48–50]) (**Figures 8** and **9C**). The buoyant crust of the Tiburon Ridge that accreted within the past 3.5 m.y. forms a seaward-dipping backstop in contact with the lower half of the accretionary wedge [48]. As much as half of the sediments underthrust at the toe of the accretionary wedge

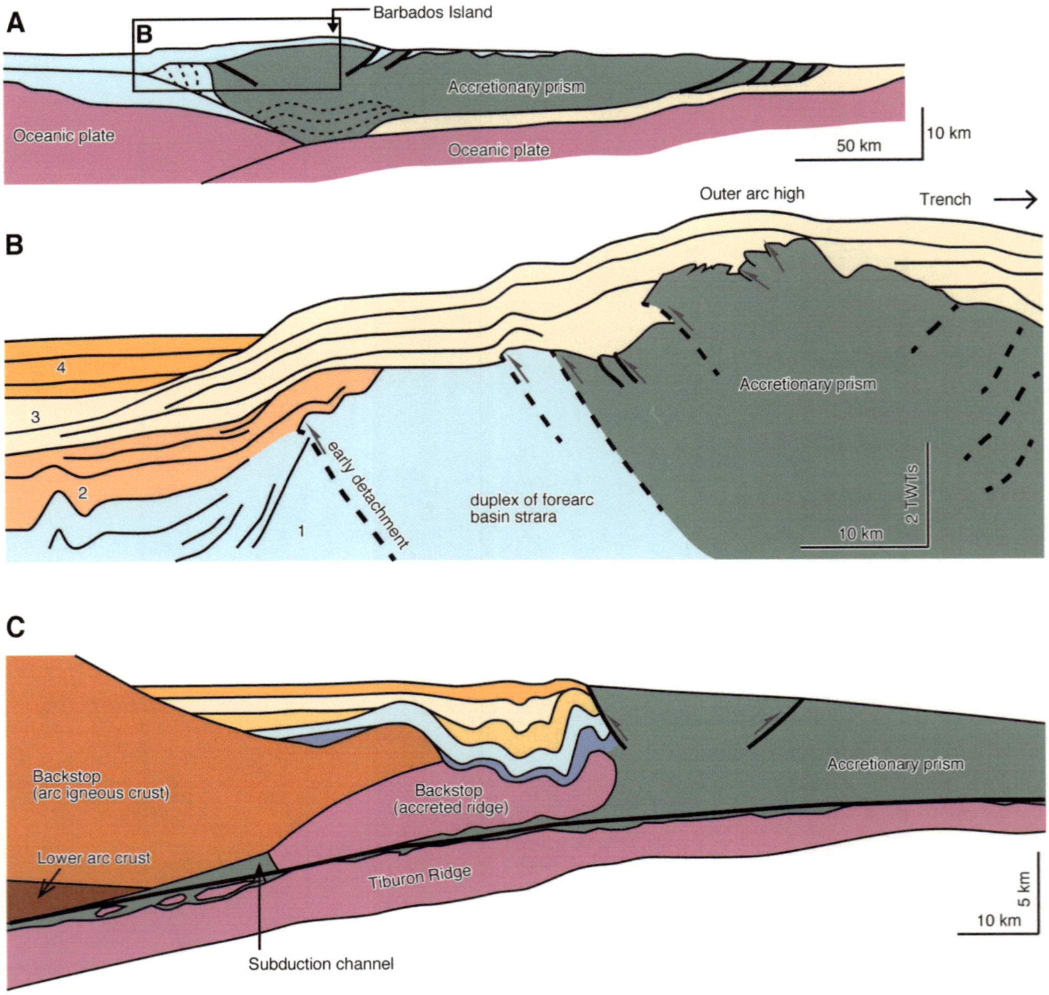

Figure 9. (A) Interpretations of the seismic profiles across the Tobago Trough in the Lesser Antilles forearc. (B) Details of seaward margin of the Tobago forearc basin. Modified from Ref. [45]. (C) A cross section across a forearc basin off Guadeloupe (modified from Ref. [48]). Seaward margins of both basins are deformed and uplifted by backthrusts. Locations of the profiles are shown in **Figure 8**.

appears to be subducted to the toe of the backstop, which were transported through a subduction channel between the subducting oceanic crust and the accretionary wedge.

6. South American margin (Columbia-Ecuador-Peru-Chile)

6.1. Plate configuration

The South American margin is a very long (~8000 km) subduction zone where the oceanic Nazca (6–7 cm/yr) and Antarctic (<2 cm/yr) plates subducts beneath the continental South

American plate (**Figure 10**). Due to its large length, a wide variety of styles of subduction is recognized, in terms of tectonic erosion/accretion [26, 51], oblique subduction [52, 53], and flat-slab and ridge subduction [54]. The width between the shelf and the trench is relatively small, and there are no large accretionary complex that does not exist throughout the margin.

6.2. Forearc basins along the Columbia-Ecuador margin

Along the Columbia margin, double forearc basins elongate parallel to the trench similar with the southwestern Alaskan margin. The inner basin is mainly onshore, being synclinal sags from Paleogene with very large (~10 km) depth (e.g., Tumaco Basin [55]) (**Figure 11A**). On

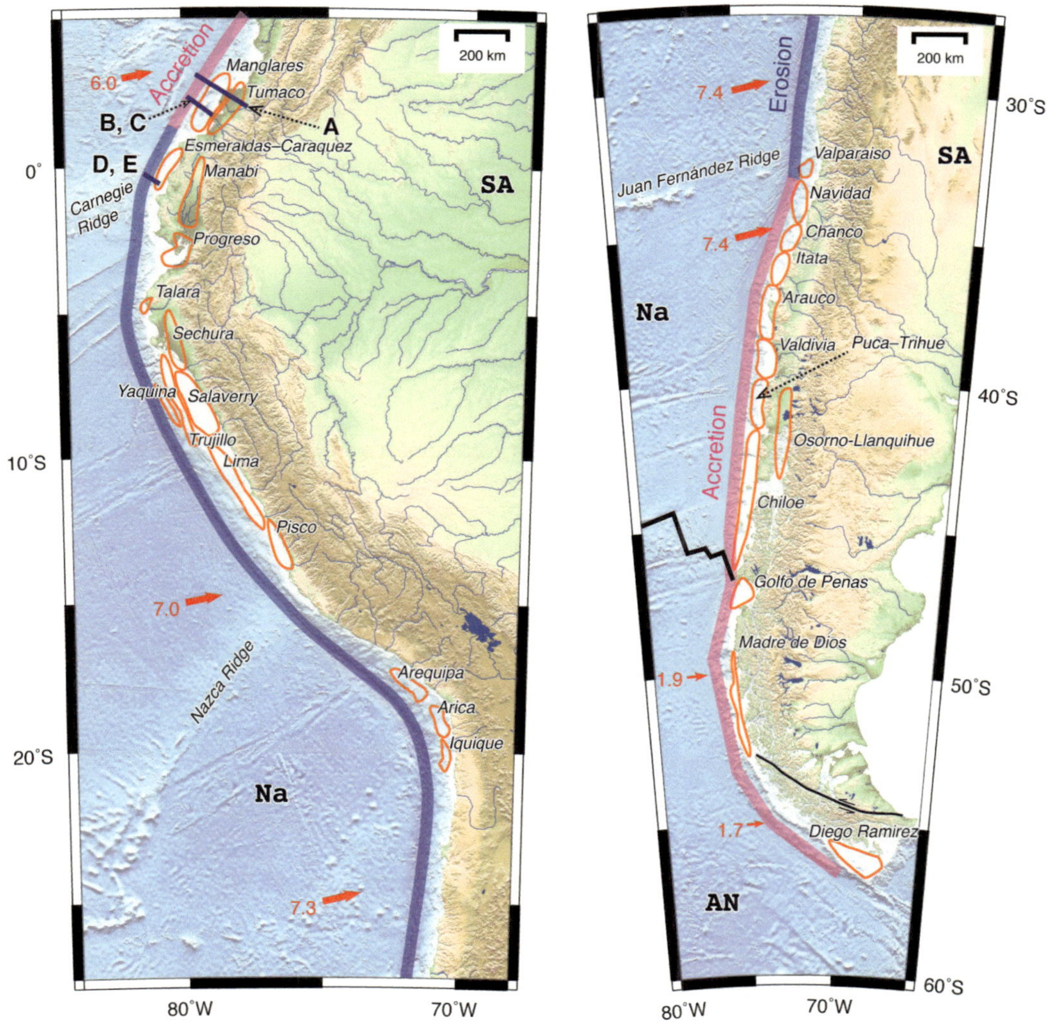

Figure 10. Index map of the South American margin (Colombian-Ecuador-Peru-Chile). Blue lines with labels A–E indicate locations of survey lines shown in **Figure 11**. Red arrows indicate the direction and velocity [cm/yr] of plate motion relative to the South American plate based on [8].

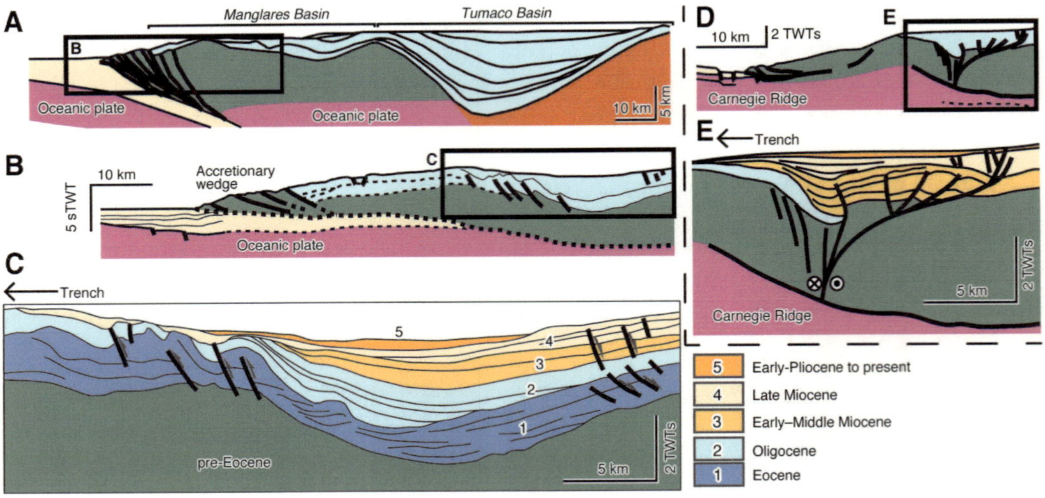

Figure 11. (A) Schematic cross section across the Tumaco and Manglares basins along the north Ecuador and south Colombia margin (modified from Ref. [58]). (B) Enlargement of the box in A. (C) Interpretation of seismic profile of the Manglares Basin (modified from Ref. [56]). Seismic units; 1—Eocene, 2—Oligocene, 3—Early-Middle Miocene, 4—Middle-Late Miocene, 5—Pliocene-Pleistocene. (D) A cross-section across the Ecuador margin (modified from Ref. [57]). (E) Interpretation of seismic profile for the boxed area in (C). Flower-structures are conspicuous. The depositional ages of the seismic units are not determined. Locations of the profiles are shown in **Figure 10**.

the other hand, offshore forearc basins are smaller and shallower than the inner basins (e.g., Manglares Basin [56]) (**Figure 11B** and **C**).

The Manglares basin is fronted by the Colombian accretionary wedge. The substrate of the basin is possibly accreted mass of an oceanic plateau. The basal unit (Unit 1 in **Figure 11C**) is composed of deep-water turbidite of the middle to late Eocene. The constant thickness suggests that they are slope (bathyal) deposits. In the Oligocene to the middle Miocene (Units 2 and 3), the outer arc high began to uplift in response to the plate kinematic reorganization during 40–35 Ma, inferred from landward propagation and onlap of the deposition. Unit 4 (late Miocene) corresponds to the overfilled sediments deposited on the truncated outer arc high, resulting in seaward progradation of the sediments. Sediments in Unit 5 fill the central part of the basin during a relax-phase of subsidence. Seaward-vergent reverse faults have tilted these forearc basin sediments landward.

On the other hand, a forearc basin along the Ecuador margin (southern part of the Esmeraldas-Caraquez Basin) is highly deformed by strike-slip faults, where Carnegie ridge crest is being subducted [57]. Conspicuous flower-structures popped up sediments in the basin. The strike-slip fault system terminates at depth against Carnegie ridge beneath the margin.

7. Tonga-Kermadec-Hikurangi margin

7.1. Margin configuration

Tonga-Kermadec-Hikurangi margin is a more than 3000 km-long boundary between the Pacific and Indo-Australian plates. Along the Tonga-Kermadec margin, a thin and less buoyant oceanic crust with deep water depth subducts beneath island arc crusts of Tonga-Kermadec ridges (**Figure 12A**). The Tonga-Kermadec margin is a typical example of tectonically erosive margins [4]. Along the Hikurangi margin, a thick (10–15 km) and buoyant oceanic crust (Hikurangi Plateau composed of Cretaceous large igneous province) with shallow water depth underthrusts beneath North Island of New Zealand (**Figure 12**). The Hikurangi margin is strongly controlled by westward oblique subduction of the Pacific plate.

At the Hikurangi margin, the subduction began at 25 Ma, when the plate motion reorganized from transtension to convergent. The Axial and Coastal ranges are composed of Mid-Cretaceous-Paleogene passive margin sedimentary rocks and Late Cretaceous accretionary

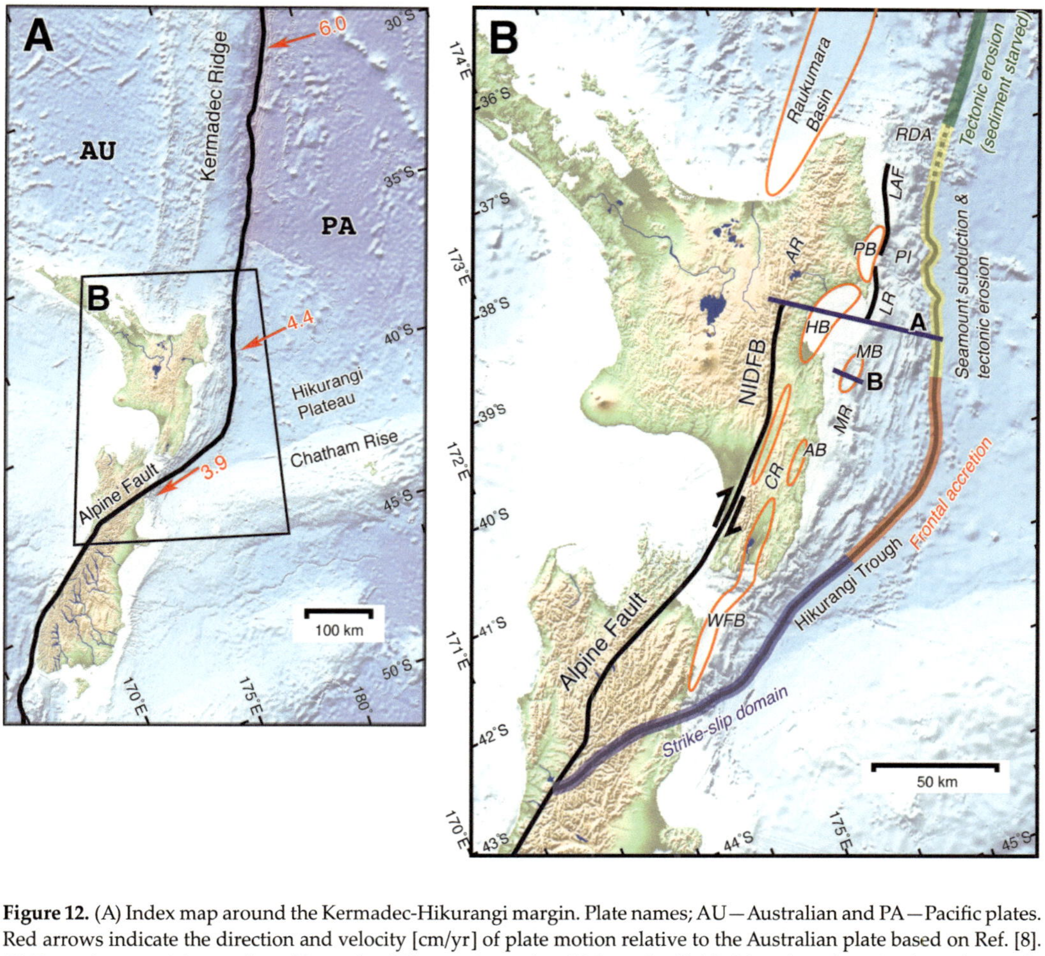

Figure 12. (A) Index map around the Kermadec-Hikurangi margin. Plate names; AU—Australian and PA—Pacific plates. Red arrows indicate the direction and velocity [cm/yr] of plate motion relative to the Australian plate based on Ref. [8]. (B) Tectonic map of the southern Kermadec-Hikurangi margin which can be divided into four domains from the north; tectonic erosion (southern Kermadec), seamount subduction and tectonic erosion (northern Hikurangi), frontal accretion (central Hikurangi), strike-slip (southern Hikurangi). Blue lines with labels A–B indicate locations of the profiles shown in **Figure 13**. Forearc basins; PB—Poverty Bay, HB—Hawke Bay, MB—Motu-o-Kura, AB—Akitio, WFB—Wairarapa-Flaxbourne basins. Abbreviations; RDA—Ruatoria debris avalanche, PI—Poverty Indentation, LAF—Lachlan-Ariel faults, LR—Lachlan Ridge, MR—Motu-o-Kura Ridge, CR—Coastal Range, AR—Axial Range, and NIDFB—North Island Dextral fault belt.

rocks, which have been affected by trench-parallel dextral strike-slip fault (the North Island Dextral Fault Belt) due to oblique subduction and has role of backstops of forearc deformation.

The southern Kermadec-Hikurangi margin can be separated into four sections (**Figure 13B**). The southern Kermadec section is a tectonically erosive (sediment starved) margin where a relatively thin subducting crust and thin trench fill sediments (<1 km) with large orthogonal convergent rate are dominated. The northern Hikurangi section is characterized by not only tectonic erosion but also submarine landslides and debris avalanches, such as Ruatoria debris avalanche and Poverty Indentation.

In the central Hikurangi section, subducting crust of the Hikurangi Plateau is thicker (more buoyant) with smoother surface topography than the north section. Frontal accretion formed a wide (>100 km) accretionary wedge represented by seaward-verging imbricated thrusts and ridges (e.g., [59, 60]), because of thick (~4 km) trench fill sediments supplied through the Hikurangi Trough from an area of high mountains in South Island [61]. The deformation front migrated seaward at 30–100 km/m.y. [59]. In the southern Hikurangi section, the orthogonal

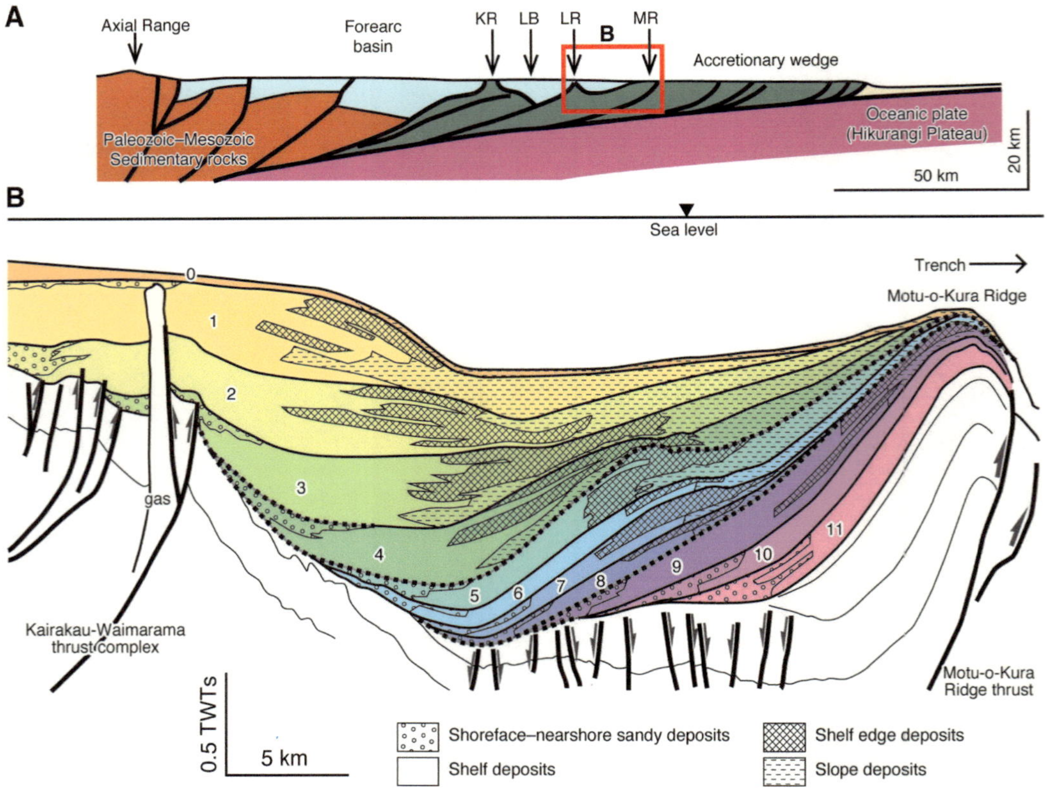

Figure 13. (A) A generalized cross section across the Hikurangi margin (modified from Ref. [64]). *Abbreviations*; MR—Motu-o-Kura Ridge, LR—Lachlan Ridge, LB—Lachlan Basin, KR—Kidnappers Ridge. (B) Details of the sequence units and depositional faces of the Motu-o-Kura Basin (modified from Ref. [63]). This basin is one of the upper trench-slope basins on the accretionary wedge, bounded by Motu-o-Kura Ridge at the seaward margin and Kidnappers Ridge at the landward margin.

convergent rate becomes slow (2.5–3.0 cm/yr). The plate boundary merges into the Alpine Fault strike-slip fault systems at the southern end. Accretionary wedge becomes narrower from central to this section (<50 km).

7.2. Motu-o-Kura basin

The main forearc basin is located inland in the central and southern Hikurangi section and offshore in the northern Hikurangi section (Poverty Bay basin and Hawke Bay basin). Number of trench-slope basins develop offshore (Akitio basin [62] and Motu-o-Kura basin [63]) on the actively growing accretionary wedge (**Figure 13A**). Because some of the basins were emerged several times during their evolutions, the deposition was highly affected by eustatic sea-level changes in addition to tectonics. In addition, the basin depocenters migrated three dimensionally with diachronous unconformities, due to the oblique subduction and segmented geometries of the trench-slope basins.

The Motu-o-Kura basin [63] includes sediments from ca. 1.1 Ma. The lower sequences (Units 9–11 in **Figure 13B**) show constant thickness in the basin and even on the Motu-o-Kura Ridge, which were interpreted as slope deposits before the ridge activated. The middle and upper sequences (Units 0–8) represent a tectonic growth sequence reflecting sedimentation contemporaneous with thrust faulting and uplift of the ridge. The strata in the basin exhibit gentle synclinal folding associated with normal faults in the center of the basin. The Motu-o-Kura Ridge thrust increased the displacement rate in the middle sequences (Units 5–8) are apparent at ca. 800 and ca. 430 ka as revealed by growth strata.

8. Summary

This review showed deposition and deformation styles of accretionary-type forearc basins were much variable. Especially, deformations in seaward margins of forearc basins are highly influenced by evolutional histories of accretionary wedge composed of a prowedge developing on the subducting plate and retro-wedges (e.g., [65–67]). Sandbox analog experiments showed two stages of the wedge evolutions [68]; Stage I was characterized by dominant retro-wedge thrusting and conjugate prowedge kinking and folding and by rapid uplift of the axial zone (outer arc high), and Stage II was characterized by progressive stacking of thrust sheets in the prowedge, by slow retro-wedge thrusting, and by slow to very slow uplift in the axial zone. Transition from stage I to stage II occurred when the growing wedges reached the critical height at which they behaved as a backstop for further prowedge accretion. In addition, the ratio of thickness of backstop (H_b) to thickness of accreted sediments (H_c) was considered as a factor for vergence partitioning in the wedge [69], suggesting the high and low (<1) ratios of H_b/H_c could resulted in proward vergence and doubly (pro- and retro-ward) vergence, respectively.

Deformation of accretionary-type forearc basins can be similarly classified into two models. We propose the following models for accretionary-type forearc basins (**Figure 14**), which are: (A) single-wedge (asymmetrically doubly vergent), (B) two-wedge (symmetrically doubly vergent), and (C) strike-slip models.

Single-wedge model: Forearc basins in this model are less contracted but tilted landward, because the outer arc highs move seaward with displacements of the fore-thrusts (splay faults) in the accretionary wedges. The basin widths are extended by this seaward movement of the outer-arc high as well as landward migration of the depocenter. Subsidence of the basin may be ascribed to an isostatic subsidence due to thickened wedge or tectonic erosion beneath the basin. Examples of this model are Lombok Basin, Nankai Trough (Kumano and Enshu troughs), Hikurangi Trough (Motu-o-Kura basin), and Colombia margin (Manglares basin).

Two-wedge model: The seaward margins of forearc basins in this model are more contracted with folds and faults by landward-vergent thrusts in the retro-wedges. The basin widths enlarge landward with migration of the depocenters. Basin subsidence can be caused by tectonic loading of thickened wedges or basal erosion. Examples of this model are Sunda (Western Java basins), Alaskan (Tugidak basin), and Lesser Antilles (Tobago basin) margins. Backthrusts-related deformations in these margins are restricted in the early stages of the basin formations. The basin strata in the recent stage are generally undeformed.

Strike-slip model: Obliquity of subducting plates can be an additional factor to modify the basin formations. High obliquity causes a strain-partitioning along the seaward margin of the basins with strike-slip faults. Transpression and transtension lead to compressional and extensional deformations, respectively. They may further overprint the preexisting deformations of single- and two-wedge models. Examples of this model are Sumatra (Simuelue and Mentawai basins), Ecuador, and southern Chile margins.

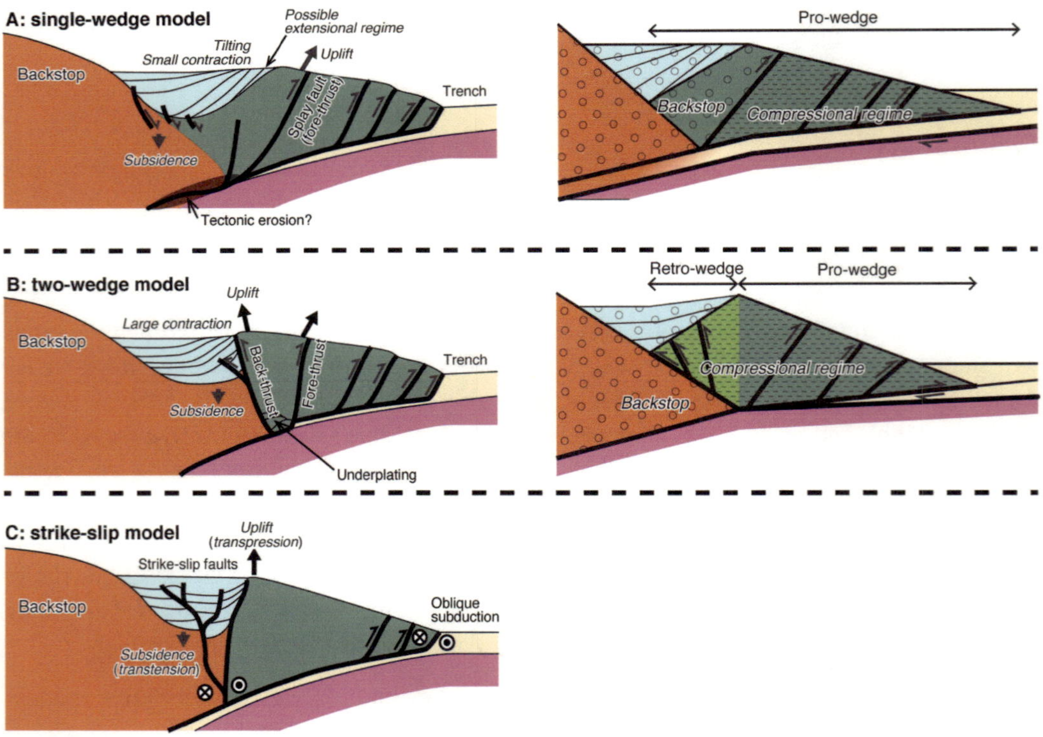

Figure 14. Schematic models of three models of deformation styles in forearc basins. (A) Single-wedge model with asymmetrical doubly vergent (singly vergent) uplift. (B) Two-wedge model with symmetrical doubly vergent uplift. (C) Strike-slip model.

These single- and two-wedge models may be comparable with P-C and P-U-C modes of [70]. The P-C mode, where pro-wedge (P) and conduit of subducting materials (C) are active, is characterized by landward-dipping backstops. Whereas the P-U-C mode (U means uplifted plug) creates an apparent seaward-dipping backstops. Amounts of materials added by underplating (basal accretion) or removed by tectonic erosion (basal erosion) can influence degree of uplift or subside the axes of the wedges (outer arc highs), respectively; the former grows uplifted plug between the pro- and retro-wedges. Therefore, material flux between the subducting and overriding plates can be one of the major causes of the differences between the two models.

Furthermore, we speculate that the retro-wedge is overthrusted on the mechanical backstop, when the cross-sectional widths of the pro-wedge are too small to absorb the total strain caused by underthrusting of the subducting plate. Internal strength (friction and cohesion) and total thickness (tectonic load) of the wedges may be other factors to determine the thrust polarity. For example, basins of the two-wedge models, such as the Western Java and Tobago basins, were deformed by landward-vergent thrusts on the retro-wedges. However, these deformations are recognized only in the lower sequences, indicating the thrust activities waned as the accretionary wedges grew.

Changes of subduction direction could cause oblique subduction, which overwrite the preexistence deformations in single-wedge (e.g., Nankai Trough) and two-wedge (e.g., Mentawai Basin) models by transpressional and transtensional deformations (strike-slip model). Spatial and temporal variations of mechanical conditions at the subduction zones might be recorded as styles of deposition and deformation of forearc basins. Forearc basin is a potentially powerful tool for unraveling histories and mechanisms of subduction zone processes.

Author details

Atsushi Noda* and Ayumu Miyakawa

*Address all correspondence to: a.noda@aist.go.jp

Geological Survey of Japan, National Institute of Advanced Industrial Science and Technology, Tsukuba, Ibaraki, Japan

References

[1] Dickinson WR. Forearc basins. In: Busby CJ, Ingersoll RV, editors. Tectonics of Sedimentary Basins, chap. 6. Oxford, United Kingdom: Blackwell Science;1995. pp. 221-261.

[2] Noda A. Forearc basins: Types, geometries, and relationships to subduction zone dynamics. Geological Society of America Bulletin. 2016;**128**(5-6):879-895

[3] Stern RJ. Subduction zones. Reviews of Geophysics. 2002;**40**(4):1012. DOI:10.1029/2001RG000108.

[4] Clift PD, Vannucchi P. Controls on tectonic accretion versus erosion in subduction zones: Implications for the origin and recycling of the continental crust. Reviews of Geophysics. 2004;**42**(RG2001). DOI: 10.1029/2003RG000127.

[5] Draut AE, Clift PD. Basins in arc-continent collisions. In: Busby C, Azor A, editors. Tectonics of Sedimentary Basins: Recent Advances, chap. 17. Chichester, UK: Wiley-Blackwell; 2012. pp. 347-368

[6] Wells RE, Blakely RJ, Sugiyama Y, Scholl DW, Dinterman PA. Basin-centered asperities in great subduction zone earthquakes: A link between slip, subsidence, and subduction erosion? Journal of Geophysical Research. 2003;**108**(B10, 2507). DOI: 10.1029/2002JB002072

[7] Song TRA, Simons M. Large trench-parallel gravity variations predict seismogenic behavior in subduction zones. Science. 2003;**301**(5633):630-633

[8] Argus DF, Gordon RG, DeMets C. Geologically current motion of 56 plates relative to the no-net-rotation reference frame. Geochemistry, Geophysics, Geosystems. 2011;**12**(Q 11001):444. DOI: 10.1029/2011GC003751.

[9] McCaffrey R. The tectonic framework of the Sumatran subduction zone. Annual Review of Earth and Planetary Sciences. 2009;**37**(1):345-366

[10] Moore GF, Curray JR, Moore DG, Karig DE. Variations in geologic structure along the Sunda Fore Arc, northeastern Indian Ocean. In: Hayes DE, editor. The Tectonic and Geologic Evolution of Southeast Asian Seas and Islands, Geophysical Monograph, no. 23. Washington, DC, United States: American Geophysical Union;1980. pp. 145-160.

[11] Müller RD, Sdrolias M, Gaina C, Roest WR. Age, spreading rates, and spreading asymmetry of the world's ocean crust. Geochemistry, Geophysics, Geosystems. 2008;**9**(Q04006). DOI: 10.1029/2007GC001743.

[12] Berglar K, Gaedicke C, Franke D, Ladage S, Klingelhoefer F, Djajadihardja YS. Structural evolution and strike-slip tectonics off north-western Sumatra. Tectonophysics. 2010; **480**:119-132

[13] van der Werff W. Variation in forearc basin development along the Sunda Arc, Indonesia. Journal of Southeast Asian Earth Sciences. 1996;**14**(5):331-349

[14] Grevemeyer I, Tiwari VM. Overriding plate controls spatial distribution of megathrust earthquakes in the Sunda–Andaman subduction zone. Earth and Planetary Science Letters. 2006;**251**(3-4):199-208

[15] Singh SC, Carton H, Tapponnier P, Hananto ND, Chauhan APS, Hartoyo D, Bayly M, Moeljopranoto S, Bunting T, Christie P, Lubis H, Martin J. Seismic evidence for broken oceanic crust in the 2004 Sumatra earthquake epicentral region. Nature Geoscience. 2008;**1**(11):777-781

[16] van der Werff W, Prasetyo H, Kusnida D, van Weering TCE. Seismic stratigraphy and Cenozoic evolution of the Lombok Forearc Basin, Eastern Sunda Arc. Marine Geology. 1994;**117**:119-134

[17] Berglar K, Gaedicke C, Lutz R, Franke D, Djajadihardja YS. Neogene subsidence and stratigraphy of the Simeulue forearc basin, Northwest Sumatra. Marine Geology. 2008; **253**(1-2):1-13

[18] Mukti MM, Singh SC, Deighton I, Hananto ND, Moeremans R, Permana H. Structural evolution of backthrusting in the Mentawai Fault Zone, offshore Sumatran forearc. Geochemistry, Geophysics, Geosystems. 2012;**13**(Q12006). DOI: 10.1029/2012GC004199.

[19] Kopp H, Hindle D, Klaeschen D, Oncken O, Reichert C, Scholl D. Anatomy of the western Java plate interface from depth-migrated seismic images. Earth and Planetary Science Letters. 2009;**288**(3-4):399-407

[20] Lüschen E, Müller C, Kopp H, Engels M, Lutz R, Planert L, Shulgin A, Djajadihardja YS. Structure, evolution and tectonic activity of the eastern Sunda forearc, Indonesia, from marine seismic investigations. Tectonophysics. 2011;**508**:6-21

[21] De Smet MEM, Barber AJ. Tertiary stratigraphy. In: Barber AJ, Crow MJ, Milsom JS, editors. Sumatra: Geology, Resources and Tectonic Evolution, *Memoir*, vol. 31, chap. 7. Geological Society, London; 2005. pp. 86-97

[22] Singh SC, Hananto N, Mukti M, Robinson DP, Das S, Chauhan A, Carton H, Gratacos B, Midnet S, Djajadihardja Y, Harjono H. Aseismic zone and earthquake segmentation associated with a deep subducted seamount in Sumatra. Nature Geoscience. 2011;**4**(5):308-311

[23] Berglar K, Gaedicke C, Ladage S, Thöle H. The Mentawai forearc sliver of Sumatra: A model for a strike-slip duplex at a regional scale. Tectonophysics. 2016; DOI: 10.1016/j.tecto.2016.09.014.

[24] Kopp H, Klaeschen D, Flueh ER, Bialas J, Reichert C. Crustal structure of the Java margin from seismic wide-angle and multichannel reflection data. Journal of Geophysical Research. 2002;**107**(B2). DOI: 10.1029/2000JB000095.

[25] Planert L, Kopp H, Lueschen E, Mueller C, Flueh ER, Shulgin A, Djajadihardja Y, Krabbenhoeft A. Lower plate structure and upper plate deformational segmentation at the Sunda–Banda arc transition, Indonesia. Journal of Geophysical Research. 2010;**115**(B08107). DOI: 10.1029/2009JB006713.

[26] von Huene R, Lallemand S. Tectonic erosion along the Japan and Peru convergent margins. Geological Society of America Bulletin. 1990;**102**(6):704-720

[27] Kimura G, Hashimoto Y, Kitamura Y, Yamaguchi A, Koge H. Middle Miocene swift migration of the TTT triple junction and rapid crustal growth in southwest Japan: A review. Tectonics. 2014;**33**(7):1219-1238

[28] Taira A. Tectonic evolution of the Japanese Island arc system. Annual Review of Earth and Planetary Sciences. 2001;**29**:109-134

[29] Park JO, Tsuru T, Kodaira S, Cummins PR, Kaneda Y. Splay fault branching along the Nankai subduction zone. Science. 2002;**297**(5584):1157-1160

[30] Strasser M, Moore GF, Kimura G, Kitamura Y, Kopf AJ, Lallemant S, Park JO, Screaton EJ, Xin S, Underwood MB, Zhao X. Origin and evolution of a splay fault in the Nankai accretionary wedge. Nature Geoscience. 2009;**2**(9):648-652

[31] Takano O, Itoh Y, Kusumoto S. Variation in forearc basin configuration and basin-filling depositional systems as a function of trench slope break development and strike-slip movement: Examples from the Cenozoic Ishikari–Sanriku-Oki and Tokai-Oki–Kumano-Nada Forearc Basins, Japan. In: Itoh Y, editor. Mechanism of Sedimentary Basin Formation: Multidisciplinary Approach on Active Plate Margins. Rijeka, Croatia: InTech; 2013. pp. 3-25

[32] Moore GF, Boston BB, Strasser M, Underwood MB, Ratliff RA. Evolution of tectono-sedimentary systems in the Kumano Basin, Nankai Trough forearc. Marine and Petroleum Geology. 2015;**67**:604-616

[33] Ramirez SG, Gulick SPS, Hayman NW. Early sedimentation and deformation in the Kumano forearc basin linked with Nankai accretionary prism evolution, southwest Japan. Geochemistry, Geophysics, Geosystems. 2015;**16**:1616-1633

[34] Martin KM, Gulick SPS, Bangs NLB, Moore GF, Ashi J, Park JO, Kuramoto S, Taira A. Possible strain partitioning structure between the Kumano fore-arc basin and the slope of the Nankai Trough accretionary prism. Geochemistry, Geophysics, Geosystems. 2010;**11**(Q0AD02). DOI: 10.1029/2009GC002668.

[35] Moore GF, Boston BB, Sacks AF, Saffer DM. Analysis of normal fault populations in the Kumano Forearc Basin, Nankai Trough, Japan: 1. Multiple orientations and generations of faults from 3-D coherency mapping. Geochemistry, Geophysics, Geosystems. 2013;**14**(6):1989-2002

[36] Tsuji T, Ashi J, Ikeda Y. Strike-slip motion of a mega-splay fault system in the Nankai oblique subduction zone. Earth, Planets and Space. 2014;**66**(1):120. doi: 10.1186/1880-5981-66-120

[37] Goto S, and others. Fuel Resource Geology Map: Eastern Nankai Trough. FR–2. Tsukuba, Japan: Geological Survey of Japan, AIST 2010, p. 10

[38] Scholl DW, Vallier TL, Stevenson AJ. Geologic evolution and petroleum geology of the Aleutian Ridge. In: Scholl DW, Grantz A, Vedder JG, editors. Geology and Resource Potential of the continental margin of western North America and adjacent ocean basins: Beaufort Sea to Baja California, Earth Science Series, vol. 6, chap. 7. Houston, TX, United States (USA): Circum-Pacific Council for Energy and Mineral Resources; 1987. pp. 123-155.

[39] von Huene R, Miller JJ, Weinrebe W. Subducting plate geology in three great earthquake ruptures of the western Alaska margin, Kodiak to Unimak. Geosphere. 2012;**8**(3):628-644

[40] Clendenen WS, Sliter WV, Byrne T. Tectonic implications of the Albatross sedimentary sequence, Sitkinak Island, Alaska. In: Bradley DC, Ford AB, editors. Geologic studies in Alaska, U.S. Geological Survey Bulletin, vol. 1999. U.S. Geological Survey;1992. pp. 52-70.

[41] Jadamec MA, Billen MI, Roeske SM. Three-dimensional numerical models of flat slab subduction and the Denali fault driving deformation in south-central Alaska. Earth and Planetary Science Letters. 2013;**376**:29-42

[42] Fisher MA. Structure and tectonic setting of continental shelf southwest of Kodiak Island, Alaska. American Association of Petroleum Geologists Bulletin. 1979;**63**(3):301-310

[43] von Huene R, Fisher M, Hampton M, Lynch M. Petroleum potential, environmental geology, and the technology for exploration and development of the Kodiak lease sale area #61, Open-File Report, vol. 80-1082. U. S. Geological Survey 1980, 70 p.

[44] Westbrook GK, Ladd JW, Buhl P, Bangs N, Tiley GJ. Cross section of an accretionary wedge: Barbados Ridge complex. Geology. 1988;**16**(7):631-635

[45] Torrini R, Speed RC. Tectonic wedging in the forearc basin: Accretionary prism transition, Lesser Antilles forearc. Journal of Geophysical Research. 1989;**94**(B8):10,549-10,584

[46] Speed R, Torrini R, Smith PL. Tectonic evolution of the Tobago Trough forearc basin. Journal of Geophysical Research. 1989;**94**(B3):2913-2936

[47] Aitken T, Mann P, Escalona A, Christeson GL. Evolution of the Grenada and Tobago basins and implications for arc migration. Marine and Petroleum Geology. 2011;**28**(1):235-258

[48] Bangs NL, Christeson GL, Shipley TH. Structure of the Lesser Antilles subduction zone backstop and its role in a large accretionary system. Journal of Geophysical Research. 2003;**108**(B7, 2358). DOI: 10.1029/2002JB002040.

[49] Laigle M, Becel A, de Voogd B, Sachpazi M, Bayrakci G, Lebrun JF, Evain M. Along-arc segmentation and interaction of subducting ridges with the Lesser Antilles Subduction forearc crust revealed by MCS imaging. Tectonophysics. 2013;**603**:32-54

[50] Evain M, Galve A, Charvis P, Laigle M, Kopp H, Bécel A, Weinzierl W, Hirn A, Flueh ER, Gallart J. Structure of the Lesser Antilles subduction forearc and backstop from 3D seismic refraction tomography. Tectonophysics. 2013;**603**:55-67

[51] von Huene R, Pecher IA, Gutscher MA. Development of the accretionary prism along Peru and material flux after subduction of Nazca Ridge. Tectonics. 1996;**15**(1):19-33

[52] Collot JY, Agudelo W, Ribodetti A, Marcaillou B. Origin of a crustal splay fault and its relation to the seismogenic zone and underplating at the erosional north Ecuador–south Colombia oceanic margin. Journal of Geophysical Research. 2008;**113**(B12102). DOI: 10.1029/2008JB005691.

[53] Polonia A, Torelli L, Brancolini G, Loreto MF. Tectonic accretion versus erosion along the southern Chile trench: Oblique subduction and margin segmentation. Tectonics. 2007;**26**(TC3005). DOI: 10.1029/2006TC001983.

[54] Gutscher MA, Spakman W, Bijwaard H, Engdahl ER. Geodynamics of flat subduction: Seismicity and tomographic constraints from the Andean margin. Tectonics. 2000;**19**:814-833

[55] Borrero C, Pardo A, Jaramillo CM, Osorio JA, Cardona A, Flores A, Echeverri S, Rosero S, Garca J, Castillo H. Tectonostratigraphy of the Cenozoic Tumaco forearc basin (Colombian Pacific) and its relationship with the northern Andes orogenic build up. Journal of South American Earth Sciences. 2012;**39**:75-92

[56] Marcaillou B, Collot JY. Chronostratigraphy and tectonic deformation of the North Ecuadorian–South Colombian offshore Manglares forearc basin. Marine Geology. 2008;**255**(1-2):30-44

[57] Collot JY, Marcaillou B, Sage F, Michaud F, Agudelo W, Charvis P, Graindorge D, Gutscher MA, Spence G. Are rupture zone limits of great subduction earthquakes controlled by upper plate structures? Evidence from multichannel seismic reflection data acquired across the northern Ecuador–southwest Colombia margin. Journal of Geophysical Research. 2004;**109**(B11103). DOI: 10.1029/2004JB003060.

[58] López ER. Evolution tectono-stratigraphique du double bassin avant: arc de la marge convergente Sud Colombienne: Nord Equatorienne pendant le Cénozoïque [Ph.D. thesis]. Nice, France: Université Nice Sophia Antipolis; 2009.

[59] Barnes PM, de Lépinay BM. Rates and mechanics of rapid frontal accretion along the very obliquely convergent southern Hikurangi margin, New Zealand. Journal of Geophysical Research. 1997;**102**(B11):24,931-24,952

[60] Ghisetti FC, Barnes PM, Ellis S, Plaza-Faverola AA, Barker DHN. The last 2 Myr of accretionary wedge construction in the central Hikurangi margin (North Island, New Zealand): Insights from structural modeling. Geochemistry, Geophysics, Geosystems. 2016;**17**:2661-2686

[61] Lewis KB. The 1500-km-long Hikurangi Channel: Trench-axis channel that escapes its trench, crosses a plateau, and feeds a fan drift. Geo-Marine Letters. 1994;**14**:19-28

[62] Bailleul J, Robin C, Chanier F, Guillocheau F, Field B, Ferriere J. Turbidite systems in the inner forearc domain of the Hikurangi convergent margin (New Zealand): New constraints on the development of trench-slope basins. Journal of Sedimentary Research. 2007;**77**(4):263-283

[63] Paquet F, Proust JN, Barnes PM, Pettinga JR. Controls on active forearc basin stratigraphy and sediment fluxes: The Pleistocene of Hawke Bay, New Zealand. Geological Society of America Bulletin. 2011;**123**(5-6):1074-1096

[64] Barnes PM, Nicol A, Harrison T. Late Cenozoic evolution and earthquake potential of an active listric thrust complex above the Hikurangi subduction zone, New Zealand. Geological Society of America Bulletin. 2002;**114**(11):1379-1405

[65] Willett SD, Beaumont C, Fullsack P. Mechanical model for the tectonics of doubly vergent compressional orogens. Geology. 1993;**21**:371-374

[66] Willett SD. Orogeny and orography: The effects of erosion on the structure of mountain belts. Journal of Geophysical Research. 1999;**104**(B12):28,957-28,981

[67] Naylor M, Sinclair HD, Willett S, Cowie PA. A discrete element model for orogenesis and accretionary wedge growth. Journal of Geophysical Research. 2005;**110**(B12403). DOI: 10.1029/2003JB002940.

[68] Storti F, Salvini F, McClay K. Synchronous and velocity-partitioned thrusting and thrust polarity reversal in experimentally produced, doubly-vergent thrust wedges: Implications for natural orogens. Tectonics. 2000;**19**:378-396

[69] Storti F, Marin RS, Faccenna C, Sainz AC. Role of the backstop-to-cover thickness ratio on vergence partitioning in experimental thrust wedges. Terra Nova. 2001;**13**(6):413-417

[70] Beaumont C, Ellis S, Pfiffner A. Dynamics of sediment subduction-accretion at convergent margins: Short-term modes, long-term deformation, and tectonic implications. Journal of Geophysical Research. 1999;**104**:17573-17601

Chapter 2

Cenozoic Fault Zone Activity and Geologic Evolution of the Offshore Regions of Fukushima and Miyagi Prefectures, Northeastern Japan, Based on Petroleum Exploration Data

Hiroyuki Arato

Additional information is available at the end of the chapter

http://dx.doi.org/10.5772/67391

Abstract

Three major fault zones were identified in the offshore regions of Fukushima and Miyagi prefectures, northeastern Japan, from petroleum exploration seismic and well data acquired by the Ministry of Economy, Trade and Industry (METI) and its predecessor: the Kesennuma tectonic line (KTL), the Ishinomaki tectonic line (ITL), and the Joban tectonic line (JTL). The stratigraphic relationships indicate that these tectonic lines were activated during the Middle Miocene, and that their activity was likely closely connected with the opening of the Japan Sea and the drifting of the northeast Japan arc to its current position. Parts of these tectonic lines were reactivated during the Late Pliocene and Quaternary in the Sendai Bay area.

Keywords: northeast Japan arc, tectonic lines, strike-slip fault, Fukushima, Miyagi

1. Introduction

The East Asian Japanese Islands are located on the convergent margin of the western Pacific Ocean. These islands are tectonically characterized by backarcs (e.g., the Japan Sea, Okinawa Trough, and Okhotsk Basin), island arcs (e.g., the Kuril, northeast Japan, Izu-Bonin, southwest Japan, and Ryukyu arcs), and trenches (e.g., the Japan, Izu-Ogasawara, Kuril, and Ryukyu trenches, and the Sagami and Nankai troughs). The various tectonic settings of the Japanese islands were determined by onshore geological research, as well as offshore oceanic geological surveys and oceanic geophysical measurements. The tectonic transformation around the Japanese islands is therefore thought to have occurred as follows:

(i) clockwise rotation of the southwest Japan arc (e.g., [1]),

(ii) deformation of the northeast Japan arc by NNW-SSE-trending strike-slip faults (e.g., [2–5]),

(iii) ocean floor spreading in the northern Japan Sea (e.g., [6–8]),

(iv) extension of the continental crust in the southern Japan Sea (e.g., [7, 9, 10]),

(v) inversion tectonics at the plate boundary on the eastern margin of the northern Japan Sea (e.g., [11]), and

(vi) rotation of the Philippine Sea Plate (e.g., [12, 13]).

Geotectonic research has predominantly focused on tectonic structure, deep crust processes, and the interaction and internal structure of oceanic and overriding plates. In contrast, scientific drilling and seismic surveys are generally not planned and designed to observe the thick Cretaceous and Cenozoic sediments on the overriding plate of arc-trench systems. Consequently, causes of the formation and deformation processes of the Japanese offshore basins are rarely discussed. Counter to the general research trend, the Japanese government has been acquiring geological and geophysical data for petroleum exploration for many years. The Ministry of International Trade and Industry (MITI) and its successor the Ministry of Economy, Trade and Industry (METI) have been conducting *MITI/METI Exploratory Survey for Petroleum Resources* programs through several iterations of a *Five-Year Plan for Domestic Oil and Natural Gas Resources*. The database covers the sedimentary basins occupying the shelf around the Japanese islands and primarily consists of data collected from more than 30 offshore exploratory wells, two-dimensional (2D) seismic surveys from an area >100,000 km, and three-dimensional (3D) seismic surveys over approximately 50,000 km^2. This database facilitates discussion of the formation/deformation of the sedimentary basins, because the surveys were originally designed to investigate thick Cretaceous and Cenozoic sediments for petroleum exploration [14]. This study focuses on the sedimentary basins in the eastern offshore regions of Fukushima and Miyagi prefectures, northeastern Japan and uses seismic structural interpretation to investigate the sedimentary and deformation patterns of the basins. These sedimentary and deformation patterns are compared with the already known tectonic framework of the northeast Japan arc and the arc-trench system. The results of this study will contribute to the on-going discussions of the active structures and/or long-term tectonic movement of northeastern Japan and may help elucidate the mechanisms of large earthquakes and disasters. The study results will also provide background knowledge for assessing global environmental variation.

2. Database

Data sets of the MITI/METI seismic survey results and geological information from three MITI exploratory wells were used to assess the tectonic framework of the forearc basins of the northeast Japan arc. The study area encompasses approximately 36,000 km^2 of the shelf adjacent to the Fukushima and Miyagi prefectures, Japan.

2.1. MITI/METI seismic surveys

The Reconnaissance Geophysical Survey "Minami Sanriku-Kashima Oki" (MITI Seismic-MSK) was conducted by MITI in 1986 using the survey vessel *Kaiyomaru*. A survey of approximately 2500 km of 2D seismic lines was performed in the area from Miyagi to the Fukushima Pacific coast using a 15–20 km grid [15].

The Reconnaissance Geophysical Survey "Abukuma Ridge Nanbu 3D" (METI 3D Seismic-ARN) was conducted by METI in 2009 using the 3D geophysical survey vessel *Shigen*.

The line locations for the 2D seismic survey and area covered by the 3D seismic survey are shown in **Figure 1**, with detailed specifications for those seismic surveys provided in **Table 1**.

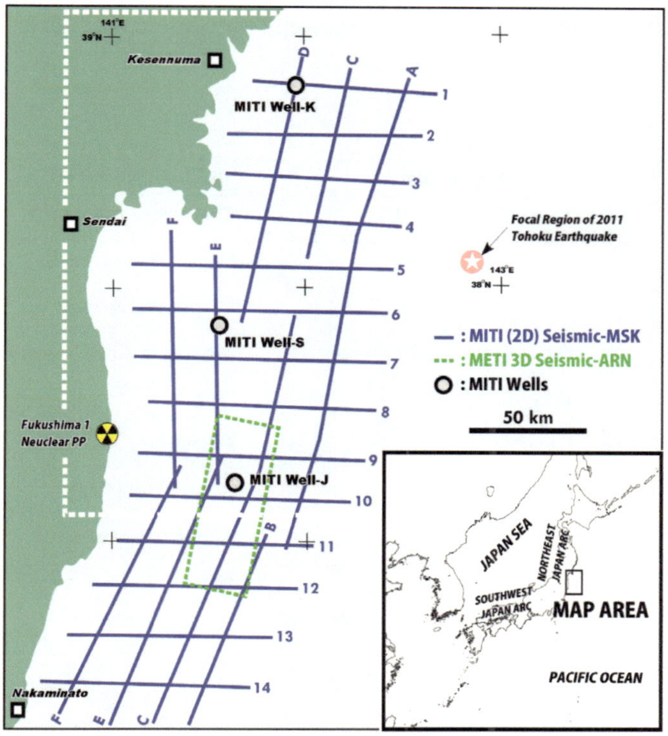

Figure 1. Index map showing the seismic lines, area, and wellsite locations. The blue lines are line location of MITI Reconnaissance Geophysical Survey "Minami Sanriku-Kashima Oki" (MITI Seismic-MSK). The numbers and alphabets indicate line name. The green dashed open box indicates the area of METI Reconnaissance Geophysical Survey "Abukuma Ridge Nanbu 3D" (METI 3D Seismic-ARN). The open circles are the location of MITI Kesennuma-oki well (MITI Well-K), MITI Soma-oki well (MITI Well-S), and MITI Joban-oki well (MITI Well-J).

2.2. MITI exploratory wells

The exploratory well "MITI Kesennuma Oki" (MITI Well-K) was drilled by MITI in 1985, approximately 30 km ESE from Kesennuma City in Miyagi prefecture, Japan. A Cretaceous granitic rock body was observed at a depth of 1843 m below mean sea level (MSL), and drilling was stopped at a depth of 2027 m MSL [16].

The exploratory well "MITI Joban Oki" (MITI Well-J) was drilled by MITI in 1990 approximately 60 km ENE from Onahama City in Fukushima prefecture, Japan. Cretaceous sediments overlain by Quaternary and Neogene strata were observed at a depth of 1345 m MSL, and drilling was continued for approximately 3000 m through the Cretaceous succession to a total depth of 3170 m MSL [17].

The exploratory well "MITI Soma Oki" (MITI Well-S) was drilled by MITI in 1991 approximately 50 km east of Soma City in Fukushima Prefecture, Japan. A Paleogene unit overlain by Quaternary and Neogene successions was observed at a depth of 1625 m MSL, and Cretaceous clastics were observed at a depth of 2450 m MSL, with a total well depth of 3500 m MSL [18].

The locations of these three MITI exploratory wells are shown in **Figure 1**, and the stratigraphic details are listed in **Table 2**.

	MITI Seismic-MSK	**METI 3D Seismic-ARN**
	Two-dimensional (2D)	**Three-dimensional (3D)**
Formal survey name	MITI Reconnaissance	METI Reconnaissance
	Geophysical Survey	Geophysical Survey
	"Minami Sanriku-Kashima Oki"	"Abukuma Ridge Nanbu 3D"
Year acquired	1986	2009
Vessel	Kaiyo Maru	Shigen
Data volume	2500.50 km	1950 km^2 (full fold)
Energy source (size)	Airgun (3223 cu.in.)	Airgun (3090 cu.in. × 2)
Streamer	2400 m	4800 m × 10
Group number	96	—
Fold	48	—
CMP bin size	—	inline: 12.5 m, xline: 25 m
Sample interval	4 ms	4 ms
Record length	5000 ms	8000 ms
Line number or area shape	EW: 14	EW: 25 km
	NS: 6	NS: 80 km

Table 1. Specifications of the MITI seismic surveys.

		MITI Well-K	MITI Well-S	MITI Well-J
Specification	Formal well name	MITI Kesennuma-oki	MITI Soma-oki	MITI Joban-oki
	Owner	Ministry of International Trade and Industry		
	Operator	Teikoku oil	Japex	Teikoku oil
	Spud in date	1984.12.04.	1990.06.21.	1991.07.13.
	Rig released date	1985.02.13.	1990.09.25.	1991.10.17.
	Water depth	240 m	137.2 m	268 m
	Total depth	2027 m	3500 m	3170 m
	Vertical/deviated	Vertical	Vertical	Vertical
	Well head location	38°48′27″ N	37°51′07″ N	37°14′28″ N
		141°58′01″ E	141°34′45″ E	141°38′30″ E
Stratigraphy	Base Quaternary	—	—	402 m
	Base Pliocene	—	525 m	588 m
	Base Upper Miocene	—	582 m	—
	Base Middle Miocene	—	825 m	625 m
	Base Lower Miocene	508 m (base Tertiary)	1562 m	1345 m
	Base Paleogene	—	2450 m	—
	Base Upper Cretaceous	1843 m	(3500 m TD)	(3170 m TD)
	TD horizon	Lower Cretaceous (granitic rock)	Upper Cretaceous	Upper Cretaceous (Turonian)

Table 2. Specifications and stratigraphy of the MITI wells. All depth from mean sea level (MSL).

3. Methods

This study was performed as follows:

(i) Stratigraphic horizons and stratigraphic units were recognized on the seismic profiles from MITI Seismic-MSK based on a general seismic interpretation method (**Table 3**).

(ii) The approximate stratigraphic position of each horizon was designated based on the stratigraphic information from MITI Well-K, Well-S, and Well-J [16–18].

(iii) Notable evidence of deformation was observed and interpreted as tectonic lineations based on planar continuity.

(iv) A discontinuity (an unconformity and its correlative conformity) was recognized in the northwestern part of the study area based on a seismic sequence interpretation method [19, 20].

(v) The relationship between these structural and stratigraphic features and the known characteristics of the adjacent onshore region was assessed.

Seismic unit	Horizon name of unit top	Geological system
A	-	Quaternary
B	Horizon 1	Pliocene
C	Horizon 2	Upper Miocene
D	Horizon 3	Middle Miocene
E	Horizon 4	Lower Miocene
F	Horizon 5	Paleogene
G	Horizon 6	Cretaceous
H	Horizon 7	Paleozoic/Mesozoic/Cretaceous Granite

Table 3. Stratigraphic position of the seismic units and their bounding horizons detected on MITI seismic MSK by using the seismic stratigraphic method.

4. Stratigraphic control of seismic units

Unit H, recognized below 1843 m in measured depth (MD) of MITI Well-K, is granite [16]. An radiometric age determination on the cored granite indicates an Early Cretaceous age, and the rock body is inferred to be correlative to intrusive granites in the South Kitakami Belt consisting of Mesozoic and Paleozoic strata [21].

Unit G is recognized below 2450 mMD of MITI Well-S [18], the interval from 508 to 1843 mMD at MITI Well-K [16], and in the interval from 1345 mMD to the total depth of 3170 mMD at MITI Well-J [17]. Cretaceous calcareous nannnofossils, planktic foraminifer fossils, and radioralian fossils have been reported from the intervals.

Unit F, found in the interval from 1562 to 2450 mMD at MITI Well-S, is correlative to Paleogene based on pollen biostratigraphic analysis. Its coaly lithofacies suggest that Unit F is equivalent to the coal-bearing Paleogene strata outcropping at the adjoining onshore area [18, 22].

From Unit E, recognized in the interval from 625 to 1345 mMD at MITI Well-J and the interval from 82 to 1562 mMD at MITI Well-S, Early Miocene diatom fossils and calcareous nannofossils and radioralian fossils have been reported [17, 18, 22].

Unit D is confirmed in the interval from 582 to 825 mMD and the interval from 588 to 625 mMD at MITI Well-S and Well-J, respectively. The Middle Miocene diatom fossils have been reported from those successions [17, 18].

http://dx.doi.org/10.5772/67391

Unit C, found in the interval from 525 to 582 mMD at MITI Well-S, includes Late Miocene radioralian fossils [18].

Unit B is correlative to Pliocene based on calcareous nannofossil biostratigraphic analysis [17].

Unit A is correlative to Quaternary based on the calcareous nannofossil and diatom biostratigraphic analyses [17, 18].

5. Results

The positions of faults and their direction of extension were estimated using seismic interpretations of the heteromorphic and structural gaps of the traced horizons. Although several smaller fault zones were identified, three relatively large fault zones were designated as tectonic lines.

5.1. The Kesennuma tectonic line

This tectonic line is located in the northern part of the study area and is typically recognized at position 1 on Line 1 (p1-1) of MITI Seismic-MSK (**Figure 2**).

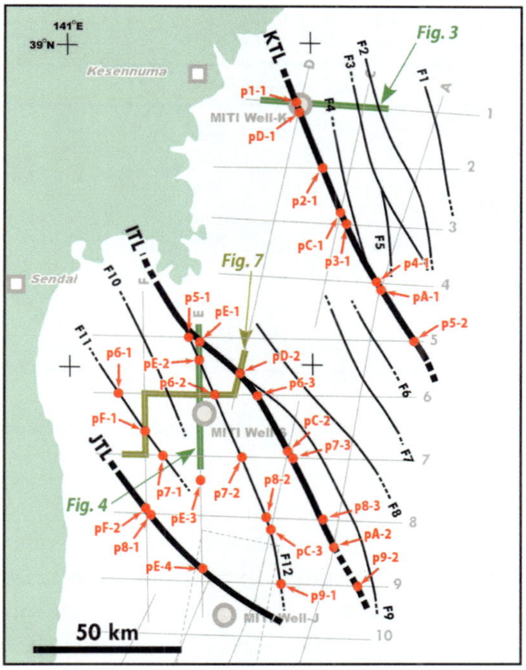

Figure 2. Map showing the positions of faults and deformed strata observed on the seismic sections (red solid circles), and traced lines of the tectonic lines (heavy black lines) and other fault zones (thinner black lines). In addition, the locations of seismic section in **Figures 3**, **4**, and **7** are shown (colored heavy lines).

Horizon 7, the upper limit of Unit H (Cretaceous granite), is recognized in MITI Well-K at a depth of 1843 m MSL (corresponding to two-way travel time = 1.43 s). An apparent west-dipping normal fault and its derived faults are located west of the MITI Well-K location on the Line 1 section (**Figure 3**). A rock unit with a two-way travel time upper limit of 1.5–2.0 s is recognized west of the fault zone, and is characterized by irregularly shaped, low frequency reflectors. The seismic facies characteristics of this unit resemble those of Unit H east of the fault zone. Furthermore, this unit decreases in depth toward a western shoreline and changes into a seismically transparent facies. Paleozoic and Mesozoic strata and granitic rocks of the southern Kitakami Belt are exposed at the surface in the region of extension. Therefore, the upper limit of this rock unit is correlative with Horizon 7, which is the top of Unit H. These spatial relationships were also observed at position pD-1 and indicate an apparent south-dipping normal fault. Therefore, the fault zone consists of a southwest-dipping normal fault with a NNW-SSE trend. The author tentatively designates this fault zone the Kesennuma tectonic line (KTL).

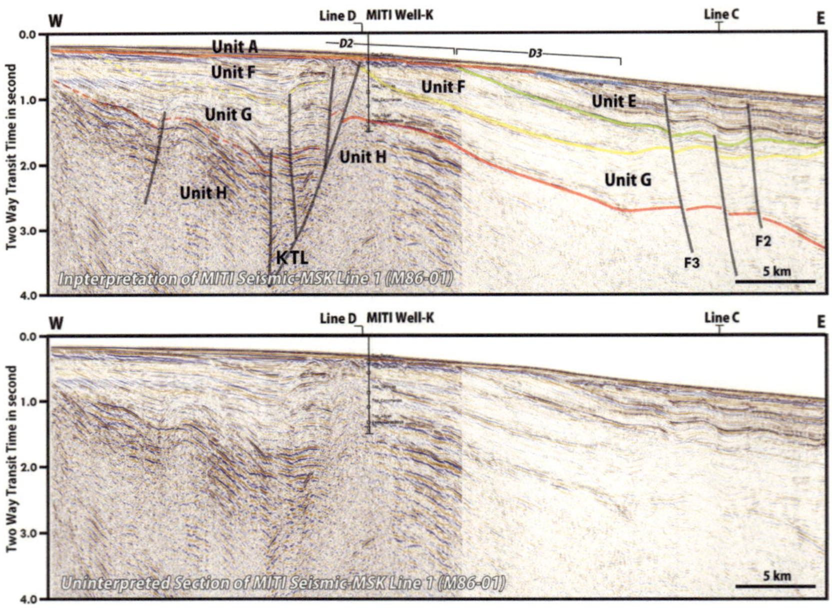

Figure 3. Uninterpreted section of MITI Seismic-MSK Line 1 (below), and its structural and stratigraphic interpretation (above).

The KTL, bounding the eastern edge of the wedge-shaped sedimentary body above Horizon 7 on Line 1, is traceable to positions p2-1, p3-1, and pC-1 with similar fault characteristics. It can also tentatively be traced to positions p4-1, pA-1, and p5-2 because some evidence of deformation is recognized, although the fault characteristics are not an exact match.

Despite a lack of direct stratigraphic control, based on a comprehensive regional seismic stratigraphic interpretation the wedge-shaped sedimentary body may consist largely of Cretaceous, Paleogene, and Lower to Middle Miocene sediments. The Upper Miocene and older strata are deformed by the KTL; however, the deformation does not extend into the overlying Units B (Pliocene) and A (Quaternary).

5.2. Ishinomaki tectonic line

This fault zone is located in the middle and southern parts of the study area. An apparent south-dipping normal fault is recognized at position pE-1 on the north-south Line E (**Figure 2**). In addition, an apparent west-dipping normal fault is observed at position p5-1 crossing the east-west Line 5 (**Figure 4**). These are judged to be different appearances of a single southwest-dipping and NW-SE trending normal fault zone. The author tentatively designates this fault zone the Ishinomaki tectonic line (ITL).

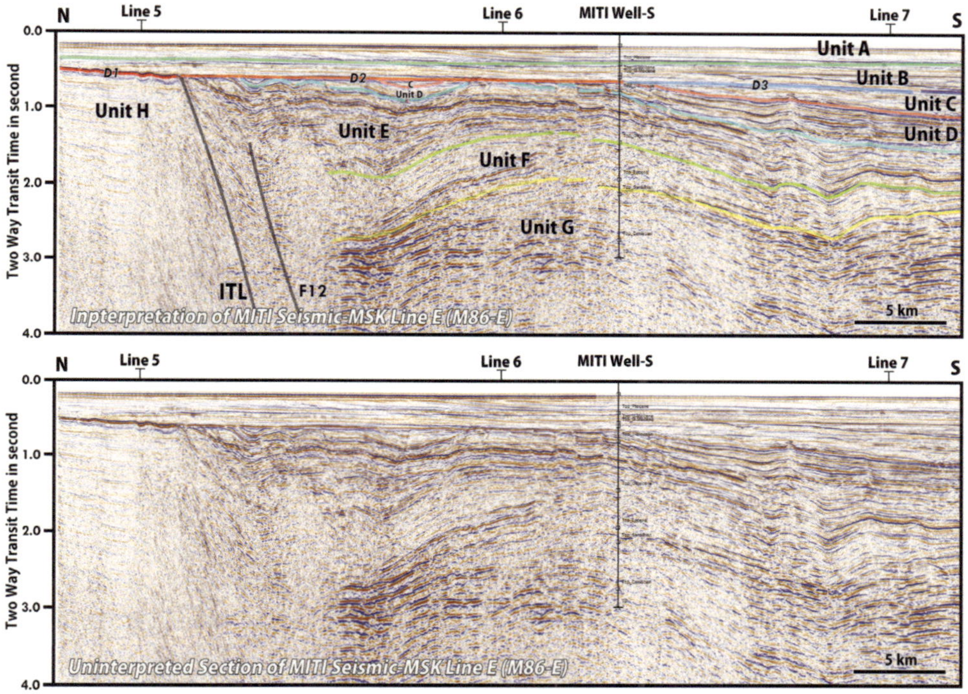

Figure 4. Uninterpreted section of MITI Seismic-MSK Line E (below), and its structural and stratigraphic interpretation (above).

The footwall block of the ITL primarily consists of a seismically transparent layer, estimated to be the seaward extension of the southern Kitakami Belt, composed of Unit H (Paleozoic and Mesozoic strata, and granitic rocks). Horizon 7, the upper limit of Unit H, increases in depth east of position p5-1, and is overlain by the wedge-shaped sedimentary body seen on Lines 1–4. Based on stratigraphic data from MITI Well-S, the hanging wall block of the ITL is interpreted to consist of Units B to G (Paleogene and Neogene, and likely also Cretaceous). The northwestern extension of the ITL is not found at the north end of Line F, although the units making up the hanging wall block are recognized. The southeastern extension of the ITL can be traced to positions pD-2, p6-3, pC-2, p7-3, and p8-3. It is characterized by deformed strata, although the vertical fault displacement is not clearly detected because it passes through the wedge-shaped sedimentary body consisting of Cretaceous and Cenozoic strata seen near MITI Well-K, and then extends southward. F12, a branched fault zone of the ITL, is also identified on the southwest side, avnd is observed at positions pE-2, p6-2, p7-3, p8-2, pC-3, and p9-1. The deformation caused by the ITL does not generally extend into the overlying Units B (Pliocene) and A (Quaternary); however, the northwest portion of the ITL slightly displaces the Pliocene and Quaternary units.

5.3. Joban tectonic line

This fault zone is located in the southwestern part of the study area. An apparent north-dipping normal fault is recognized at position pF-2 on the north-south trending Line F (**Figure 2**). A flexural structure of Unit E (Lower Miocene), differentiated from the overburden, is found at position p8-1. Based on the spatial relationships, this fault zone is an apparent northeast-dipping normal fault, with a left-lateral strike-slip sense. The southeastern extension of this fault zone is found at position pE-4. The author tentatively designates this fault zone the Joban tectonic line (JTL).

F11 is another fault zone similar to the JTL, and is recognized at positions p6-1, p7-1, and pF-1. Unit C (Upper Miocene) and older strata are deformed by the JTL; however, the deformation does not extend into the overlying Units B (Pliocene) and A (Quaternary). A set of the JTL and F11 faces the ITL across the graben that occupies Sendai Bay, and shows a NW to SE extensional trend.

5.4. Other faults

Fault F4, derived from the KTL, is found on the northeast side of the KTL. Faults F1, F2, F3, and F5 are recognized northeast of the KTL and F4, and are characterized by a direction and dip similar to the KTL and F4. Faults F6, F7, F8, and F9 are found in the area between the ITL and the KTL/F11. F6 and F7 are northeast-dipping normal faults. In contrast, F8 and F9 are southwest-dipping similar to the ITL. F10 is found within the graben between the ITL and the JTL/F11. All of these fault zones and the three tectonic lines show a NW-SE to NNW-SSE extensional trend.

5.5. Discontinuity

As a distinctive feature of basin development, a discontinuity is recognized in the seismic sections in the northwestern and western near-coast part of the study area. The strata above the discontinuity are Units B (Pliocene) and A (Quaternary). This discontinuity is subdivided into three zones based on the ages of the underlying strata and seismic sequence configurations (**Figure 5**).

The zone where eroded Unit H (Paleozoic, Mesozoic, and granitic rocks) unconformably underlies Unit B (Pliocene) is designated D1 (**Figure 6**). D1 is discernible at the western ends of Lines 2, 3, 4, 6, and 7 by an irregular or rough surface and its undersurface truncation, which is a type of reflection termination (Vail [19]). The eroded top surfaces of horsts are unconformably overlain by Unit B (Pliocene) at the western end of Lines 5 and 6 and the northern end of Line F (**Figure 7**).

http://dx.doi.org/10.5772/67391

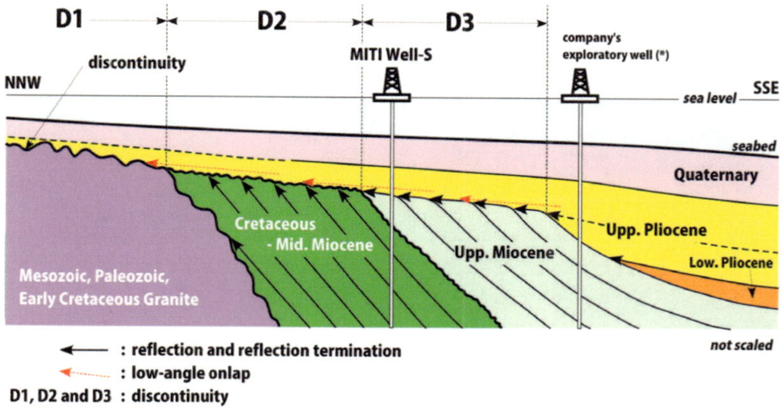

Figure 5. Schematic dip sectional image of the discontinuity and its subdivision showing stratigraphic relationship between its undersurface and overlying strata.

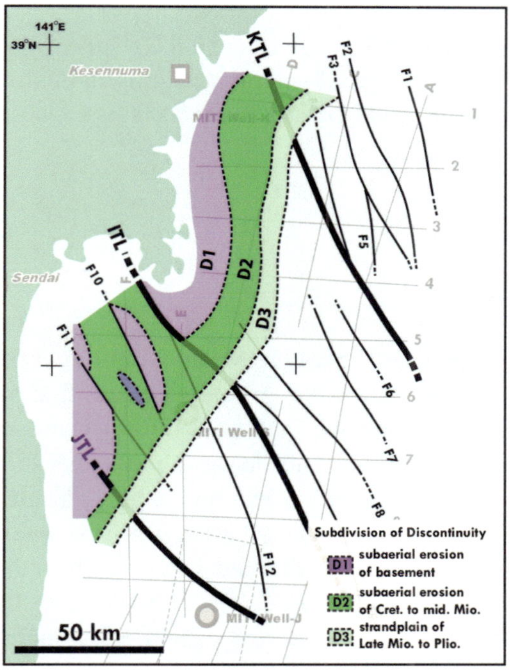

Figure 6. Map showing the distribution of discontinuity and its subdivisional zones.

The zone where eroded units G (Cretaceous) and F (Paleogene) underlie Unit B (Pliocene) is designated D2. D2 is found in the areas east and southeast of D1, and can be detected by low angle onlaps on a relatively smooth surface, its undersurface truncation, and east- or southeast-dipping reflections.

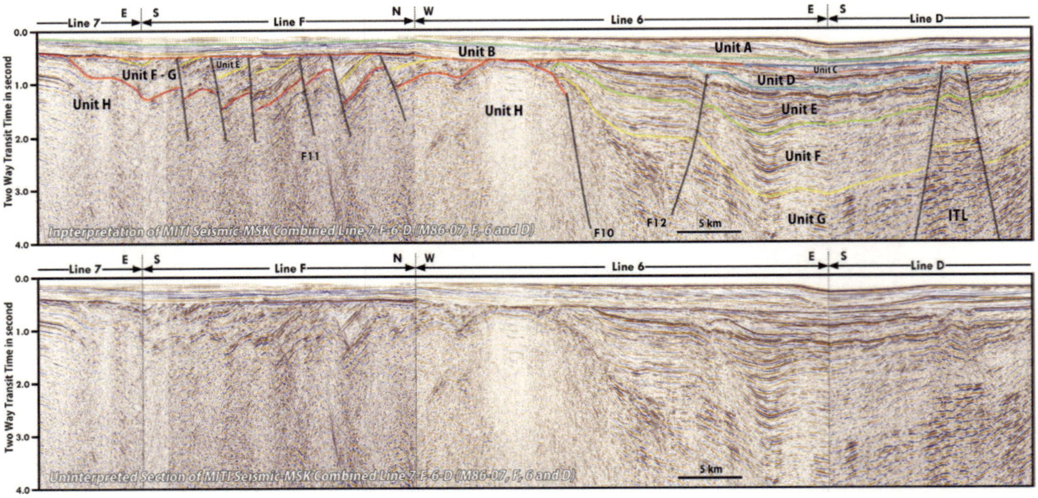

Figure 7. Uninterpreted composite Section of MITI Seismic-MSK Lines 7, F, 6, and D (below), and its structural and stratigraphic interpretation (above).

The zone where the top surface of the prograding Unit C (Upper Miocene) is overlain by Unit B (Pliocene) is designated D3. There is no seismic stratigraphic evidence of erosion in D3, and undersurface toplap of east- or southeast-dipping reflections are observed.

6. Discussion

6.1. Strike of the tectonic lines

The KTL, ITL, JTL and other fault zones are observed in the 2D MITI Seismic-MSK data acquired using an approximately 15–20 km grid. This means that the existence of a fault or deformed strata cannot be confirmed within the grid boxes lacking seismic lines. This is common for regional 2D seismic data sets, and does not mean that they are not useful for the interpretation of fault strike. With sufficiently high intensity, length, and a relatively stable extension direction, a tectonic line identified on a single seismic line may extend to any position on three other seismic lines surrounding the grid box. Thus, the observed deformation points are located on the lines extending NW-SE to NNW-SSE. This interpretation of the strike direction of the tectonic lines is supported by marine geological surveys and an active fault survey as described below.

Marine geological maps covering the northwestern part of the study area were compiled using a high resolution and high density shallow 2D seismic data set acquired by the Geological Survey of Japan. A shallow part of the northwestern portion of the ITL is indicated on the map off "Kinkazan" as the Ishinomaki Bay Fault, which extends NNW-SSE [23]. A shallow, northwestern portion of KTL is shown on the map off "Kamaishi", which also extends NNW-SSE [24].

Tohoku Electric Power Co., Inc. conducted the active fault survey around the northern part of Sendai Bay using high resolution and high density shallow 2D seismic data set. The results of the survey are documented in handouts from a meeting held by the Nuclear Regulation Authority of Japan on April 16, 2014 [25]. The handout suggests that a NNW-SSE trending fault system identified by the survey may correspond to the Ishinomaki Bay Fault.

6.2. Sense of the tectonic lines

The tectonic lines are estimated to have a relatively straight extensional trend. However, the sense of displacement is not as constant. For example, a tectonic line that appears as a normal fault with relatively large vertical displacement in one area can be identified in other area as local deformation of strata without vertical displacement. Generally, a fault with a vertical component of displacement is relatively recognizable on a 2D seismic profile compared with strike-slip or wrench faults lacking vertical displacement, because seismic surveys use vertical elastic waves and measure their reflections.

The following model is proposed to further the discussion (**Figure 8 (1)**):

(i) there is a syncline located west of a north-south trending anticline,

(ii) a northwest-southeast trending sinistral strike-slip fault cuts both the syncline and anticline diagonally,

(iii) the fault plane dips southwest,

(iv) the displacement of the strike-slip fault is half of the wavelength of the syncline-anticline cycle, and

(v) some east-west 2D seismic lines are available in this area.

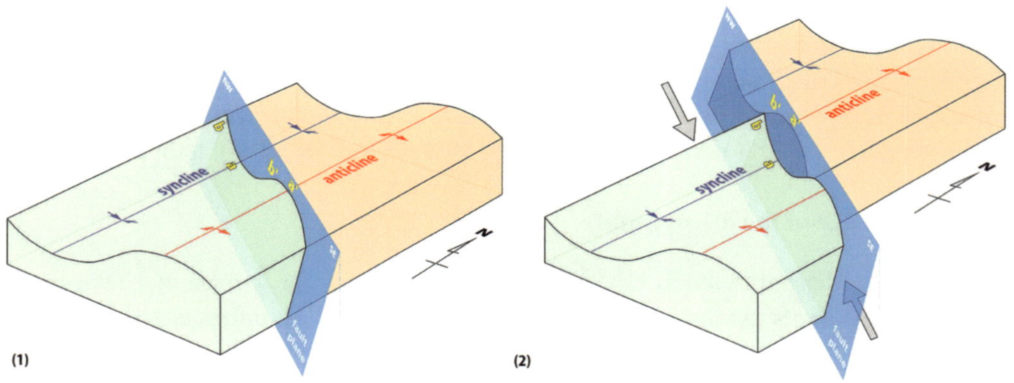

Figure 8. Conceptual model illustrating a north-south trending syncline and anticline set cut by a diagonal southwest dipping sinistral strike-slip fault.

Within this model, because the southwest block of the fault slips toward the southeast, the axis of the syncline (a) may adjoin the axis of the anticline (a′) on the northeast block (**Figure 8 (2)**). The fault will appear as a west-dipping normal fault with significant vertical displacement on the east-west seismic line at this position (**Figure 9**).

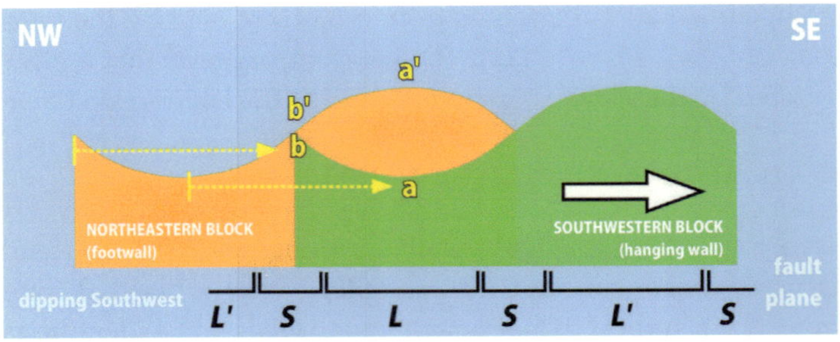

Figure 9. The fault plane view of the conceptual model in **Figure 8** showing the positional relationship between the northeastern block and southwestern block after sinistral strike-slip. [L] and [L'] represent the area with apparent significant vertical displacement, and [S] represents the area with less vertical displacement. Furthermore, [L] will appear apparent southwest-dipping normal fault and [L'] will be reverse fault.

In contrast, with the same fault activity, the western flank of the syncline (b) on the southwest block may be adjacent to the western flank of the anticline (b') on the northeast block. The less vertically displaced deformation will be observed in the seismic section at this position.

This model suggests that the strike-slip sense causes tectonic lines to appear as an apparently normal fault, with vertical displacement in one area and as deformed strata lacking vertical displacement in other areas. Flower-like faulting and local folding without significant vertical displacement are frequently observed on the lines of the MITI Seismic-MSK survey. With this evidence, it is conceivable that the tectonic lines in the study area may have a strike-slip sense. If it is possible to acquire enough data, based on the above-mentioned model the horizontal displacement and direction of movement should be discussed in detail. However, the 2D MITI Seismic-MSK survey, with its rough grid data, is not worthy of further sophisticated discussion.

In contrast, 3D seismic data can be used for such discussions if the conditions are suitable. For example, the so-called "Abukuma Ridge" [26], lying subsurface along the edge of the continental shelf in the study area, is cut by the southeastern extension of the JTL. Based on METI 3D Seismic-ARN data this deformation is caused by a sinistral strike-slip fault [27].

6.3. Formation age of the discontinuity

D2, part of the discontinuity, is characterized by an eroded flat surface of Cretaceous to Middle Miocene strata overlain by Pliocene units with a low angle onlap. Based on a microbiostratigraphic study using a company's exploratory well samples, this surface at position pE-3 is covered by Upper Pliocene sediments and lacks the Lower Pliocene units [28]. This suggests that the low angle onlap above D2 must be a coastal onlap at the base of the Upper Pliocene, indicating a landward shift of the coastline (transgression) in the Late Pliocene.

The landward portion of the discontinuity, D1, is characterized by an eroded undulating surface of Paleozoic to Mesozoic sediments or granitic rocks overlain by Upper Pliocene sediments. This surface is interpreted as a drowned coast caused by further transgression in the Late Pliocene. This interpretation is supported by the landward connection of D1 to the recent Sanriku coast ria.

The basinward portion of the discontinuity, D3, is characterized by toplap or asymptotic truncation of Upper Miocene strata overlain by low angle onlap of Upper Pliocene sediments. This surface is relatively flat, and in some areas, lowers stepwise basinward. This suggests that D3 was formed as a Late Miocene strand plain. In the eastern and southeastern basinward directions, D3 is correlative with Horizon 2 at the top of Upper Miocene. Farther southeast, the Lower Pliocene conformably overlies the Upper Miocene [29]. Therefore, the relative sea level fall that formed the discontinuity, which consists of D1, the D2 erosional surface, and the D3 strand plain is estimated to have occurred during the Middle Pliocene.

Upper Miocene strata with a vertical thickness of 57 m were detected in the interval between depths 525 and 582 m MSL below D2 at MITI Well-S, and are unconformably underlain by Middle Miocene units [18]. Based on the stratigraphic data from the well drilled near position pE-3 this interval corresponds to the lower portion of the Upper Miocene [22, 28]. The unconformity between the Middle Miocene and the lower part of the Upper Miocene is also truncated by the erosion that occurred on the D2 surface. This stratigraphic relationship indicates that the erosion began after the deposition of the lower part of the Upper Miocene strata. Therefore, the discontinuity is interpreted to have formed between the early Late Miocene and Late Pliocene.

6.4. Active age of the tectonic lines

The discontinuity is not folded or faulted by the KTL or the JTL/F11. In addition, the ITL does not deform the discontinuity and its overlying sediments except in the Sendai Bay area. The strata above the discontinuity are generally flat and stable despite the existence of tectonic lines like the KTL, JTL, and ITL, as well as the undulating topography of the discontinuity itself. This likely indicates that the most active period of the tectonic lines had ended before the formation of the discontinuity, terminating before the Late Miocene. Furthermore, based on the interpretation of the METI Seismic ARN data, the Lower Miocene strata are cut at the Abkukuma Ridge by the southeast portion of the JTL [27]. Therefore, it is inferred that the tectonic lines were most active at least during the early Middle Miocene.

Those tectonic lines resemble generally Tanakura Tectonic Line (TTL), Hatakawa Tectonic Line (HTL), or Futaba Fault (FF) in sense [3, 4], which are the most important tectonic elements of the southern part of northeastern Japan. The activity of TTL had been active until the Middle Miocene; however, the activity is devided into several stages after the Middle Cretaceous. TTL is considered to connect the east margin of the Japan Sea Transform (JET) bounding the eastern limit of spreading area of the northern Japan Sea [4]. The Middle Miocene is the age of final stage for the opening of Japan Basin and Yamato Basin [6, 7]. The characteristics and sense of KTL, ITL, and JTL should be considered from a wider viewpoint of large scale tectonic framework around the northeastern Japan and northern Japan Sea.

6.5. Subsurface graben in Sendai Bay

The ITL is a fault with an apparent southwest dip in the Sendai Bay area. In contrast, the JTL/F11 is an apparent northeast-dipping fault. Therefore, the area bounded in the northeast by the ITL and in the southwest by the JTL/F11 is a graben that extends NNW, landward to the Kitakami Lowland that is characterized by a graben-like topographic feature [30]. Based on the active age of the boundary-limiting tectonic lines, this graben is inferred to have formed prior to the Middle Miocene.

The ITL does not generally deform the discontinuity and its overburden in the study area; however, it does displace shallow sediments within the Sendai Bay area. This deformation is possibly due to an active structure, because the deformed overburden consists of Upper Pliocene and Quaternary sediments. The footwall of the ITL in the deformed area is interpreted to consist of rigid rocks within the Southern Kitakami Belt, such as Mesozoic sediments, Paleozoic sediments, or granitic rocks. This suggests that the Sendai Bay portion of the ITL is reactivated, or that the relatively unrigid Cretaceous to Middle Miocene hanging wall strata are differentially compacted.

The MITI Seismic-MSK data show an upward swelling of the hanging wall block along the tectonic line that seems to indicate the occurrence of inversion tectonics during the Pliocene or Pleistocene. Although no displacement seems to reach the ocean bottom, because the Pliocene and Pleistocene deposits are thin their internal seismic facies and reflection termination patterns are not easily observed or recognized. The existence of several multiples beneath the ocean bottom further complicates the proper observation of seismic characters in this series of seismic data.

The Ishinomaki Bay Fault (the northwestern shallow portion of the ITL) is a reverse fault that cuts the discontinuity and reaches the overlying uAK supersequence that is roughly correlative with the Upper Pliocene; however, no evidence for the deformation of the Pleistocene AK2r subsequence has been noted [23]. Based on the interpretation of the Tohoku Electric Power survey results, some active faults displace not only the discontinuity and its overlying Pliocene strata, but also influence the lower part of the Pleistocene sediments.

Based on the evidence presented, the deformation of the discontinuity and its overlying sediments is inferred to reflect inversion tectonics, with a reactivation of the ITL.

7. Conclusion

The tectonic deformation history of the offshore regions of the Fukushima and Miyagi prefectures, Japan, was studied using MITI Seismic-MSK, METI 3D Seismic-ARN, and MITI well data. Three major tectonic lines and several medium-scale fault zones were recognized in the study area.

Despite the limitations of the large grids of the 2D seismic configuration, based on a general interpretation method the strike direction of the tectonic lines is inferred to be NNW-SSE to NW-SE.

The sense of the tectonic lines is not constant for each line. This can occur in the case of oblique cuts of anticlinal-synclinal trends by a strike-slip fault. In this case a sinistral strike-slip fault displaces the axis of the Abukuma Ridge, as determined from the METI 3D Seismic-ARN data.

http://dx.doi.org/10.5772/67391

Based on this investigation, the most active period of the tectonic lines is estimated to be after the Early Miocene, during which time Abukuma Ridge was formed. The activity ceased prior to the formation of the discontinuity that eroded the strata deformed by the activity of the tectonic lines. The discontinuity is interpreted to have formed between the early Late Miocene to Late Pliocene. Therefore, the tectonic lines must have been active during the early Middle Miocene at least. Based on the stratigraphic interpretation of MITI Seismic-MSK sections, an angular unconformity between the Lower Miocene succession and its overlying Middle Miocene is recognized in the part of this study area. The sedimentary basin including this study area has a two-story layered configuration bounded by the top Lower Miocene unconformity. The second story is obviously the succession in the forearc basin of the island arc after drifted to its current position. The lower story, in contrast, may represent succession within a basin of a continental margin arc in front of the Eurasian Continent. Not only the top Lower Miocene unconformity but also the tectonic lines detected in this study must have close relationship with such major tectonic deformation process of East Asia.

The tectonic lines bounding the subsurface graben in the Sendai Bay Area affected the deformation of the discontinuity and its overlying strata. The deformation must therefore be related to Quaternary inversion tectonics with a reactivation of the ITL. This graben is inferred to extend landward to the Kitakami Lowland.

Acknowledgements

The author wishes to express his gratitude to the Ministry of Economy, Trade and Industry (METI) and the Japan Oil, Gas and Metals National Corporation (JOGMEC), who provided their exploration data for use in this study and permitted publication.

Author details

Hiroyuki Arato

Address all correspondence to: h_arato@gipc.akita-u.ac.jp

Graduate School of International Resource Sciences, Akita University, Japan

References

[1] Otofuji Y, Matsuda T. Paleomagnetic evidence for the clockwise rotation of Southwest Japan. Earth and Planetary Science Letters 1983; 62: 349-359.

[2] Ehiro M, Okami K, Kanisawa S. Recent progress and further subjects in studies on the "Hayachine Tectonic Belt" in the Kitakami Massif, Northeast Japan. Earth Science 1988; 42: 317-335.

[3] Otsuki K, Ehiro M. Cretaceous left-lateral faulting in Northeast Japan and its bearing on the origin of geologic structure of Japan. Journal of Geological Society of Japan 1992; 98: 1097-1112. (in Japanese with English abstract)

[4] Yanai S, Aoki K, Akahori Y. Opening of Japan Sea and major tectonic lines of Japan: MTL, TTL and Fossa Magna. Journal of Geography 2010; 119: 1079-1124 (in Japanese with English abstract)

[5] Takahashi Y, Mao J, Zhao X. Timing of mylonitization in the Nihonkoku Mylonite Zone of north Central Japan: Implications for Cretaceous to Paleogene sinistral ductile deformation in the Japanese Islands. Journal of Asian Earth Sciences 2012; 47: 265-280.

[6] Tamaki K. Geological structures of the Japan sea and its tectonic implications. Bulletin of the Geological Survey of Japan 1988; 39: 269-365.

[7] Jolivet L, Tamaki K. Neogene kinematics in the Japan Sea region and volcanic activity of the northeast Japan arc. In: Tamaki K, Suyehiro K, Allan J, MacWilliams M. editors. Proceedings of the Ocean Drilling Program, Scientific Results 127/128 Part 2 1311-1331; College Station, Texas (Ocean Drilling Program); 1992.

[8] Tamaki K, Suyehiro K, Allan J, Ingle Jr J C, Pisciotto K A. Tectonic synthesis and implications of Japan Sea ODP drilling. In: Tamaki K, Suyehiro K, Allan J, MacWilliams M. editors. Proceedings of the Ocean Drilling Program, Scientific Results 127/128 Part 2 1333-1348; College Station, Texas (Ocean Drilling Program); 1992.

[9] Itoh Y, Uno K, Arato H. Seismic evidence of divergent rifting and subsequence deformation in the southern Japan Sea, and a Cenozoic tectonic synthesis of the eastern Eurasian margin. Journal of Asian Earth Sciences 2006; 27: 933-942.

[10] Itoh Y, Tsutsumi H, Yamamoto H, Arato H. Active right-lateral strike-slip fault zone along the southern margin of the Japan Sea. Tectonophysics 2002; 351: 301-314.

[11] Okamura Y. Relationships between geological structure and earthquake source faults along the eastern margin of the Japan Sea. Journal of Geological Society of Japan 2010; 116: 337-353.

[12] Takahashi M, Saito K. Miocene intra-arc bending at an arc-arc collision zone, central Japan. The Island Arc 1997; 6: 168-182.

[13] Takahashi M. Tectonic boundary between Northeast and Southwest Japan Arcs during Japan Sea opening. Journal of the Geological Society of Japan 2006; 112: 14-32 (in Japanese with English abstract)

[14] Arato H. The future aspect of energy resource development: What do we learn after exploration and development histories of the oil and gas fields in Akita province? Journal of the Japanese Association for Petroleum Technology 2016; 81: 271-280 (in Japanese with English abstract)

[15] Kanamori K. Part I. Comprehensive Outcomes of Phases 5 and 6 Five Year Plans for Domestic Oil and Natural Gas Resources Development. In: Japan Natural Gas Association, Japan Offshore Petroleum Development Association, editors. Petroleum Resources of Japan, Revised Version 1992; 3-34 (in Japanese)

[16] Sato K. 3.6.4 MITI "Kesennuma Oki" Well. In: Japanese Association of Petroleum Technology, editor. Recent Japanese Petroleum Exploration and Development 1993; 133-136 (in Japanese)

[17] Sato K. 3.6.6 MITI "Joban Oki" Well. In: Japanese Association of Petroleum Technology, editor. Recent Japanese Petroleum Exploration and Development 1993; 133-136 (in Japanese)

[18] Minowa H. 3.6.5 MITI "Off Soma" Well. In: Japanese Association of Petroleum Technology, editor. Recent Japanese Petroleum Exploration and Development 1993; 141-144 (in Japanese)

[19] Mitchum RM Jr, Vail P R, Sangree J B. Seismic stratigraphy and global changes of sea level, Part 6: Stratigraphic interpretation of seismic reflection patterns in depositional sequences. In: Payton C E, editor. Seismic Stratigraphy - application to hydrocarbon exploration. Memoir of American Association of Petroleum Geologists 1977; 26: 117-133.

[20] Arato H, Takano O. Significance of sequence stratigraphy in petroleum exploration. In: Saito Y, Hoyanagi K, Ito M, editors. Sequence Stratigraphy - Toward a New Dynamic Stratigraphy. The Memoirs of the Geological Society of Japan 1995; 45: 43-60 (in Japanese with English abstract)

[21] Shibata K. Isotopic ages of granit core from the MITI-Kesennuma Oki Well. Bulletin of the Geological Survey of Japan 1986; 37: 467-470.

[22] Kato S, Tazawa K, Nakano T. Cenozoic strontium isotope stratigraphy in the MITI Soma-oki well, northeastern Japan. Journal of the Geological Society of Japan 1997; 103: 1046-1052. (in Japanese with English Abstract)

[23] Nakamura, K. Geological map off Kinkazan. Marine Geological Map Series 33: Geological Survey of Japan, 1990.

[24] Okamura Y, Nishimura K, Tanahashi M. Explanatory notes of geological map off Kamaishi. Marine Geology Map Series 22, Geological Survey of Japan 1983; 25 p (in Japanese with English abstract)

[25] Meeting Handouts for the Nuclear Regulation Authority of Japan [Internet]. 2014. Available from: https://www.nsr.go.jp/disclosure/committee/yuushikisya/tekigousei/power_plants/h26hy/20140616.thml [Accessed: 2016-12-01].

[26] Iwazaki T. Part IV, Chapter 2. Pacific Coast Northesat Japan. In: Japan Natural Gas Association, Japan Offshore Petroleum Development Association, editor. Petroleum Resources of Japan, Revised Version 1992; 192-211 (in Japanese)

[27] Chiyonobu S, Arato H. Fault geometry and its characteristics in the southern part of Abukuma ridge, offshore Fukushima Prefecture, Japan [Internet]. 2015. Available from http://www2.jpgu.org/meeting/2015/session/PDF/S-GL37/SGL37-P03_E.pdf [Accessed: 2016-12-01]

[28] Kameo K, Sato T. Recent development of calcareous nannofossil biostratigraphy and its application- Neogeone and Quaternary stratigraphy of offshore Joban based on calcareous nannofossils. Journal of the Japanese Association for Petroleum Technology 1999; 64: 16-27 (in Japanese with English abstract)

[29] Iwata T, Hirai A, Inaba T, Hirano M. Petroleum system in the Offshore Joban Basin, northeast Japan. Journal of the Japanese Association for Petroleum Technology 2002; 67: 62-71 (in Japanese with English abstract)

[30] Minoura K. The tectonic development of northeast Japan. Journal of the Geodetic Society of Japan 1985; 31: 97-105 (in Japanese with English abstract)

Section 2

Special Section: Median Tectonic Line - Multidisciplinary Dissection of a Longstanding Crustal Break on the Eurasian Margin

Prologue

Yasuto Itoh

Additional information is available at the end of the chapter

At the beginning of the special section, "Median Tectonic Line: Multidisciplinary Dissection of a Long-Standing Crustal Break on the Eurasian Margin," the author attempts to provide readers with a tectonic overview of the world's largest class of intraplate faults. Figure 1 summarizes the regional tectonic context and geologic features together with study areas of the contributors.

The incipient stage of the Median Tectonic Line (MTL) dates back to the Cretaceous, when a remarkably rapid northward motion of the Izanagi Plate provoked the sinistral wrenching and eventual breakup of the eastern Eurasian margin [1]. At that time, the MTL was connected to the Central Sikhote-Alin fault to the north and constituted a regional transcurrent fault [2]. The propagating termination of the MTL was the site of a pull-apart basin formation that was buried by an enormous amount of clastic sequences of the Izumi Group [3].

During most of the Paleogene, the MTL is considered to have been dormant because basin formation along the fault became inactive. Exceptionally, rapid back-arc opening in the Japan Sea during the early Miocene resulted in the reduction of the structure, after which the MTL acted as an arc-bisecting fault of southwest Japan. The episodic rifting and rapid migration of the island arc provoked the formation of marine sedimentary basins along the fault [4].

Since the end of the Miocene, the activity of the MTL has been governed by the convergent mode of the Philippine Sea Plate. The Pliocene was heralded by an intensive inversion event related to the north–south compressive regime [5]. Although the crustal shortening is most obviously observed on the back-arc shelf, coeval contraction features can also be identified on the forearc shelf [6]. Watershed mountain ranges emerged on the northern bank of the MTL, as was clarified by provenance studies of basin-filling coarse clasts [7, 8]. The Philippine Sea Plate shifted its converging direction counterclockwise at c.a. 2 to 1 Ma [9] and urged the MTL to produce frequent dextral slips.

These long-standing tectonic episodes are closely linked to remarkable differences in the evolutionary processes of geologic terranes on both sides of the fault. Its structural features are superficially self-contradictory, reflecting temporal fluctuations in tectonic stress regimes and resultant overlaid image of crustal failures. It is also marked by an anomalously broad damage zone accompanied by the active migration of formation fluids.

Multidisciplinary research is necessary to obtain a comprehensive view of such a complex development of an arc-dividing fault. In this special section, the three-dimensional architecture of the fault is presented by Itoh et al. (Chapter 3) and Kusumoto et al. (Chapter 4) based on a reflection seismic survey and gravity analysis combined with numerical modeling, respectively. The thermal history around the MTL is discussed by Itoh et al. (Chapter 5) through the thermochronological analysis of apatite and zircon grains obtained from sedimentary rocks (Cretaceous Izumi Group) that buried the fault-related pull-apart basin. Then, the elemental transition related to the fluid circulation along the brecciated fault damage zone during the earliest Paleogene is described by Kaneko et al. (Chapter 6) by means of a geochemical study. Finally, the timeline of tectonic deformation on the fault is presented by Itoh and Iwata (Chapter 7) based on a detailed structural analysis of the Izumi Group.

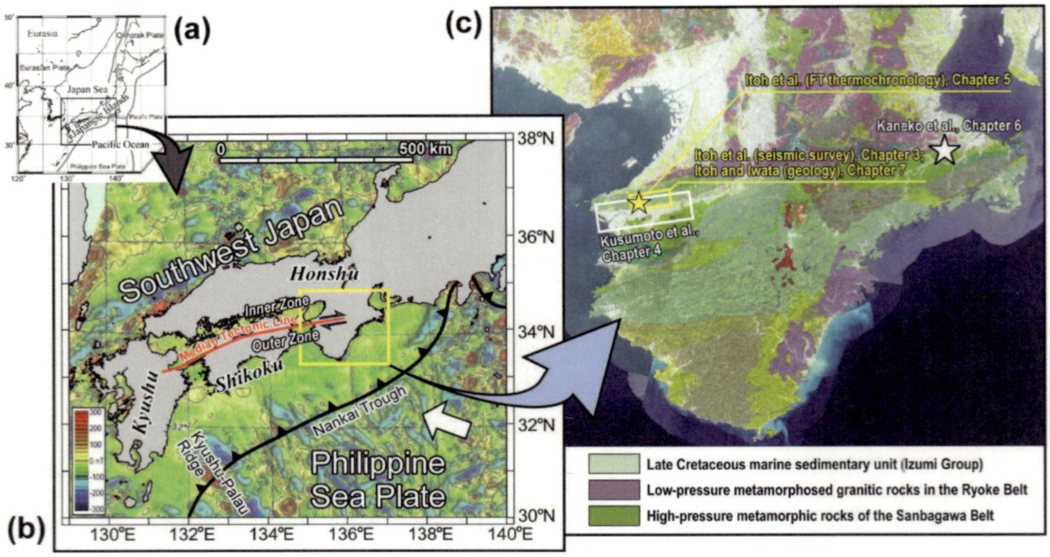

Figure 1 (a, b) Regional tectonic context and (c) geologic features together with the study areas of the contributors. The base maps of (b) the geomagnetic anomaly and (c) the geology are adapted from Ref. [10] and [11], respectively.

Author details

Yasuto Itoh[1]*, Tomotaka Iwata[2] and Keiji Takemura[3]

*Address all correspondence to: yasutokov@yahoo.co.jp

1 Graduate School of Science, Osaka Prefecture University, Osaka, Japan

References

[1] Fitch TJ. Plate convergence, transcurrent faults, and internal deformation adjacent to southeast Asia and the western Pacific. Journal of Geophysical Research. 1972; 77: 4432-4460

[2] Itoh Y, Uno K, Arato H. Seismic evidence of divergent rifting and subsequent deformation in the southern Japan Sea, and a Cenozoic tectonic synthesis of the eastern Eurasian margin. Journal of Asian Earth Sciences. 2006; 27: 933-942

[3] Noda A, Toshimitsu S. Backward stacking of submarine channel-fan successions controlled by strike-slip faulting: The Izumi Group (Cretaceous), southwest Japan. Lithosphere. 2009; 1: 41-59

[4] Itoigawa J. Miocene series in the Setouchi Province - Overview. In: Yamashita N, Kaseno Y, Itoigawa J, editors. Regional Geology of Japan Part 5 Chubu II. Tokyo: Kyoritsu Shuppan Co., Ltd.; 1988. p. 116-117

[5] Itoh Y, Nagasaki Y. Crustal shortening of Southwest Japan in the Late Miocene. The Island Arc. 1996; 5: 337-353

[6] Takano O, Nishimura M, Fujii T, Saeki T. Sequence stratigraphic distribution analysis of methane-hydrate-bearing submarine-fan turbidite sandstones in the eastern Nankai Trough area: relationship between turbidite facies distributions and BSR occurrence. Journal of Geography. 2009; 118: 776-792

[7] Oka Y. The formation of the Izumi range and the Osaka Group. The Quaternary Research. 1978; 16: 201-210

[8] Oka Y, Sangawa A. The formation of the sedimentary basin in the east of Inland Sea and the uplift of the Awaji Island, Japan. Journal of Geography. 1981; 90: 393-409

[9] Nakamura K, Renard V, Angelier J, Azema J, Bourgois J, Deplus C, Fujioka K, Hamano Y, Huchon P, Kinoshita H, Labaume P, Ogawa Y, Seno T, Takeuchi A, Tanahashi M, Uchiyama A, Vigneresse J-L. Oblique and near collision subduction, Sagami and Suruga Troughs - preliminary results of the French-Japanese 1984 Kaiko cruise, Leg 2. Earth and Planetary Science Letters. 1987; 83: 229-242

[10] Geological Survey of Japan, AIST, CCOP (Coordinating Committee for Coastal and Offshore Geoscience Programmes in East and Southeast Asia), editors. Magnetic Anomaly Map of East Asia 1:4,000,000 CD-ROM Version (2nd Edition), Digital Geoscience Map P-3. Tsukuba: Geological Survey of Japan, National Institute of Advanced Industrial Science and Technology; 2002

[11] Geological Survey of Japan, AIST, editor. Seamless Digital Geological Map of Japan 1:200,000 (July 3, 2012 Version), Research Information Database DB084. Tsukuba: Geological Survey of Japan, National Institute of Advanced Industrial Science and Technology; 2012

Chapter 3

Three-Dimensional Architecture of the Median Tectonic Line in Southwest Japan Based on Detailed Reflection Seismic and Drilling Surveys

Yasuto Itoh, Tomotaka Iwata and Keiji Takemura

Additional information is available at the end of the chapter

http://dx.doi.org/10.5772/67434

Abstract

The subsurface morphology of an arc-bisecting tectonic zone has been unraveled by means of well-organized seismic investigation with the aid of borehole stratigraphic control. The Median Tectonic Line (MTL) active fault system in southwestern Japan, one of the world's largest intraplate transcurrent faults, is driven by the recent oblique subduction of the Philippine Sea Plate. Six tied seismic profiles covering the mountainous range, the southern foothill of which is truncated by the MTL, were used to visualize the Quaternary basins on both feet of the Izumi Mountains. North- and east-trending basement deformation was confirmed on the northern and southern sides of the watershed, respectively; this deformation reflects the spatial diversity in tectonic stress. Seismic data on the southern Izumi flank revealed a low-angle fault parallel to the MTL active fault system; this fault may be interpreted as a dormant structure that developed from 6 to 2 Ma under the intermittent increases of the compressive regime. A kink zone in the upthrown block of the thrust was identified on seismic profiles and continuously traced through field geological survey. This zone confirms the prevailing contractional phase related to the transient convergence mode of the oceanic plate.

Keywords: active fault, seismic survey, borehole, southwest Japan, Median Tectonic Line

1. Introduction

On convergent margins, the deformation and rearrangement of island arcs are often provoked by the oblique subduction of oceanic plates [1]. Located on the eastern Eurasian margin, the

Japanese Archipelago has been a site of vigorous tectonic events for more than hundreds of millions of years. Huzita contended that recent crustal deformation has been accompanied by frequent motions on basin-fringing active faults in his pioneering papers [2, 3]. He placed an emphasis on the role of the Median Tectonic Line (hereafter abbreviated as MTL), dextral motions on which have recurred during the Quaternary and bisect the southwestern Japan arc. Evaluation of its neotectonic activity has been established based on its geomorphological features [4, 5]. In the present chapter, the authors concentrate on the eastern part of the MTL running through the Kii Peninsula, where a comprehensive fault study was recently executed by the Japanese Ministry of Education, Culture, Sports, Science and Technology (MEXT). Before the national project, detailed geologic surveys including the excavation of surface ruptures [6] have been used to describe the superficial deformation pattern along the fault, and a series of offshore [7] and onshore [8] seismic studies have been presented as useful profiles across the fault. As for the western half of the peninsula, a tectonic model has been proposed to strike a balance between the fault structure and its dominant motion senses [9]. However, the three-dimensional architecture of the MTL, one of the largest intraplate faults in the world, has never been addressed by means of detailed seismic data with a reliable stratigraphic control on deep boreholes; this theme was the focus of the present study.

2. Seismic surveys

The dataset used in this study consists of six tied seismic lines, two of which cross control boreholes. **Figure 1** delineates the study area, the reflection seismic array, and the boring locations.

In 1996, the National Research Institute for Earth Science and Disaster Resilience (NIED) conducted a seismic survey of the approximately 5300 m long K96-1 line along the southwestern coast of the Osaka Plain (**Figure 2**; [10]). During the shooting, 240 channels of geophones at 25 m intervals recorded the energy released from two vibrators (Y-2400) shot at 50 m intervals. The raw seismic data were stacked and then subjected to a poststack processing sequence to enhance the resolution.

In 1996, NIED conducted a seismic survey of the approximately 4800 m long K96-2 line that is connected to the K96-1 line at its northwestern end (**Figure 3**; [10]). During the shooting, 240 channels of geophones at 25 m intervals recorded the energy released from two vibrators (Y-2400) shot at 50 m intervals. The raw seismic data were stacked and then subjected to a poststack processing sequence to enhance the resolution.

In 1996, NIED conducted a seismic survey of the approximately 17,600 m long N96-1 line that runs through the Izumi Mountains and crosses the MTL (**Figure 4**; [11]). During the shooting, 240 channels of geophones at 25 m intervals recorded the energy released from four vibrators (Y-2400) shot at 50 m intervals. The raw seismic data were stacked and then subjected to a poststack processing sequence to enhance the resolution.

In 2006, MEXT et al. conducted a seismic survey of the approximately 22,200 m long Izumi line that runs through the Izumi Mountains and crosses the MTL (**Figure 5**; [12]). During the shooting, 555 channels of geophones at 40 m intervals recorded the energy released from four

http://dx.doi.org/10.5772/67434

vibrators (HEMI-50) shot at 10 m intervals. The raw seismic data were stacked and then subjected to a poststack processing sequence to enhance the resolution.

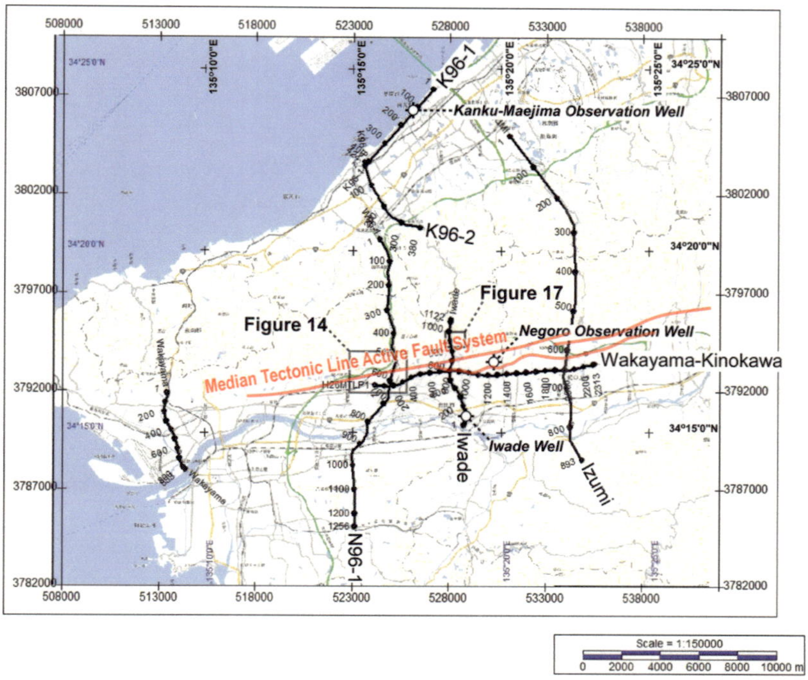

Figure 1. Index map of the seismic and drilling survey area. Common midpoint (CMP) numbers are attached to the seismic lines. Refer to the prologue of this section for regional tectonics.

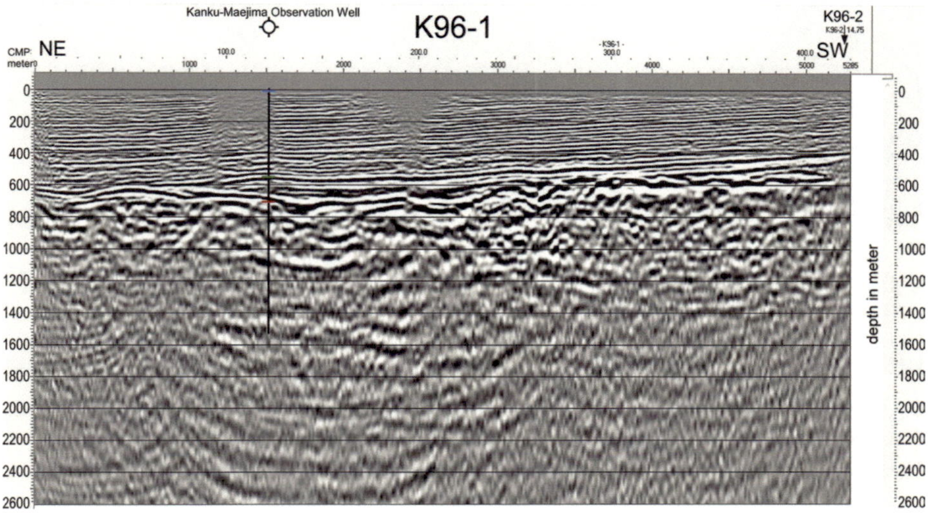

Figure 2. Raw seismic profile of the K96-1 line without vertical exaggeration [10]. See **Figure 1** for line location.

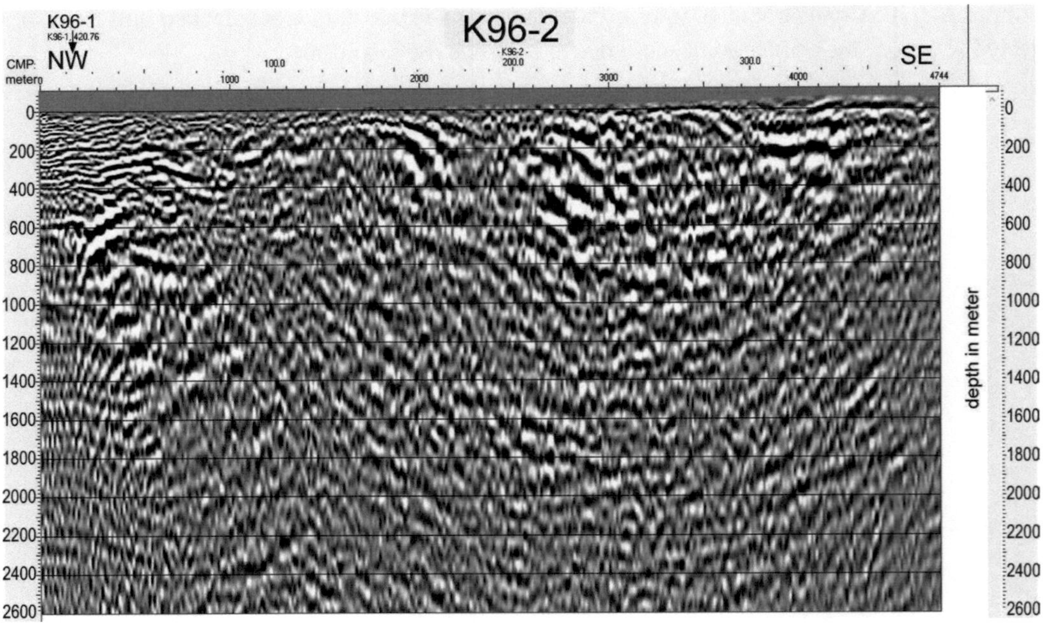

Figure 3. Raw seismic profile of the K96-2 line without vertical exaggeration [10]. See **Figure 1** for line location.

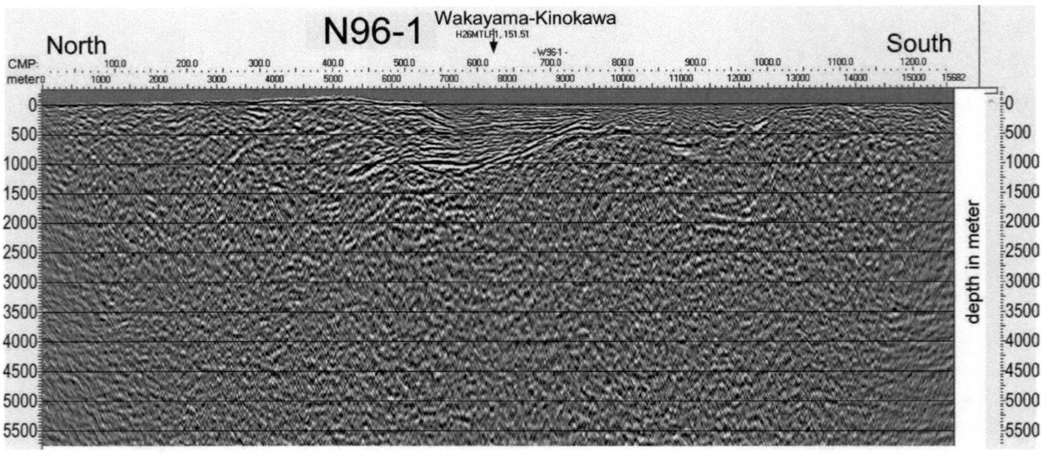

Figure 4. Raw seismic profile of the N96-1 line without vertical exaggeration [11]. See **Figure 1** for line location.

In 2014, MEXT and the Disaster Prevention Research Institute (DPRI) conducted a seismic survey of the approximately 5760 m long Iwade line that is located on the southern flank of the Izumi Mountains and crosses the MTL (**Figure 6**; [13]). During the shooting, a minimum of 100 channels of geophones at 10 m intervals recorded the energy released from one vibrator (EnviroVibe) shot at 10 m intervals. The raw seismic data were stacked and then subjected to a poststack processing sequence to enhance the resolution.

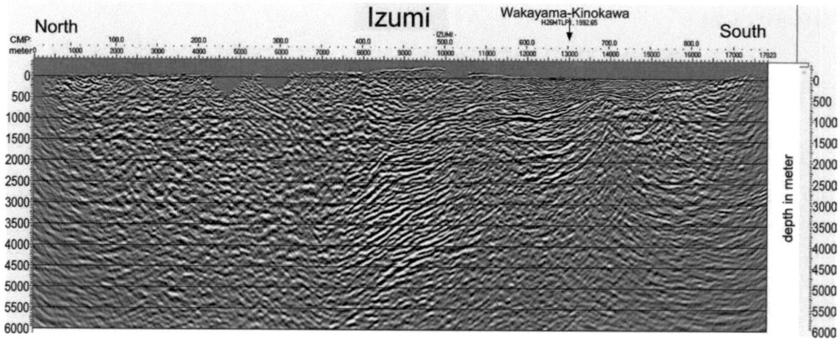

Figure 5. Raw seismic profile of the Izumi line without vertical exaggeration [12]. See **Figure 1** for line location.

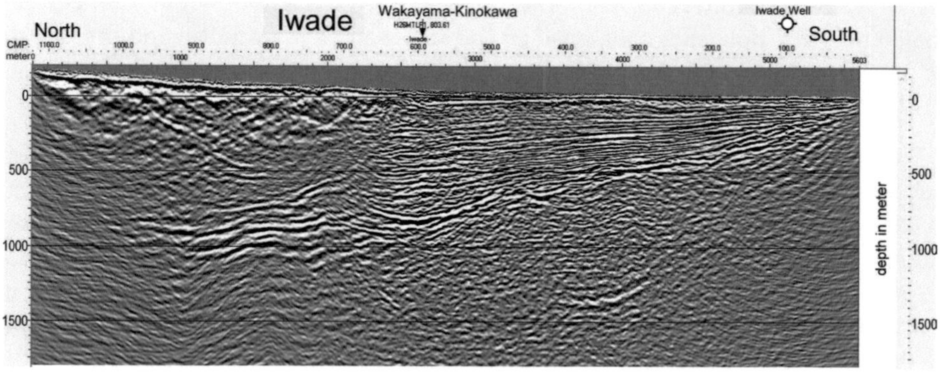

Figure 6. Raw seismic profile of the Iwade line without vertical exaggeration [13]. See **Figure 1** for line location.

In 2015, MEXT and DPRI conducted a seismic survey of the approximately 11,774 m long Wakayama-Kinokawa line, which runs parallel to the MTL and crosses the N96-1, Iwade, and Izumi lines (**Figure 7**; [14]). During the shooting, a minimum of 120 channels of geophones at 10 m intervals recorded the energy released from one vibrator (EnviroVibe) shot at 10 m intervals. The raw seismic data were stacked and then subjected to a poststack processing sequence to enhance the resolution.

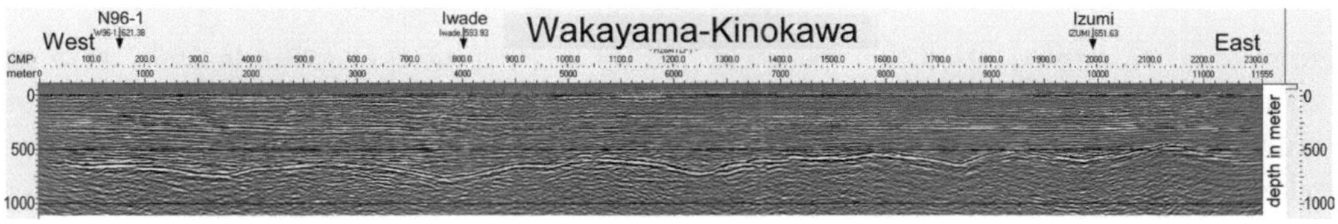

Figure 7. Raw seismic profile of the Wakayama-Kinokawa line without vertical exaggeration [14]. See **Figure 1** for line location.

3. Drilling surveys

3.1. Negoro Observation Well

The Negoro Observation Well was drilled vertically on the southern foothill of the Izumi Mountains (**Figure 1**). Continuous core samples were recovered during drilling and were identified as sheared sandstones and unconsolidated clastics intercalating a glassy volcanic ash (V290) from the ground surface to a depth of 137 m and from 137 to 625 m (TD), respectively (**Figure 8a**; [15]).

Mizuno et al. [15] assigned the upper consolidated unit to the late Cretaceous Izumi Group, which forms the watershed Izumi Mountains, and the lower loose unit to the Plio-Pleistocene Shobudani Formation. Plant macrofossils in the lower unit contained *Liquidambar* and *Pseudolarix* with last occurrences around the late Gauss Chron and early Matuyama Chron, respectively. The stratigraphic position of the unit was concordant with tephrochronology and paleomagnetic polarities confirmed on the same samples. As for the radiometric dating of the unconsolidated unit, fission-track and U-Pb ages of a surface ash sample taken near the drilling site were determined to be 1.33 and 1.50 Ma, respectively [16].

Contact between the lower (younger) and upper (older) units was regarded as a fault based on the presence of fault breccia and fault clay. The uppermost part of the Plio-Pleistocene unit (1 m thick) is sheared and tilted and suggests the occurrence of recent slips on the unit boundary. On the assumption that the discontinuity plane is connected to the Negoro-Minami Fault fringing the southern rim of the foothill, the reverse fault is estimated to dip approximately 30° northward. Additionally, an active fault was found on an outcrop adjacent to the drilling location [6], where the high-angle ENE-trending Negoro Fault with dominant dextral motion was confirmed.

In summary, the Negoro Observation Well provides the geometry of local active faults around the southern Izumi range and the age span of the recent fluvial basin developed along the MTL active fault system (see **Figure 1**).

3.2. Kanku-Maejima Observation Well

The Kanku-Maejima Observation Well was drilled vertically on the southwestern coast of the Osaka Plain, tied with the seismic K96-1 line (**Figure 1**). The lithologies of the penetrated strata were described based on cutting observations taken at depth intervals of 10 m and classified into upper unconsolidated Quaternary clastics and partly metamorphosed igneous rocks from the ground surface to 556 m and from 556 to 1535 m (TD), respectively (**Figure 8b**; [17]).

However, it should be noted that the upper part of the metamorphosed unit (556–700 m) is characterized by obvious stratification on the K96-1 line (**Figure 2**) and the top of the unit constitutes an aggradational surface of the overlying sediments. Thus, the authors interpret the equivocal section as crushed monomictic clasts originating exclusively from the Ryoke metamorphic rocks exposed on the northern flank of the Izumi range.

http://dx.doi.org/10.5772/67434

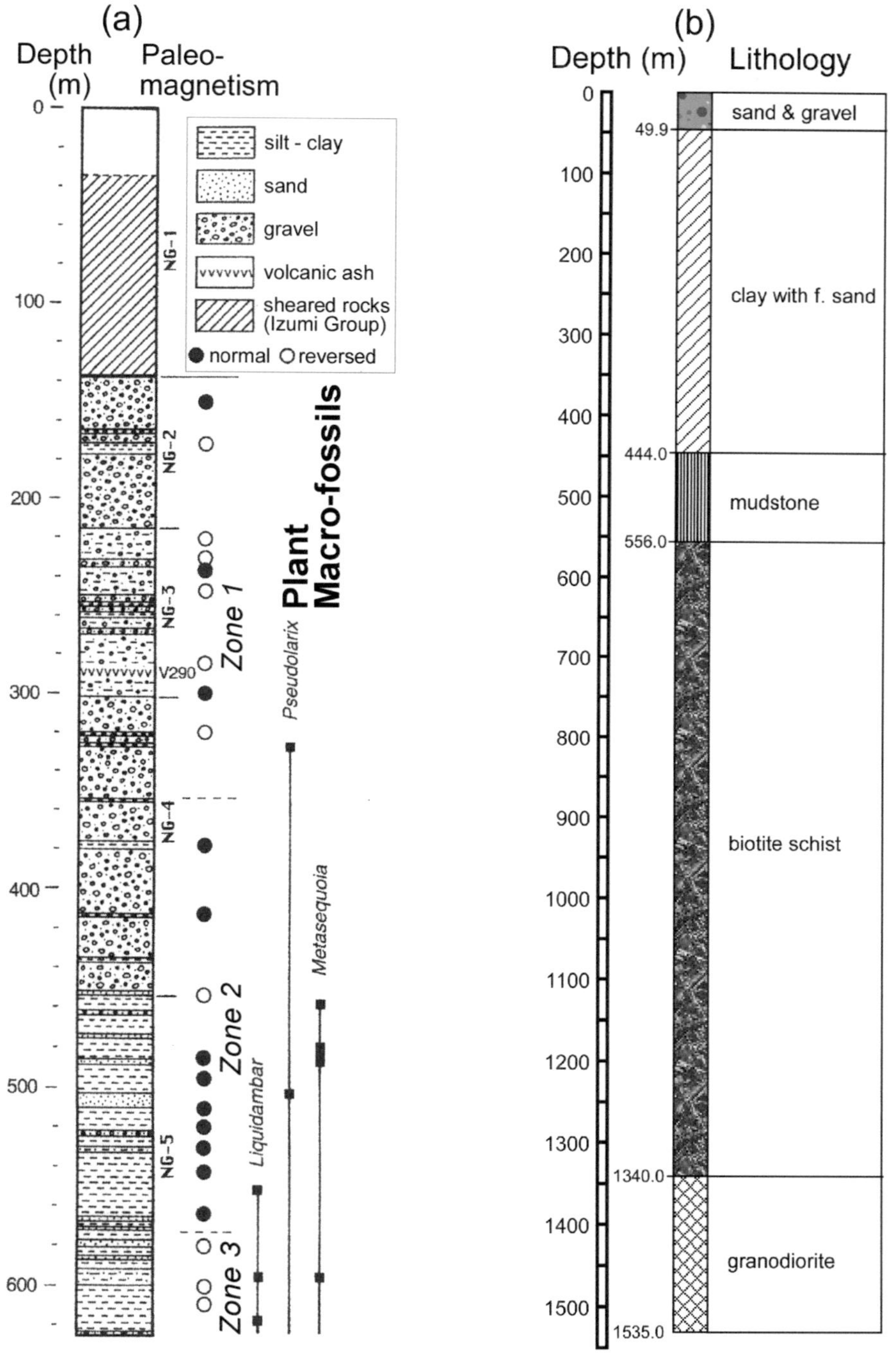

Figure 8. Geologic summaries of the (a) Negoro and (b) Kanku-Maejima Observation Wells [15, 16]. See **Figure 1** for drilling locations.

In summary, the Kanku-Maejima Observation Well provides the precise depths of the top of the acoustic basement (700 m) and the horizon of a drastic change in the clastic compositions, reflecting the progressive exhumation of the hinterland (556 m) on the K96-1.

3.3. Iwade Well

The Iwade Well was drilled vertically on the north bank of the Kinokawa River, tied with the seismic Iwade line (**Figure 1**). Continuous core samples and spot cuttings were recovered from intervals shallower and deeper than a depth of 189 m, respectively, and the strata were categorized into terrace deposits, Quaternary fluvial sediments, and Sanbagawa metamorphic rocks from the ground surface to 23 m, from 23 to 163 m, and from 163 to 300 m (TD), respectively (**Figure 9**; [18]).

A series of tephrochronological analyses that was executed for many dispersed ashes failed to correlate the ashes with widespread tephra. Although normal and reversed pDRMs were observed in the laboratory, unstable rock magnetic behavior during thermal treatments hindered the determination of polarity epochs. Fluvial sediments in the Iwade Well were correlated with the Plio-Pleistocene Shobudani Formation ad hoc [18] based on the stratigraphic study of the nearby Negoro Observation Well. However, it seems that a formal definition of the Shobudani Formation has not been established. Mizuno [19] defined the "Shobudani" unit as sediments that buried a fluvial basin along the MTL active fault system (see **Figure 1**), but the formation is assigned to three distant epochs, as he had previously suggested [15]. Recent U-Pb ages [16] obtained from volcanic ashes intercalated in the Shobudani Formation within the geologic map of Kokawa [6] showed a large scatter ranging from 0.26 to 1.50 Ma. In summary,

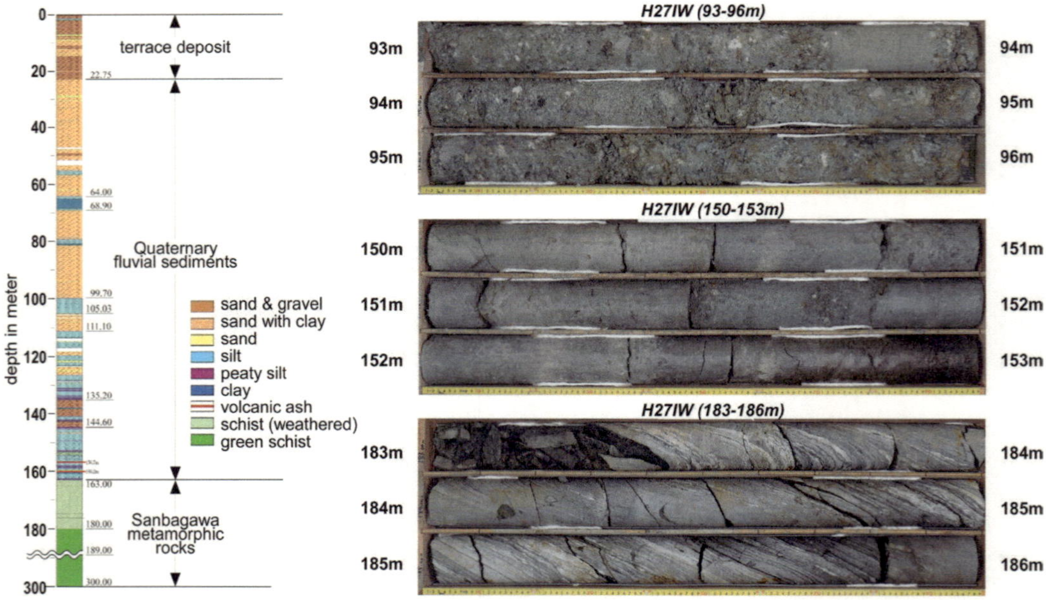

Figure 9. Geologic summary of the Iwade Well [18]. See **Figure 1** for drilling location.

http://dx.doi.org/10.5772/67434

the "Shobudani" is an artifact or a hodgepodge of fluvial sequences along the MTL, and further chronological and sedimentological analyses should be made to obtain the definitive tectonic context of the sedimentary unit. Thus, the Iwade Well is simply referenced as an indicator of depth of the top of the acoustic basement on the Iwade line.

4. Geologic interpretation of seismic profiles

After exploring the stratigraphic information of boreholes in the study area, seismic interpretation was performed to obtain the subsurface structural architecture. In this section, a series of seismic features and their tectonic implications are presented.

4.1. K96-1 line

Figure 10 shows the results of the seismic interpretation of the K96-1 line; the top of the acoustic basement and the boundary between the lower metamorphosed clasts and the upper normal sediments are delineated as red and green lines, respectively.

It is noteworthy that the reflection terminations of the normal sediments are onlapping onto the lower unit surface in the southwestward direction. Such a tendency and the gentle tilt of the sedimentary units imply the recent uplift of the southwestern basin margin. This is a north-trending basement high discovered by means of gravity, seismic, and borehole investigation (**Figure 11**; [20–22]). The homoclinal tilting of the upper unit is suggestive of the episodic and possibly ongoing uplift of the mound accompanied by basement deformation.

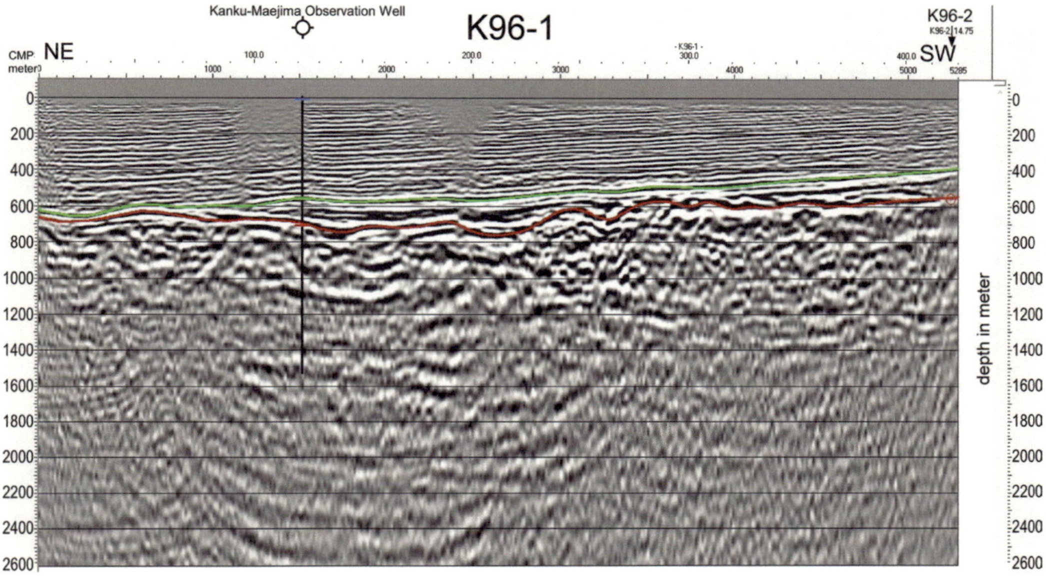

Figure 10. Interpreted seismic profile of the K96-1 line without vertical exaggeration. See **Figure 1** for line location.

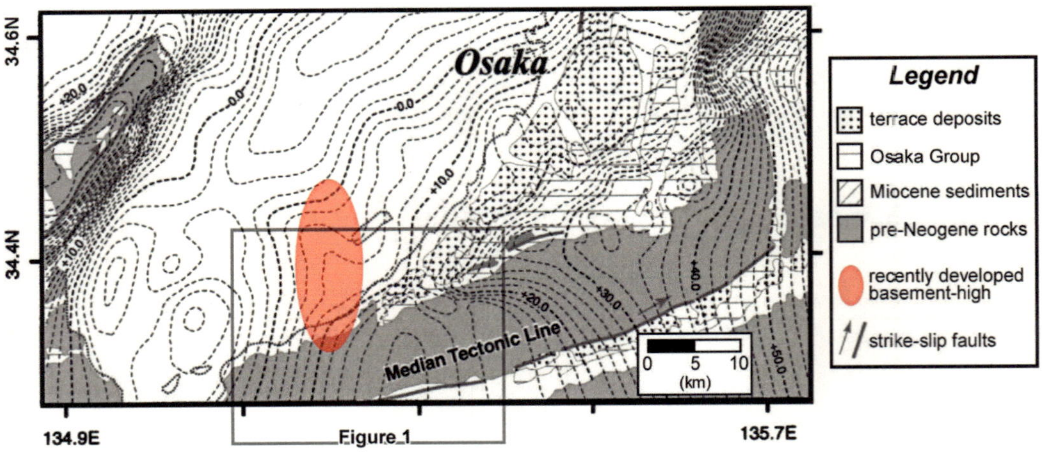

Figure 11. Regional basin morphological map of the southern Osaka Bay after Itoh et al. [20, 21]. Gravity contours are drawn at 2-mGal intervals.

4.2. K96-2 line

Regarding the K96-2 line, traceable seismic horizons are recognizable at its northwestern end, where the Cretaceous Izumi Group is covered by Quaternary sediments (**Figure 12**). The top of the acoustic basement (red in the downthrown block, maroon in the upthrown block) and the boundary between the metamorphosed clasts and normal sediments (green) are traceable from the K96-1 line. They are cut by two reverse faults, recent cumulative slips on which are unclear because the green horizon was not identified in the upthrown block.

A previous seismic interpretation [23] indicated that these faults are active faults parallel to the coastline. However, it seems that north-to-NNW-trending warping is a recent geomorphic feature around the northern foothill of the Izumi range [24], and one of the largest bumps coincides with the gentle uplift zone described on the K96-1 line. The presence of a superficial rupture on the K96-2 faults cannot be confirmed. In an effort to visualize the three-dimensional architecture of the Osaka sedimentary basin [25, 26], structural diversity that may reflect the transient convergence mode of oceanic plates (the Pacific and Philippine Sea Plates) has been recognized. The latest research [16] has presented a chronicle of the waxing and waning of neotectonic stresses and the varied structural development around the southern portion of the basin.

4.3. N96-1 line

On the southern side of the Izumi Mountains, only the top of acoustic basement is traceable through the four investigated sections. On the N96-1 line, the basement seismic horizon is shown as a red line in the downthrown block and a maroon line in the upthrown block, which is sandwiched between two faults (**Figure 13**). A kink zone in the Cretaceous Izumi Group was observed between the reverse faults.

http://dx.doi.org/10.5772/67434

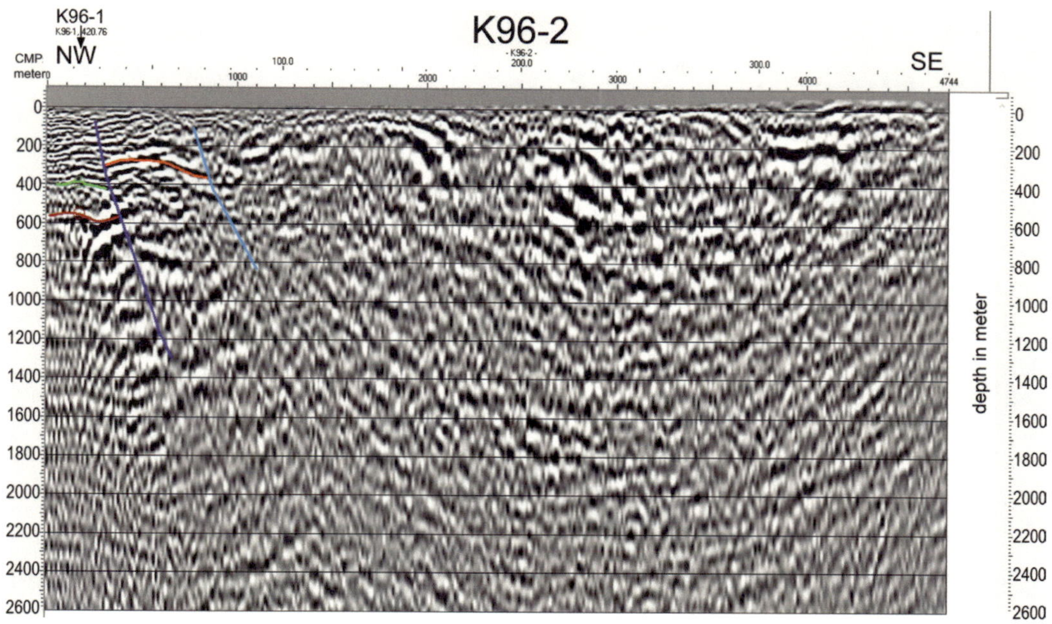

Figure 12. Interpreted seismic profile of the K96-2 line without vertical exaggeration. See **Figure 1** for line location.

The present field observation has indicated that the ENE-trending kink zone (parallel to the MTL) extends to the surrounding areas (**Figure 14**). First, it should be noted that the kink trend deviates from the original structure of a plunging syncline that developed after the initial deposition of the Izumi Group. Second, it is important that the kink is considerably distant from the active dextral trace of the MTL. In the present study, it was interpreted that the kink zone developed under the influence of compressive stress episodically raised from the Pliocene to early Pleistocene.

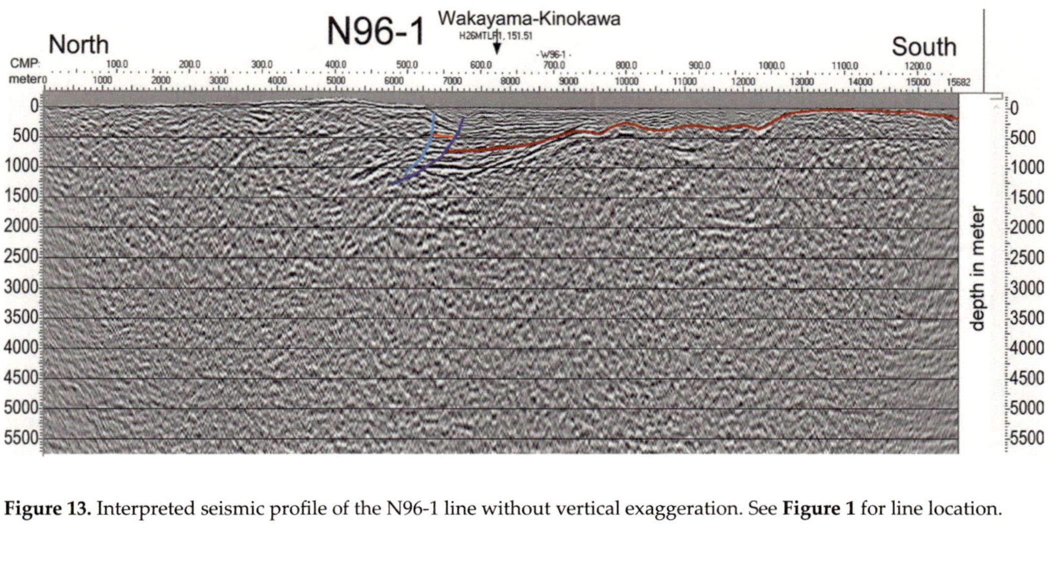

Figure 13. Interpreted seismic profile of the N96-1 line without vertical exaggeration. See **Figure 1** for line location.

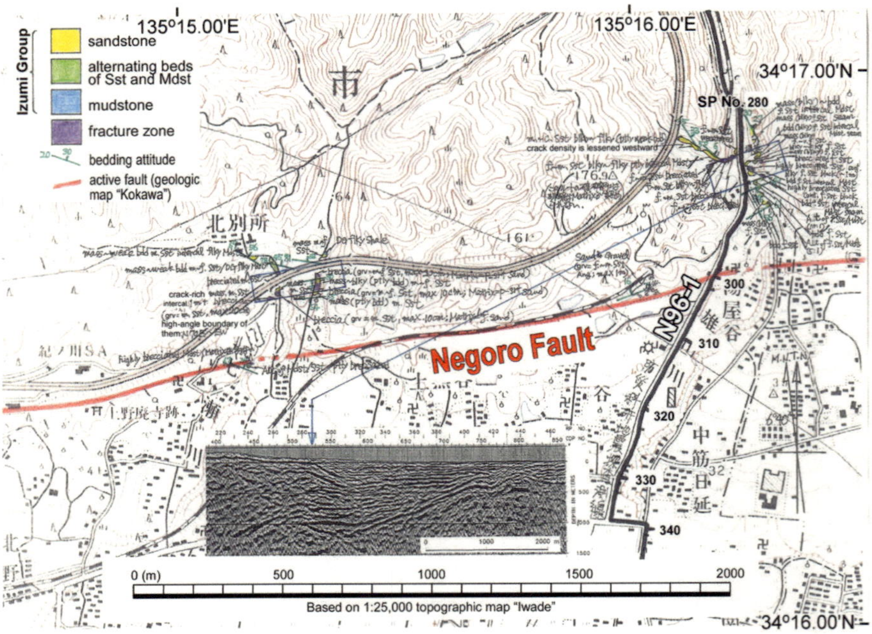

Figure 14. Field observation of a deformation zone along the N96-1 line. See Figure 1 for mapped area.

4.4. Izumi line

On the Izumi line, a single low-angle fault and the top of acoustic basement were delineated (**Figure 15**). A half-graben buried by the Quaternary sediments is clearly shown on the profile.

On this seismic line, the deformation zone of the upthrown block of the thrust was not clearly identified.

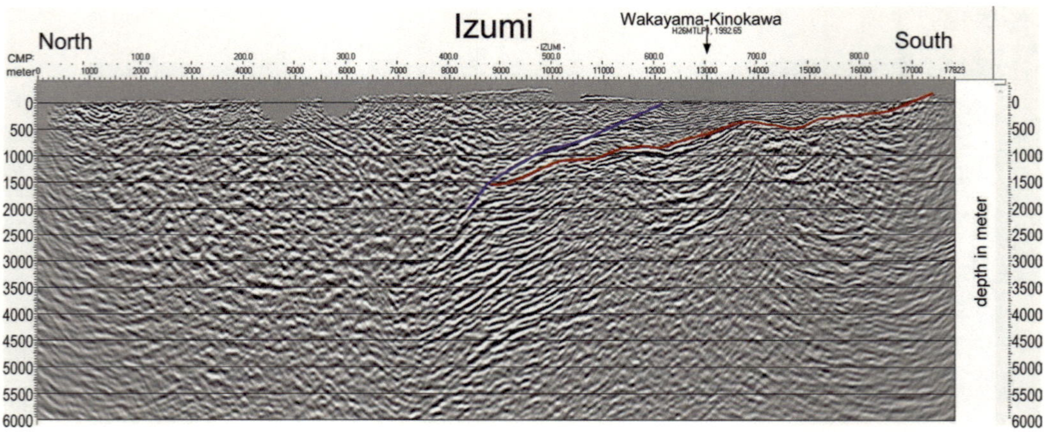

Figure 15. Interpreted seismic profile of the Izumi line without vertical exaggeration. See Figure 1 for line location.

4.5. Iwade line

On the Iwade line, a single low-angle fault and the top of acoustic basement were delineated (**Figure 16**). A half-graben buried by the Quaternary sediments is clearly shown on the profile.

On this seismic line, the structure of the upthrown block of the thrust appears to be chaotic and to not contain a clear kink zone. As shown in **Figure 17**, the present field survey has revealed a divergent deformation zone adjacent to the Iwade line.

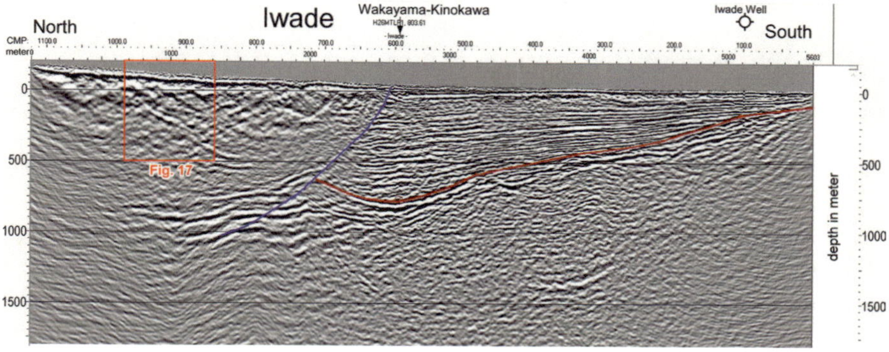

Figure 16. Interpreted seismic profile of the Iwade line without vertical exaggeration. See **Figure 1** for line location.

Figure 17. Field observation of a deformation zone along the Iwade line. See **Figure 1** for mapped area.

4.6. Wakayama-Kinokawa line

On the Wakayama-Kinokawa line, the gentle undulation of the top of the acoustic basement was delineated (**Figure 18**). As shown by the north-south seismic lines, the southern foothill of the Izumi range is characterized by a general ENE-trending structure that agrees with the topographic features. Hence, the discovered warping on the trend-parallel line implies the existence of subordinate crustal strain along the fault. Furthermore, the Quaternary sediments on the profile were observed to be nearly undeformed.

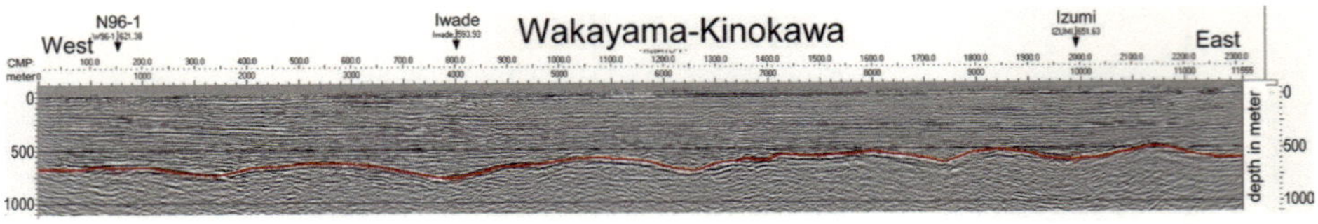

Figure 18. Interpreted seismic profile of the Wakayama-Kinokawa line without vertical exaggeration. See **Figure 1** for line location.

The basal heaves on the profile appear to correspond to uplifted areas of the southern foothill of the Izumi range. Although the flat attitudes of the recent sediments indicate that they do not have a direct relationship with the structural development processes, longstanding activities of the MTL since the Cretaceous may have developed similar wrenching trends along the fault. The gravity anomaly trend [27] shown in **Figure 19** implies that the most remarkable structural break lies under the present Kinokawa River, where a distinct gap in metamorphic grades in the Sanbagawa metamorphic complex was confirmed [28]. A similar crustal structure is also supported by the geomagnetic anomaly trend [29].

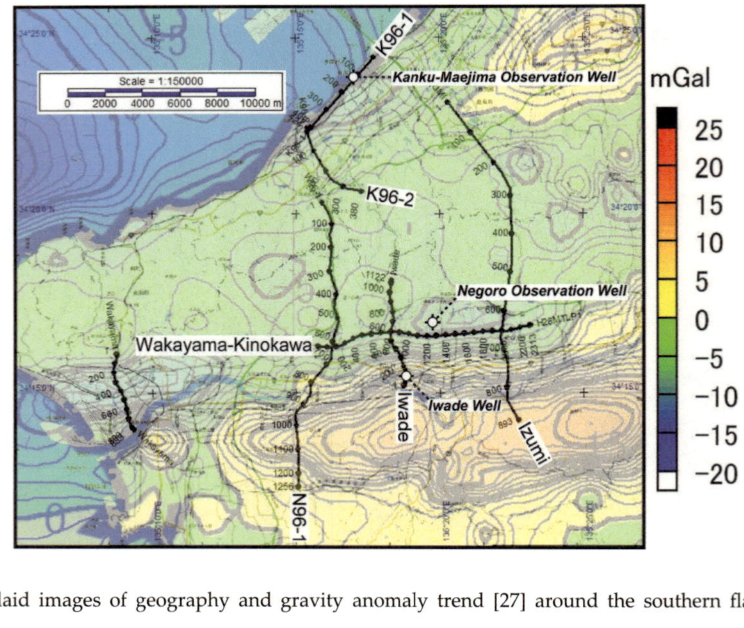

Figure 19. Overlaid images of geography and gravity anomaly trend [27] around the southern flank of the Izumi Mountains.

http://dx.doi.org/10.5772/67434

5. Tectonic context of the unraveled structure along the MTL

The present 3-year geophysical and geological exploration along the MTL active fault system (see **Figure 1**) has clarified its detailed structure. **Figure 20** presents the first-ever images of the three-dimensional architecture around the MTL.

The MTL delineated through reflection seismic study is characterized by a low-angle feature, which should have reverse motion sense considering the shape of foreland basin and deformation of the Quaternary strata burying the depression. Sato et al. [9] have argued that the MTL was initiated as a subduction surface of the Izanagi Plate during the Early Cretaceous and became reactivated as a low-angle dextrally moving thrust during the Quaternary. Their intriguing scenario appears to ignore the presence of active high-angle right-lateral fault that is separate from and runs parallel to the remarkable thrust [6, 13, 14, 18]. Although such a hypothesis should agree with geological evidence, the low-angle model has some notable discrepancies. First, a high-pressure metamorphic belt is expected to develop on the upthrown block of the down-going slab, but the Sanbagawa metamorphic complex exists exclusively on the downthrown side of the thrust [30]. Second, the Cretaceous Izumi Group resting on the thrust is a series of turbidite sequences that buried a gigantic pull-apart basin [31], which should be formed not on a thrust but on the stepping part of high-angle lateral faults. As for the strange occurrence of a metamorphosed terrane, Sato et al. [9] proposed a conciliatory plot that the plate boundary jumped toward the ocean side as a result of the normal convergence of the Pacific Plate during the Late Cretaceous. Unfortunately, this is in disagreement with the coeval deformation history of an accretionary complex of the Shimanto Belt [32], which requires a nearby transcurrent fault. A regional paleogeographic reconstruction [33] showed that the MTL constituted a Cretaceous strike-slip break along the Eurasian margin together with the Central Sikhote Alin Fault. It is most probable that the low-angle fault on the seismic profiles has a younger origin unrelated to the Mesozoic transform boundary.

The most notable suspect in the extensive contractional event is the north-south shortening of southwestern Japan in the Late Miocene [34]. Although this tectonic deformation seems to be concentrated on the margin of the Japan Sea back-arc basin, the present thermochronological analysis based on the annealing of fission tracks in apatite and zircon separated from sandstones of

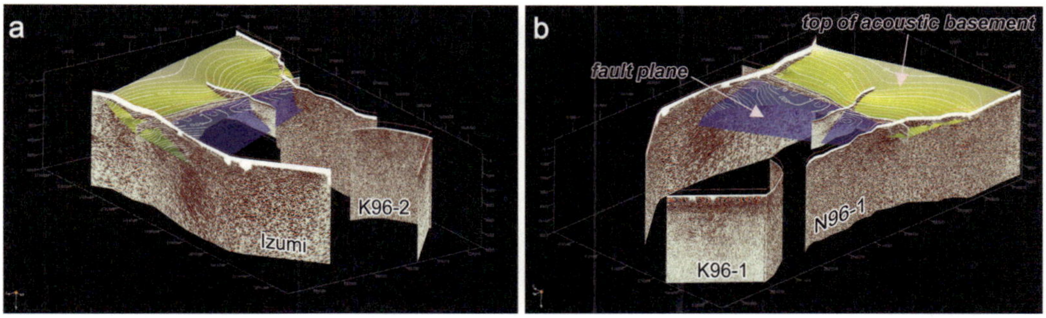

Figure 20. Bird's-eye views of the subsurface structural architecture around the Izumi Mountains from (a) the northeast and (b) the northwest based on reflection seismic interpretation. The moss green and semi-transparent blue surfaces represent the top of the acoustic basement and the most remarkable fault, respectively. See **Figure 1** for line locations.

the Izumi Group revealed that significant vertical exhumation of the sedimentary unit occurred near the end of the Miocene (Chapter 5). A sediment provenance study revealed crucial evidence of another epoch of contraction. Oka [35] noted that the lower part of the Quaternary strata exposed on the northern flank of the Izumi range contains a considerable amount of crystalline schist clasts derived from the Sanbagawa metamorphic complex, a fact that suggests that the emergence of the watershed Izumi Mountains occurred later than 3 Ma. This uplift event seems to have prevailed along the MTL [36] and to have been accompanied by the delayed deformation of the forearc [37]. The authors conclude that the remarkable thrust on the southern Izumi range developed between 6 and 2 Ma under the intermittent rise of compressive stress. Since then, the mountain-building activity has been dormant except for an episodic event that caused the uplift of the northern Izumi foothill [35] at ca. 1 Ma. This most recent incident may have been responsible for formation of the small fractures observed on the K96-2 line.

The recent reinforced dextral motion on the MTL is probably linked to a counterclockwise shift of the converging direction of the Philippine Sea Plate between 2 and 1 Ma [38], which provoked the wrench deformation of southwestern Japan [39], the migration of the forearc sliver [40], and the eventual crustal break on the back-arc shelf [41]. In an edited collection of the latest results of multidisciplinary studies on the mechanism of sedimentary basin formation [42], Itoh et al. [43] described a complicated subsurface structure in southwestern Japan related to the differential motion of fault-bounded crustal blocks. The evolutionary process of the MTL active fault system reflects the temporal shifts in the motion of the Philippine Sea Plate [44], and coming tectonic episodes (e.g., seamount subduction) may revitalize the Izumi thrust in the future.

6. Conclusions

The present integrated research has enabled the visualization of a thrust lying beneath the Izumi Mountains running parallel to the MTL active fault system, as presented in **Figure 21**.

With the aid of the stratigraphic control of three boreholes in the study area, seismic horizons were traced through the six tied reflection survey lines covering the watershed mountains, and the morphological characteristics of the Quaternary basins were successfully delineated on both flanks of the mountainous area. On the southern flank, the authors found a low-angle north-dipping fault, the up- and downthrown sides of which consist of the Cretaceous sedimentary rocks (Izumi Group) and unconsolidated fluvial sediments, respectively. Based on the subsidence pattern and the deformation mode of the infilled sediments of the recent asymmetric basin, it is clear that reverse slips are dominant on the fault. An ENE-trending continuous kink zone within the Izumi Group that has been detected on the seismic profiles and confirmed by geologic survey suggests the contractional deformation of the upthrown wedge. Considering the geomorphological features around the study area, the dextral movement on the MTL active fault system, which runs along the foothill margin to the south of the thrust, is likely to accommodate recent shear stress provoked by the oblique subduction of the Philippine Sea Plate. From the regional neotectonic context of the southwestern Japan arc, the dormant thrust likely evolved from 6 to 2 Ma, causing the uplift of the mountains under intermittent phases of north-south compression. At the present stage of tectonic history, the low- and high-angle crustal breaks act as material and mechanical boundaries, respectively.

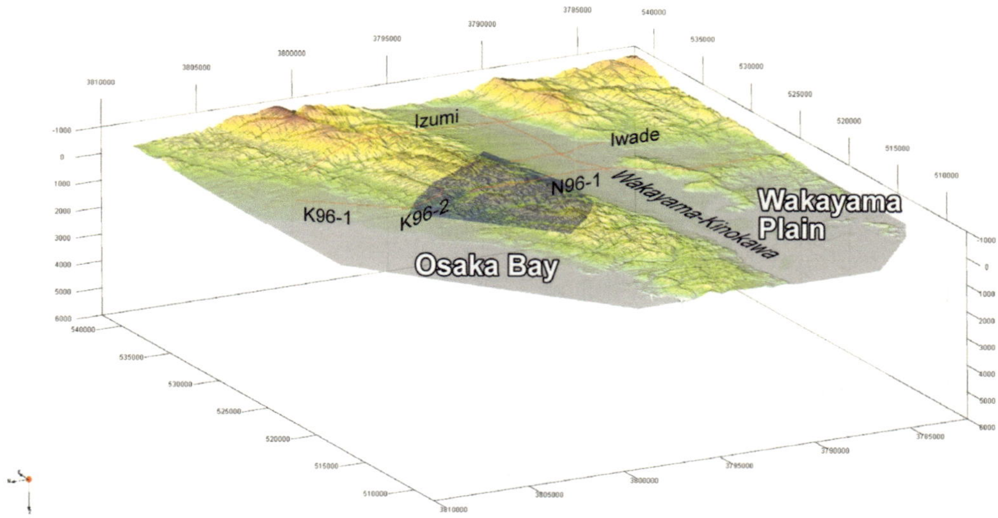

Figure 21. Bird's-eye DEM topographic map around the Izumi Mountains from the northwest. The most remarkable fault is shown as a semi-transparent blue surface. See **Figure 1** for line locations.

Among significant results of the authors' research, bird's-eye movies of three-dimensional fault architecture based on seismic interpretation and some original figures in this chapter are available at OPERA:Osaka Prefecture University Education and Research Archives (http://hdl.handle.net/10466/15058).

Acknowledgements

The authors like to thank the National Research Institute for Earth Science and Disaster Resilience for permission to publish this work. This study was executed as a part of the Comprehensive Research and Survey for the Median Tectonic Line Fault System (Eastern Margin of Kongo Mountains–Southern Margin of Izumi Mountains; FY 2013–2015) organized by the Research and Development Bureau, the Ministry of Education, Culture, Sports, Science and Technology, and the Disaster Prevention Research Institute, Kyoto University. Constructive comments by Professor Hiroyuki Arato greatly helped to improve an early version of the manuscript.

Author details

Yasuto Itoh[1]*, Tomotaka Iwata[2] and Keiji Takemura[3]

*Address all correspondence to: yasutokov@yahoo.co.jp

1 Graduate School of Science, Osaka Prefecture University, Osaka, Japan

2 Disaster Prevention Research Institute, Kyoto University, Uji, Japan

3 Institute for Geothermal Sciences, Kyoto University, Beppu, Japan

References

[1] Fitch TJ. Plate convergence, transcurrent faults, and internal deformation adjacent to southeast Asia and the western Pacific. Journal of Geophysical Research. 1972; 77: 4432-4460.

[2] Huzita K. The Quaternary tectonic stress states of Southwest Japan. Journal of Geosciences, Osaka City University. 1976; 20: 93-103.

[3] Huzita K. Role of the Median Tectonic Line in the Quaternary tectonics of the Japanese islands. Memoir of Geological Society of Japan. 1980; 18: 129-153.

[4] Research Group for Active Faults. The Active Faults in Japan: Sheet Maps and Inventories rev.ed. Tokyo: University of Tokyo Press; 1991.

[5] Okada A, Togo M, editors. Active Faults in Kinki. Tokyo: University of Tokyo Press; 2000. 395 p.

[6] Makimoto H, Miyata T, Mizuno K, Sangawa A. Geology of the Kokawa District, with Geological Sheet Map at 1:50,000. Tsukuba: Geological Survey of Japan, AIST; 2004. 89 p.

[7] Yoshikawa S, Kadosawa H, Mitsuhashi A, Iwasaki Y. Geological structure of the Median Tectonic Line in Tomogashima Strait area. Journal of the Japan Society for Marine Surveys and Technology. 1996; 8: 1-10.

[8] Yoshikawa S, Iwasaki Y, Ikawa T, Yokota H. Geological structure of the MTL in west Wakayama by reflection seismic study. Memoir of Geological Society of Japan. 1992; 40: 177-186.

[9] Sato H, Kato N, Abe S, Van Horne A, Takeda T. Reactivation of an old plate interface as a strike-slip fault in a slip-partitioned system: Median Tectonic Line, SW Japan. Tectonophysics. 2015; 644-645: 58-67.

[10] NIED (National Research Institute for Earth Science and Disaster Resilience). Report of Vibroseis and VSP Structural Survey Related to the Kanku-Maejima Boring Exploration by the Science and Technology Agency. Tsukuba: NIED; 1997.

[11] Kasahara K, Aoi S, Kimura H, Ikawa T, Kawanaka T, Mizohata S, Okamoto S. Subsurface structure derived from the seismic reflection survey around Median Tactonic Line at northwestern part of Wakayama. In: Abstracts of the Annual Meeting of Japan Geoscience Union. Tokyo: JpGU; 1998. SL-p019.

[12] MEXT (Ministry of Education, Culture, Sports, Science and Technology, Japan), ERI (Earthquake Research Institute, The University of Tokyo), DPRI (Disaster Prevention Research Institute, Kyoto University), NIED (National Research Institute for Earth Science and Disaster Resilience). Research for Crustal Structure of Metropolises (Project for Mitigation of Catastrophic Earthquake Disaster in Metropolises), Heisei 18 Fiscal Year Report. Tokyo: Research and Development Bureau, MEXT; 2007.

[13] MEXT (Ministry of Education, Culture, Sports, Science and Technology, Japan), DPRI (Disaster Prevention Research Institute, Kyoto University). Comprehensive Research and Survey for the Median Tectonic Line Fault System (Eastern Margin of Kongo Mountains – Southern Margin of Izumi Mountains), Heisei 25 Fiscal Year Report. Uji: DPRI; 2014.

[14] MEXT (Ministry of Education, Culture, Sports, Science and Technology, Japan), DPRI (Disaster Prevention Research Institute, Kyoto University). Comprehensive Research and Survey for the Median Tectonic Line Fault System (Eastern Margin of Kongo Mountains – Southern Margin of Izumi Mountains), Heisei 26 Fiscal Year Report. Uji: DPRI; 2015.

[15] Mizuno K, Tsukuda E, Takahashi M, Momohara A, Uchiyama T. Subsurface geology of the Wakayama Plain, southwestern Japan based on the deep boring survey. Journal of Geological Society of Japan. 1999; 105: 235-238.

[16] Itoh Y. Subsurface structure of Osaka sedimentary basin and its tectonic evolution. In: Itoh Y, Kusumoto S, Takemura K, editors. Research Frontiers of Sedimentary Basin Interiors: A Case Study and Methodological Review on an Oblique Convergent Margin. New York: Nova Science Publishers, Inc.; 2016. pp. 27-70.

[17] NIED (National Research Institute for Earth Science and Disaster Resilience). Report of the Kanku-Maejima Observation Well by the Science and Technology Agency. Tsukuba: NIED; 1997.

[18] MEXT (Ministry of Education, Culture, Sports, Science and Technology, Japan), DPRI (Disaster Prevention Research Institute, Kyoto University). Comprehensive Research and Survey for the Median Tectonic Line Fault System (Eastern Margin of Kongo Mountains – Southern Margin of Izumi Mountains), Heisei 27 Fiscal Year Report. Uji: DPRI; 2016.

[19] Mizuno K. Age and tectonic development of the Plio-Pleistocene sedimentary basins along the Median Tectonic Line, Southwest Japan. Memoir of Geological Society of Japan. 1992; 40: 1-14.

[20] Itoh Y, Takemura K, Ishiyama T, Tanaka Y, Iwaki H. Basin formation at a contractional bend of a large transcurrent fault: Plio-Pleistocene subsidence of the Kobe and northern Osaka Basins, Japan. Tectonophysics. 2000; 321: 327-341.

[21] Itoh Y, Takemura K, Kawabata D, Tanaka Y, Nakaseko K. Quaternary tectonic warping and strata formation in the southern Osaka Basin inferred from reflection seismic interpretation and borehole sequences. Journal of Asian Earth Sciences. 2001; 20: 45-58.

[22] Inoue N, Kitada N, Takemura K, Fukuda K, Emura T. Three-dimensional subsurface structure model of Kansai International Airport by integration of borehole data and seismic profiles. Geotechnical and Geological Engineering. 2012; DOI:10.1007/s10706-012-9568-4

[23] Kondo H, Sugito N, Yoshioka T, Tsutsumi H, Kimura H. Revisited spatial distribution of the Uemachi fault zone, Osaka, revealed by digital elevation models. Active Fault Research. 2015; 42: 1-34.

[24] Oka Y. The geomorphology and the crustal movements in the southeastern part of the Osaka Plain. Geographical Review of Japan. 1961; 34: 523-535.

[25] Itoh Y, Kusumoto S, Takemura K. Characteristic basin formation at terminations of a large transcurrent fault – basin configuration based on gravity and geomagnetic data. In: Itoh Y, editor. Mechanism of Sedimentary Basin Formation – Multidisciplinary Approach on Active Plate Margins. Rijeka: InTech; 2013. http://dx.doi.org/10.5772/56702

[26] Itoh Y, Kusumoto S, Takemura K. Tectonically controlled asymmetric basin formation and evolution: an example from an active plate margin. In: Veress B, Szigethy J, editors. Horizons in Earth Science Research, v14. New York: Nova Science Publishers, Inc.; 2015. pp. 123-141. https://www.novapublishers.com/catalog/product_info.php?products_id=57536

[27] Gravity CD-ROM of Japan, ver2, Digital Geoscience Map P-2 [CD-ROM]. Tsukuba: Geological Survey of Japan; 2004.

[28] Hirota Y. Geology of the Sambagawa metamorphic belt in western Kii Peninsula, Japan. Memoirs of Faculty of Science, Shimane University. 1991; 25: 131-142.

[29] Nakatsuka T, Okuma S. Aeromagnetic Anomalies Database of Japan, Digital Geoscience Map P-6. Tsukuba: Geological Survey of Japan; 2005.

[30] Wang CL, Maekawa H. Albite-biotite zone of the Sanbagawa metamorphic belt in the northwestern part of the Kii Peninsula, Japan. Journal of Mineralogy, Petrology and Economic Geology. 1997; 92: 43-54.

[31] Noda A, Toshimitsu S. Backward stacking of submarine channel-fan successions controlled by strike-slip faulting: The Izumi Group (Cretaceous), southwest Japan. Lithosphere. 2009; 1: 41-59.

[32] Tokiwa T. Timing of dextral oblique subduction along the eastern margin of the Asian continent in the Late Cretaceous: evidence from the accretionary complex of the Shimanto Belt in the Kii Peninsula, Southwest Japan. Island Arc. 2009; 18: 306-319.

[33] Itoh Y, Uno K, Arato H. Seismic evidence of divergent rifting and subsequent deformation in the southern Japan Sea, and a Cenozoic tectonic synthesis of the eastern Eurasian margin. Journal of Asian Earth Sciences. 2006; 27: 933-942.

[34] Itoh Y, Nagasaki Y. Crustal shortening of Southwest Japan in the Late Miocene. The Island Arc. 1996; 5: 337-353.

[35] Oka Y. The formation of the Izumi range and the Osaka Group. The Quaternary Research. 1978; 16: 201-210.

[36] Oka Y, Sangawa A. The formation of the sedimentary basin in the east of Inland Sea and the uplift of the Awaji Island, Japan. Journal of Geography. 1981; 90: 393-409.

[37] Takano O, Nishimura M, Fujii T, Saeki T. Sequence stratigraphic distribution analysis of methane-hydrate-bearing submarine-fan turbidite sandstones in the eastern Nankai Trough area: relationship between turbidite facies distributions and BSR occurrence. Journal of Geography. 2009; 118: 776-792.

[38] Nakamura K, Renard V, Angelier J, Azema J, Bourgois J, Deplus C, Fujioka K, Hamano Y, Huchon P, Kinoshita H, Labaume P, Ogawa Y, Seno T, Takeuchi A, Tanahashi M, Uchiyama A, Vigneresse J-L. Oblique and near collision subduction, Sagami and Suruga Troughs — preliminary results of the French-Japanese 1984 Kaiko cruise, Leg 2. Earth and Planetary Science Letters. 1987; 83: 229-242.

[39] Itoh Y, Takemura K. Quaternary geomorphic trends within Southwest Japan: extensive wrench deformation related to transcurrent motions of the Median Tectonic Line. Tectonophysics. 1993; 227: 95-104.

[40] Itoh Y, Takemura K. Mode of Quaternary crustal deformation of Kyushu Island, Japan. In: Shichi R, Heki K, Kasahara M, Kawasaki I, Murakami M, Nakahori Y, Okada Y, Okubo S, Ota Y, Takemoto S, editors. Proceedings of the 8th International Symposium on Recent Crustal Movement, CRCM'93, 6-11 December 1993, Kobe, Japan. Kyoto: The Local Organizing Committee for the CRCM'93; 1994.

[41] Itoh Y, Tsutsumi H, Yamamoto H, Arato H. Active right-lateral strike-slip fault zone along the southern margin of the Japan Sea. Tectonophysics. 2002; 351: 301-314.

[42] Itoh Y, editor. Mechanism of Sedimentary Basin Formation – Multidisciplinary Approach on Active Plate Margins. Rijeka: InTech; 2013. http://dx.doi.org/10.5772/50016

[43] Itoh Y, Takemura K, Kusumoto S. Neotectonic intra-arc basins within southwest Japan – conspicuous basin-forming process related to differential motion of crustal blocks. In: Itoh Y, editor. Mechanism of Sedimentary Basin Formation – Multidisciplinary Approach on Active Plate Margins. Rijeka: InTech; 2013. http://dx.doi.org/10.5772/56588

[44] Itoh Y, Kusumoto S, Takemura K. Evolutionary process of Beppu Bay in central Kyushu, Japan: a quantitative study of the basin-forming process controlled by plate convergence modes. Earth, Planets and Space. 2014; 66: 74. DOI:10.1186/1880-5981-66-74

Chapter 4

Inconsistent Structure and Motion of the Eastern Median Tectonic Line, Southwest Japan, during the Quaternary

Shigekazu Kusumoto, Keiji Takemura and
Yasuto Itoh

Additional information is available at the end of the chapter

http://dx.doi.org/10.5772/67964

Abstract

The Median Tectonic Line (MTL) is the largest tectonic line in southwest Japan, and its eastern portion has moved as a right-lateral fault with a reverse fault component during the Quaternary. Although a high dip of the MTL has been suggested from geomorphological studies, reflection surveys have indicated a low dip of 30–55°. Thus, the MTL shows contradiction between its fault dip and fault motion. In this study, we attempted to re-estimate the dip of the MTL by gravity anomaly, gravity gradient tensor, and numerical simulation, restoring topographies caused by lateral faulting. The numerical simulations suggested that a fault dip of 70–75° is a reasonable dip of the MTL. These high-dip faults are able to move as a lateral fault and have the possibility of reverse fault motion. Deformation patterns caused by faults with these dips are in harmony with the accumulated geological and geomorphological evidence for motions of the MTL. On the other hand, gravity and gravity gradient analyses showed only a material boundary with low dip. This suggested that the MTL does not have conspicuous density contrast at the boundary.

Keywords: motion boundary, material boundary, geophysical prospecting, numerical simulation, fault dip, Median Tectonic Line, Quaternary, lateral fault, reverse fault

1. Introduction

As is well known, the oblique subduction of a plate forms a strike-slip fault parallel to the trench on the continental plate side (e.g., [1]). In general, a fault of this type has vertical or subvertical dip, and these faults have been found widely at plate boundary margins of the world (e.g., [1, 2]). The Median Tectonic Line (MTL) is a lateral fault located in the plate boundary margin between the Philippine Sea plate and the Eurasian plate (**Figure 1**), and the MTL fits perfectly the class of fault mentioned above.

The MTL, which has a length of more than 1000 km, is the largest tectonic line in southwest Japan, and its eastern portion has moved as a right-lateral fault with a reverse-fault component during the Quaternary. Because the MTL is a very important tectonic line for understanding the tectonics in and around southwest Japan, many geological, geophysical, and geomorphological surveys have been conducted.

Geomorphological studies have suggested that the MTL is basically vertical (e.g., [3, 4]), and its slip rate has been estimated as 5–10 mm/yr in the Shikoku District (e.g., [4]). On the other hand, reflection surveys have indicated a dip of the MTL of between 30 and 55°N (e.g., [5–11]), although Itoh et al. [12] found high-dip faults in the Quaternary layer. In general, it is known that such low-dip faults could not move as a lateral fault, because the overburden pressure acting on the fault plane would be too large to allow lateral motion and the dips would not have optimum orientation for releasing the shear stress (e.g., [13, 14]). In order to solve this inconsistency, Sato et al. [15] suggested that the MTL, which functioned as a megathrust in the Cretaceous, was reactivated as a strike-slip fault under subsequent-phase shear stresses.

In addition to these scientific interests, there are some societal concerns or desires pertaining to this tectonic line. As mentioned above, the MTL, with a length of more than 1000 km, is the

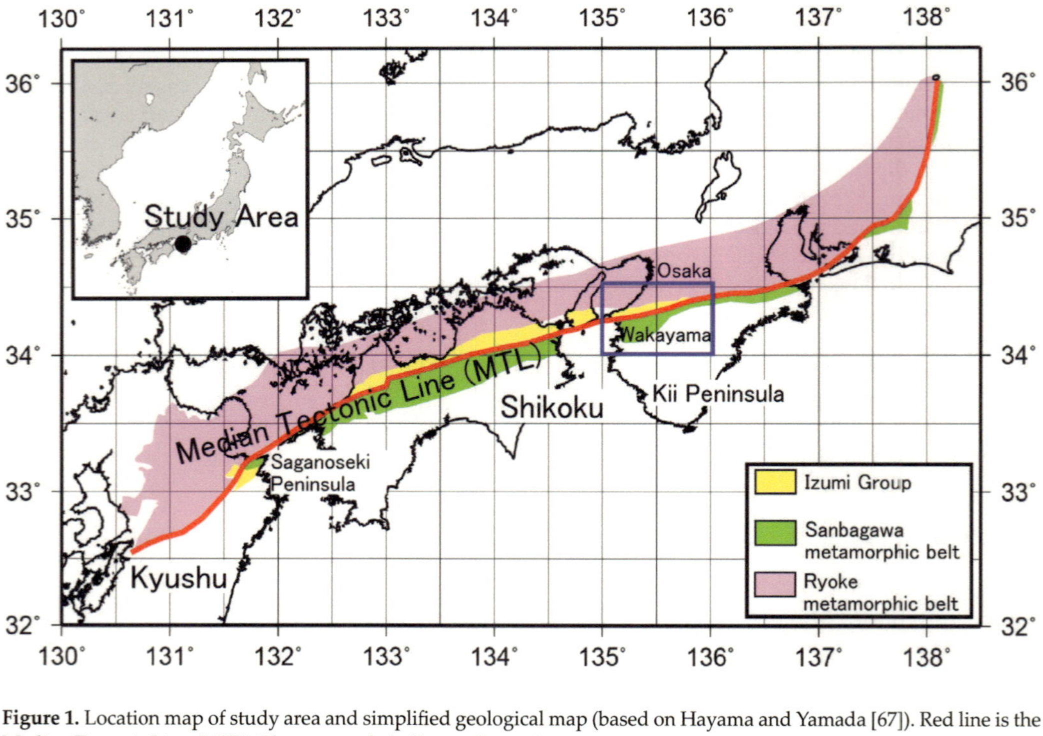

Figure 1. Location map of study area and simplified geological map (based on Hayama and Yamada [67]). Red line is the Median Tectonic Line (MTL). Blue rectangle indicates the study area.

http://dx.doi.org/10.5772/67964

largest tectonic line in southwest Japan. Consequently, several cities are in proximity to the MTL. In particular, there are large cities and international airports around the eastern end of its Quaternary activity.

The dip of a fault affects the size of its area of disaster occurrence (e.g., [16, 17]) and is an important parameter in numerical simulations for hazard map development (e.g., [18]). In general, the size of the area of disaster occurrence is wider on the hanging wall side than on the footwall side of a fault (e.g., [17]). Consequently, if the MTL was a low-dip fault, as shown in reflection surveys, and were to move, the Osaka area at the Kii Peninsula (the eastern end of Quaternary MTL activity) would sustain serious damage. On the other hand, if the MTL was a high-dip fault, as suggested by geomorphological studies, and were to move, both the Osaka area and the Wakayama area of the Kii Peninsula would sustain damage. Therefore, knowing the dip of the MTL, as a seismogenic fault, is very important not only for advancing scientific knowledge but also for societal reasons.

In this study, we attempted to estimate the dip of the MTL from the viewpoints of both a material boundary and a motion boundary. For estimation of the MTL fault dip as a material boundary, we employed gravity anomaly and gravity gradient tensor data, as excellent seismic surveys have been conducted on the MTL. For estimation of the MTL fault dip as a motion boundary, we employed numerical simulations based on dislocation theory in an elastic medium and evaluated deformation fields at the fault terminus caused by lateral fault motions.

2. The gravity anomaly and its interpretations

In this study, we used the Bouguer anomaly database of Komazawa [19]. As the Bouguer anomaly database employed here provides users with 0.5 km × 0.5 km mesh data, we could discuss structures with sizes larger than several kilometers.

Figure 2 is the residual Bouguer anomaly map of the study area. Here, we employed the Bouguer anomaly data of which the Bouguer density is 2670 kg/m^3. In order to obtain the residual Bouguer anomaly map, we estimated the first trend surface of the Bouguer anomaly in this region by the least-squares method from the Bouguer anomaly and removed it from the original Bouguer anomaly data.

Figure 2 is characterized by a distribution of gravity lows and highs. The gravity lows occur in the Izumi Mountains and on the Wakayama Plain and the gravity highs are distributed on the north and south sides of these gravity lows. By comparing **Figure 2** with **Figure 3** (the surface geological map), we found that the gravity highs correspond to outcrops of the Sanbagawa metamorphic rocks and the Ryoke plutonic rocks. The gravity lows are observed in the Izumi Mountains, consisting of gravel, sandstone, and mudstone of the late Cretaceous, and in the Wakayama Plain, consisting of Quaternary sediments, because the densities of these lithologies are not high.

As is well known, techniques for extracting the locations of structural boundaries (namely, edge), without geological and geophysical information, from a gravity anomaly map are called

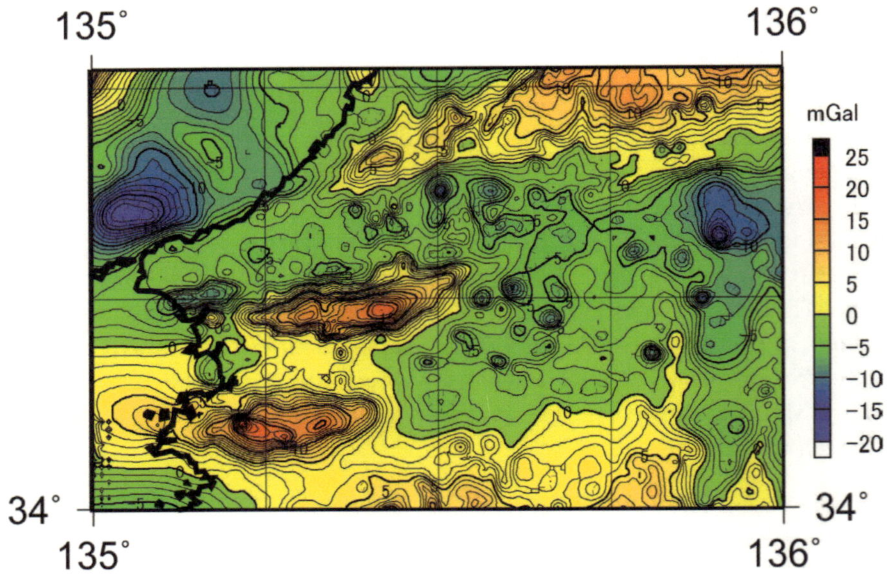

Figure 2. Residual Bouguer anomaly map of the study area and surroundings. The contour interval is 1 mGal. In the residual Bouguer anomaly map, the regional linear trend of the Bouguer anomalies, such as the effect of the subducting plate, is estimated and removed by the least-squares method from the Bouguer anomalies. The Bouguer anomaly used here is based on the gravity anomaly database (0.5 km × 0.5 km mesh data) by Komazawa [19], and a Bouguer density of 2670 kg/m^3 is employed.

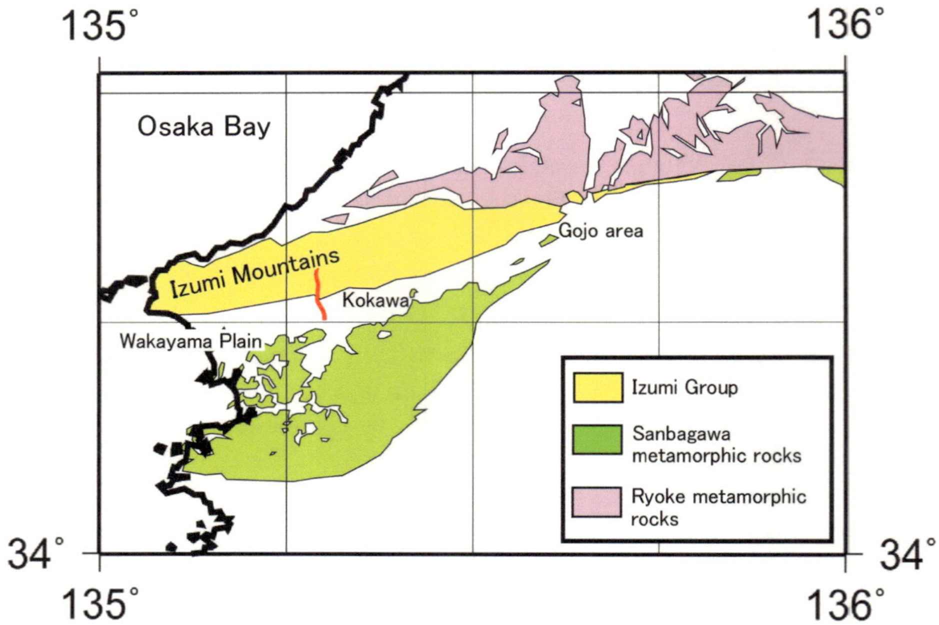

Figure 3. Simplified geological map of the study area (based on Ref. [68]). Izumi Group, Ryoke metamorphic rocks (granites and plutonic rocks), and Sanbagawa metamorphic rocks are shown. The red line in this figure indicates the seismic survey line shown in **Figure 8**.

semi-automatic interpretation methods. These techniques have played an important role in the qualitative interpretation of gravity anomaly maps. A typical technique is the horizontal gravity gradient method (e.g., [20, 21]), which uses the horizontal gravity gradient *HG* defined as follows:

$$HG = \sqrt{\left(\frac{\partial g}{\partial x}\right)^2 + \left(\frac{\partial g}{\partial y}\right)^2}, \tag{1}$$

where g is the Bouguer anomaly. The horizontal gravity gradient method has frequently been employed to find edges caused by structural boundaries such as faults or material boundaries (e.g., [22–24]).

In recent years, more sensitive and keen methods have been developed, and their extraction abilities have been investigated (e.g., [25, 26]). Li et al. [25] evaluated the extraction abilities of existing semi-automatic interpretation methods by numerical tests and pointed out following characteristics. The *HG* method (e.g., [20]), shown by Eq. (1) can determine structural boundaries of the causative body but has decreased ability with depth. The *TDR* (Eq. (2); [27]),

$$TDR = \arctan\left[\frac{\frac{\partial g}{\partial z}}{\sqrt{\left(\frac{\partial g}{\partial x}\right)^2 + \left(\frac{\partial g}{\partial y}\right)^2}}\right], \tag{2}$$

and the *THETA* (Eq. (3); [28]),

$$THETA = \arccos\left[\frac{\sqrt{\left(\frac{\partial g}{\partial x}\right)^2 + \left(\frac{\partial g}{\partial y}\right)^2}}{\sqrt{\left(\frac{\partial g}{\partial x}\right)^2 + \left(\frac{\partial g}{\partial y}\right)^2 + \left(\frac{\partial g}{\partial z}\right)^2}}\right], \tag{3}$$

are also able to extract edges, but these methods draw a phantom edge at the zero line in the case of side-by-side positive and negative anomalies. The *TDX* (Eq. (4); [29])

$$TDX = \arctan\left[\frac{\sqrt{\left(\frac{\partial g}{\partial x}\right)^2 + \left(\frac{\partial g}{\partial y}\right)^2}}{\left|\frac{\partial g}{\partial z}\right|}\right], \tag{4}$$

has the same phantom edge but can provide a clearer edge than the *TDR* or *THETA*.

Zhang et al. [26] showed that the *ILP* (Eq. (5); [30]),

$$ILP = \arcsin\left[\frac{\sqrt{\left(\frac{\partial g}{\partial x}\right)^2 + \left(\frac{\partial g}{\partial y}\right)^2}}{\sqrt{\left(\frac{\partial g}{\partial x}\right)^2 + \left(\frac{\partial g}{\partial y}\right)^2 + \left(\frac{\partial^2 g}{\partial x^2} + \frac{\partial^2 g}{\partial y^2}\right)^2}}\right], \tag{5}$$

and the *TAHG* (Eq. (6); [31]),

$$TAHG = \arctan\left[\frac{\frac{\partial HG}{\partial z}}{\sqrt{\left(\frac{\partial HG}{\partial x}\right)^2 + \left(\frac{\partial HG}{\partial y}\right)^2}}\right], \tag{6}$$

successfully extract the edges of potential field data relating to multisubsurface structures. For evaluating extraction ability, Li et al. [25] and Zhang et al. [26] employed the cube model and the rectangular parallelepiped model, which have vertical structural boundaries.

To extract the edge of the MTL trace from the residual Bouguer anomaly data shown in **Figure 2**, we applied the *HG*, *TDX*, *ILP*, and *TAHG*, which have been confirmed as excellent extraction methods. **Figures 4–7** show the results obtained by each method.

The *HG* method successfully extracted high-gradient anomalies between gravity highs and gravity lows (**Figure 4**). Additionally, the method extracted east-west trending, short, high-gradient anomalies on the Wakayama Plain and the boundary between the plain and the Izumi Mountains. However, these anomalies are not linked to one another continuously, and it is difficult to state that the *HG* method was able to extract the trace of the MTL sufficiently.

The *TDX*, *ILP*, and *TAHG* indicate structural boundary locations by their peak values. Similar to the *HG* method, the *TDX* method successfully extracted high-gradient anomalies between gravity highs and gravity lows (**Figure 5**). Although the method also extracted the east-west trending lines, they do not correspond to the trace of the MTL. The *TDX* at the northern part might indicate boundaries among the Quaternary sediments of the Osaka Plain, the Ryoke plutonic rocks, and the Izumi Group that forms the Izumi Mountains.

The *ILP* and *TAHG* methods were developed to find the edge of the potential fields related to multi subsurface structures, and they generally emphasize very small signals caused by deep structures. The *ILP* map extracted the east-west trending lines, but it is difficult to determine whether or not the map extracted the MTL trace (**Figure 6**). The *TAHG* map is similar to the *ILP* map and has similar difficulties of interpretation (**Figure 7**).

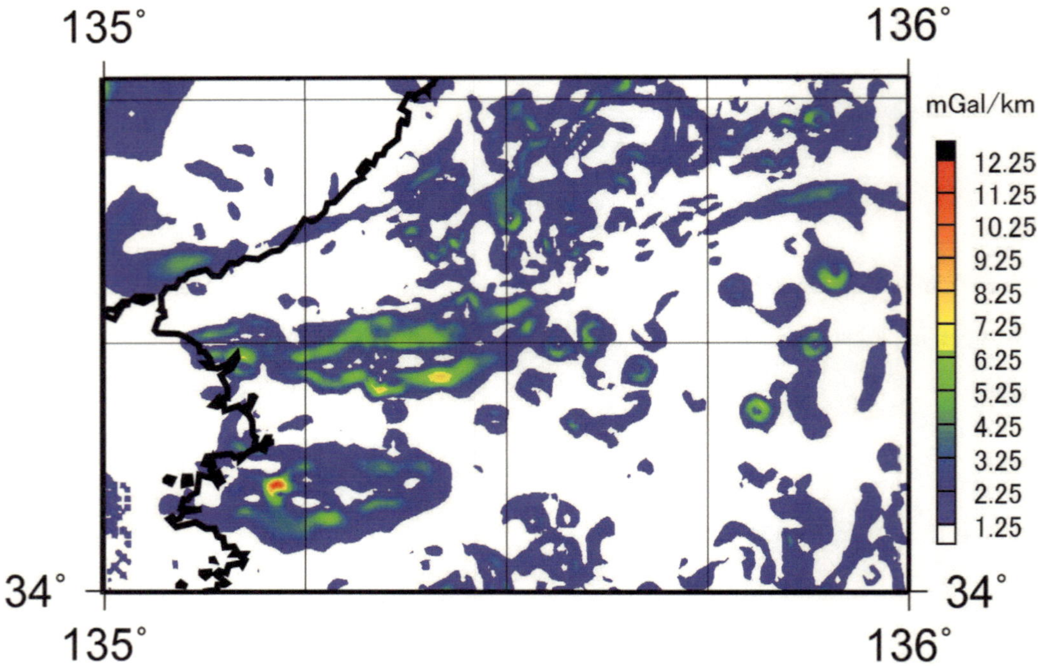

Figure 4. Horizontal gravity gradient map for values over 1.25 mGal/km. Colored areas indicate locations of structural boundaries such as faults or contacts between different materials.

From these results, we can only conclude that these extraction methods cannot extract the MTL trace sufficiently. Thus, we considered that there would be no conspicuous density contrast around the MTL. In the next section, we estimate the density contrast between the Izumi Group and the Quaternary sediments in the Wakayama Plain.

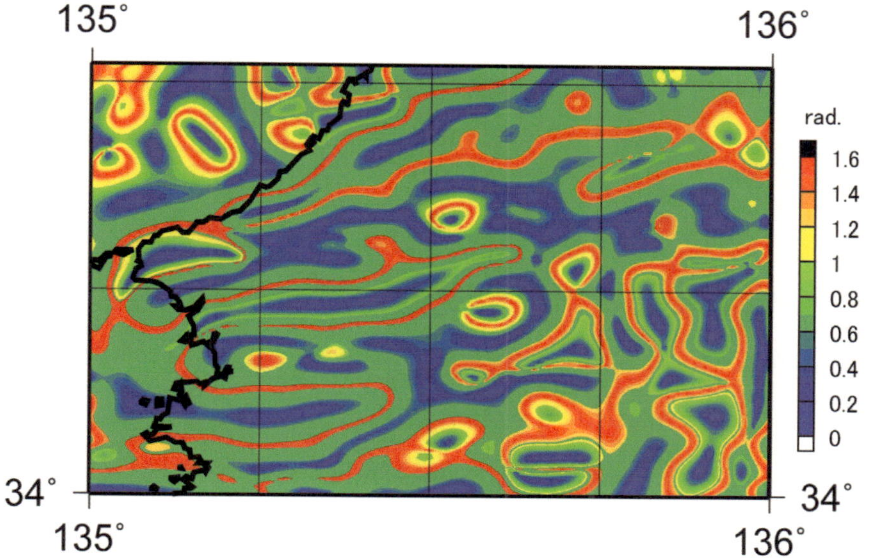

Figure 5. *TDX* map. Units are radians. This map indicates geological or structural boundary by the line *TDX* = max (here, *TDX* = 1.6).

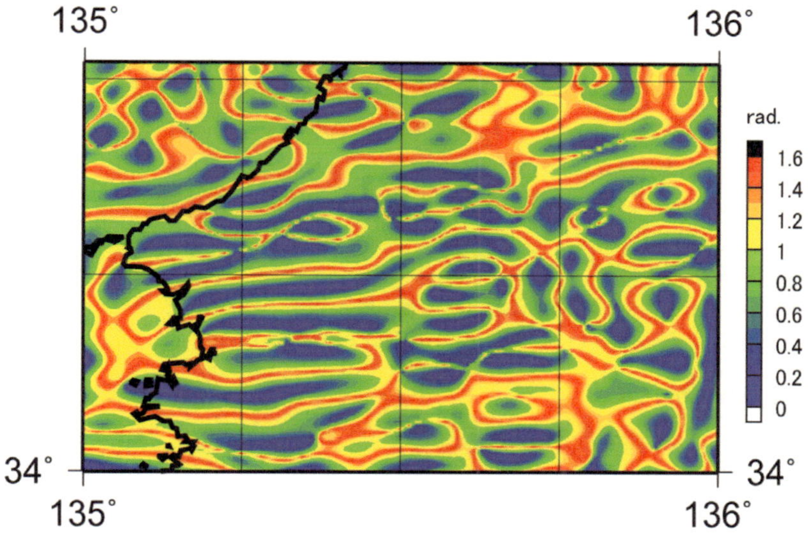

Figure 6. *ILP* map. Units are radians. This map extracts the edge of the gravity anomaly related to multisubsurface structures.

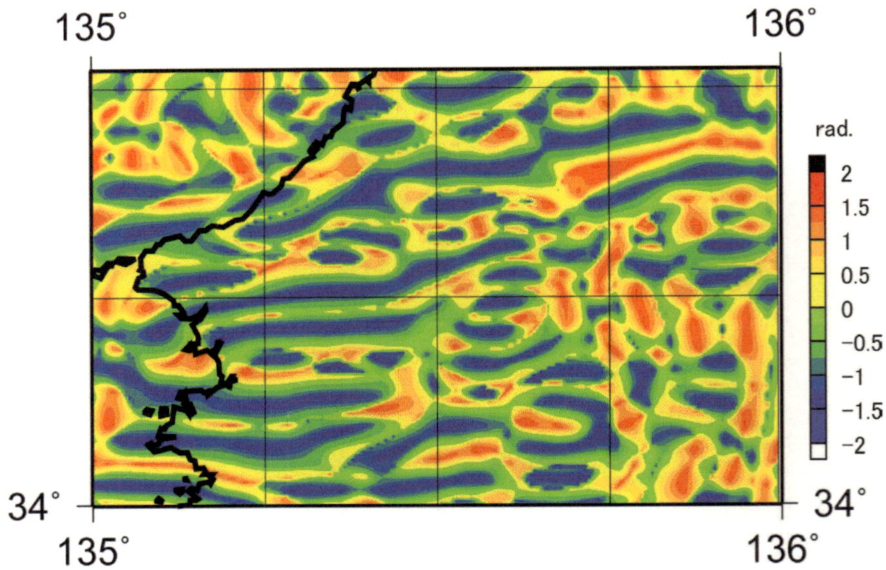

Figure 7. *TAHG* map. Units are radians. This map extracts the edge of the gravity anomaly related to multisubsurface structures.

3. Estimation of density contrast around the MTL

Figure 8 shows a depth-converted seismic profile crossing the MTL (red line shown in **Figure 3**) [32]. Here, the MTL is the boundary between the Izumi Group and the Quaternary sediment. **Figure 8** shows that the dip angle of the MTL is high near the surface but that it becomes gradually lower with depth.

Here, we simplified the subsurface structure (**Figure 8**) to a three-layer model (**Figure 9**), and we assigned densities of 2800 and 2400 kg/m^3 to the lowest layer equivalent to the Sanbagawa metamorphic rocks and the uppermost layer equivalent to the Izumi Group, respectively. The density of the Sanbagawa metamorphic rocks was referred to that at Saganoseki Peninsula, Oita, Kyushu (e.g., [33]). The density of the Izumi Group was a density typical of the solidified sediment. In this study, we changed the density of the Quaternary sediment layer within the range of 2000–2400 kg/m^3 and estimated the optimum density of the Quaternary sediment layer by comparing the calculated Bouguer anomalies with the observed ones. The two-dimensional Talwani's method [34] was employed for the calculations.

The results are summarized in **Figure 10**, and they show that the optimum density of the Quaternary sedimentary layer is 2300 kg/m^3. This supports the subsurface structures estimated by the seismic reflection survey, and it indicates that there is no conspicuous density contrast around the MTL. Specifically, the MTL is a low-dip fault as a material boundary, and its motion boundary is difficult to identify by gravity anomalies because it does not have a conspicuous density contrast.

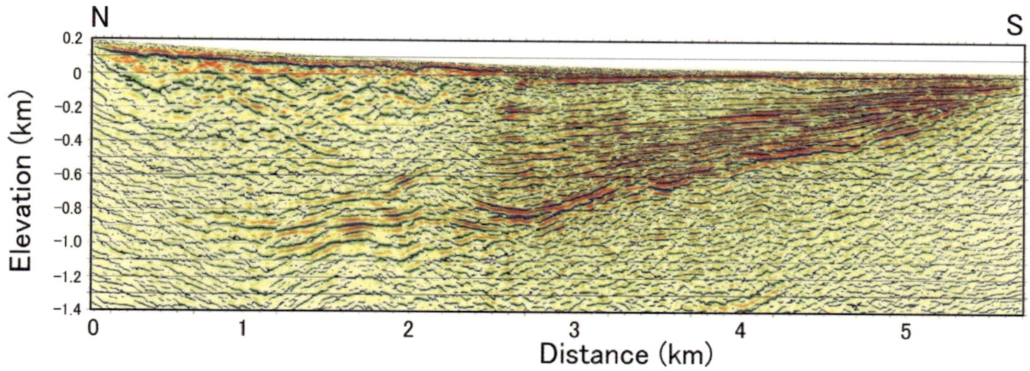

Figure 8. Depth-converted seismic profile crossing the MTL (after Ref. [32]). Location of this profile is shown in **Figure 2** by the red line.

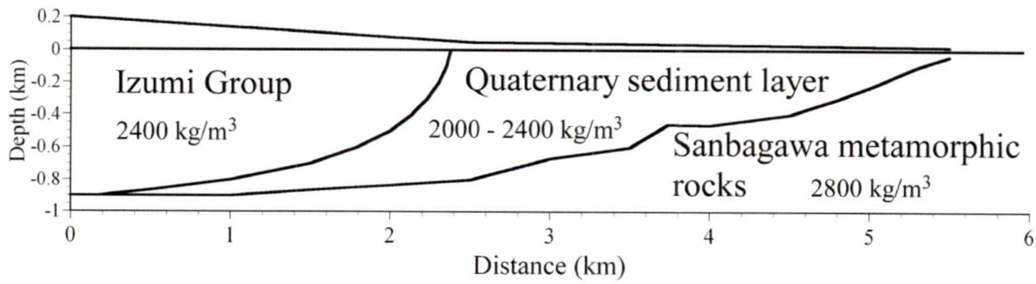

Figure 9. Simplified subsurface structure model based on the depth-converted seismic profile shown in **Figure 8**.

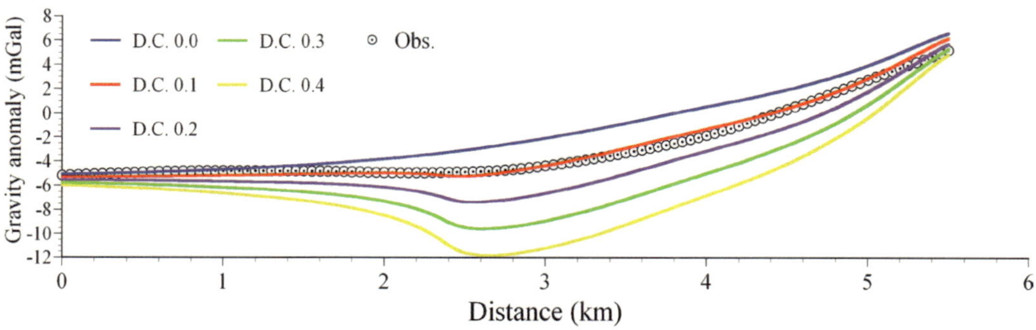

Figure 10. Estimation of the density contrast among the Izumi Group, the Sanbagawa metamorphic rocks, and the Quaternary sediment. In this figure, the circles indicate the observed gravity anomalies, and the colored lines indicate the calculated gravity anomalies from assumed subsurface density structures. Values of D.C. X.X indicate the density contrast between the Izumi Group and the Quaternary sediment in the unit of g/cm^3. For example, D.C. 0.1 means that the density contrast between the Izumi Group and the Quaternary sediment is 0.1 g/cm^3 (=100 kg/m^3).

4. Analysis using the gravity gradient tensor

In recent years, the gravity gradient tensor, whose response to subsurface structures is more sensitive than that of the gravity anomaly, has been employed for obtaining detailed subsurface structure data. The gravity gradient tensor Γ is defined by the differential coefficients of the gravitational potential W (e.g., [35, 36]) as follows:

$$\Gamma = \begin{bmatrix} \frac{\partial^2 W}{\partial x^2} & \frac{\partial^2 W}{\partial y \partial x} & \frac{\partial^2 W}{\partial z \partial x} \\ \frac{\partial^2 W}{\partial x \partial y} & \frac{\partial^2 W}{\partial y^2} & \frac{\partial^2 W}{\partial z \partial y} \\ \frac{\partial^2 W}{\partial x \partial z} & \frac{\partial^2 W}{\partial y \partial z} & \frac{\partial^2 W}{\partial z^2} \end{bmatrix}. \tag{7}$$

The gravity gradient tensor is symmetric (e.g., [35]), and the sum of its diagonal components is zero as the gravitational potential satisfies Laplace's equation:

$$\frac{\partial^2 W}{\partial x^2} + \frac{\partial^2 W}{\partial y^2} + \frac{\partial^2 W}{\partial z^2} = 0. \tag{8}$$

Various analysis techniques using the gravity gradient tensor have been suggested and discussed (e.g., [37–41]). Among these methodologies, Beiki and Pedersen [42] showed that the maximum eigenvector of the gravity gradient tensor points to the causative body (**Figure 11a**). Kusumoto [43] considered that the basement consists of an aggregate of high-density prisms (**Figure 11b**) and applied Beiki's technique [42, 44] to the estimation of the fault dip. This method provided results in which the fault dip estimated by the gravity gradient tensor was in harmony with the dip observed from seismic surveys [43, 45].

Kusumoto [46] showed that in two-dimensional analysis the maximum eigenvector of the tensor points to a high-density causative body (**Figure 11a**) and that the dip of the maximum eigenvector closely follows the dip of the normal fault (**Figure 11b**). In addition, he showed that the minimum eigenvector of the tensor points to a low-density causative body (**Figure 11c**) and that the dip of the minimum eigenvector closely follows the dip of the reverse fault (**Figure 11d**).

In two-dimensional analysis, the gravity gradient tensor Γ_{2D} on the profile is defined as follows (e.g., [38]):

$$\Gamma_{2D} = \begin{bmatrix} \frac{\partial^2 W}{\partial x^2} & \frac{\partial^2 W}{\partial z \partial x} \\ \frac{\partial^2 W}{\partial x \partial z} & \frac{\partial^2 W}{\partial z^2} \end{bmatrix}. \tag{9}$$

Because the gravity potential satisfies the Laplace's equation $\partial^2 W/\partial x^2 + \partial^2 W/\partial z^2 = 0$, we find the relationship $\partial^2 W/\partial z^2 = -\partial^2 W/\partial x^2$. Additionally, $\partial^2 W/\partial x \partial z = \partial^2 W/\partial z \partial x$, because the gravity gradient tensor is a symmetric tensor (e.g., [35]).

As gravity gradiometry surveys have not been conducted in this study area, we estimated the gravity gradient tensor from the gravity anomalies on the seismic reflection survey profile (**Figure 10**), using the Fourier transform method (e.g., [46, 47]), and calculated the dips of the maximum and minimum eigenvectors of the tensor. The dips, α, of the maximum and minimum eigenvectors of the tensor were calculated by

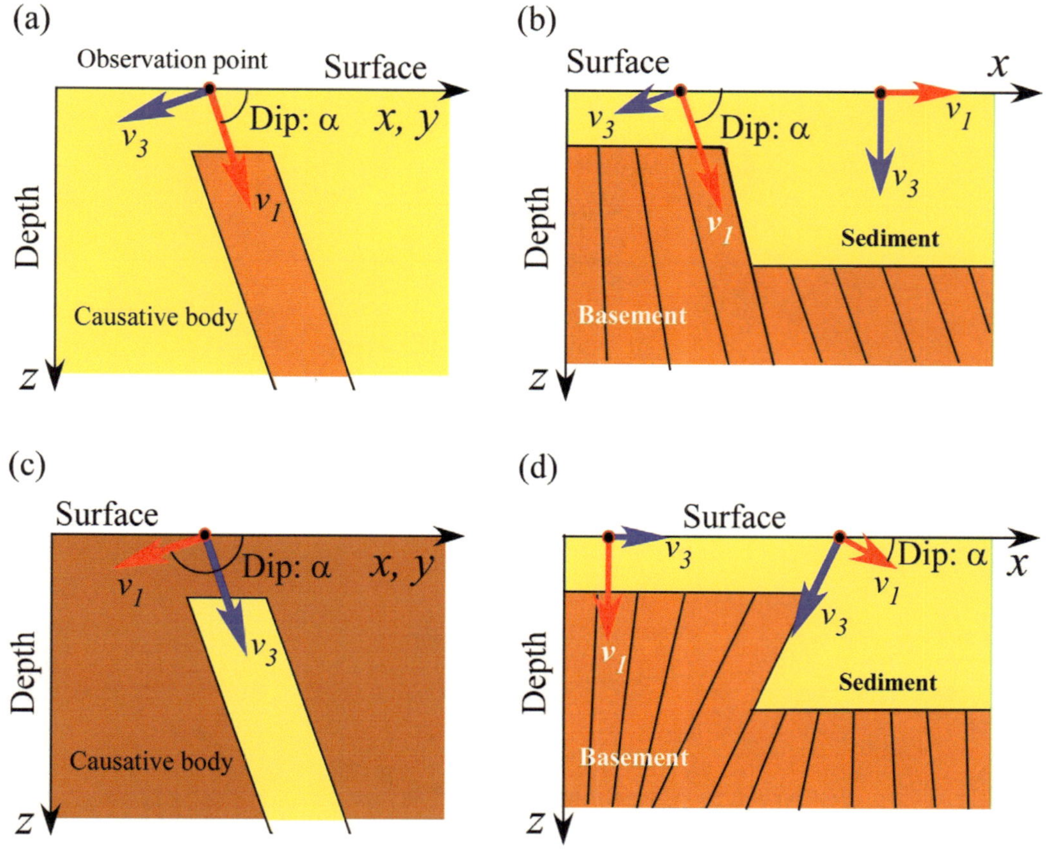

Figure 11. Schematic illustration of the maximum eigenvector and the minimum eigenvector for two-dimensional (2-D) structures. (a) Basic model (high-density causative body model). In this figure, v_1 and v_3 are the maximum eigenvector (red) and the minimum eigenvector (blue) of the gravity gradient tensor, respectively. The maximum eigenvector points to the high-density causative body (brown). (b) Normal fault model. The basement consists of an aggregate of high-density prisms (brown), and the angle, α, of the maximum eigenvector (red) of the tensor follows the fault dip. The minimum eigenvector on the thick sedimentary layer points to the vertical. (c) Low-density causative body model. The minimum eigenvector points to the low-density causative body (light yellow). (d) Reverse fault model. The basement consists of an aggregate of high density prisms (brown), and the angle, α, of the minimum eigenvector (blue) of the tensor follows the fault dip. The maximum eigenvector (red) on the basement points to the vertical.

$$\alpha = \arctan\left(\frac{v_z}{v_x}\right), \tag{10}$$

where v_x and v_z are the x and z components of each eigenvector, respectively.

Figure 12 shows each component of the gravity gradient tensor, and **Figure 13** shows the distributions of the maximum eigenvector (red) and the minimum eigenvector (blue) of the gravity gradient tensor. From **Figure 13**, it was found that the dips of the maximum eigenvectors follow the dip of the Sanbagawa metamorphic layer. On the other hand, the minimum eigenvector points vertically, and the dips of the eigenvectors do not follow the fault dip of the MTL. We considered that this occurs because the structural boundary could not be recognized due to the low-density sediment layers near the surface in the Izumi Group and in

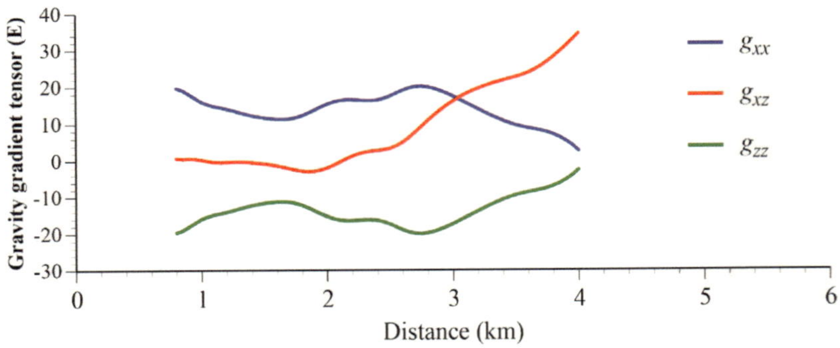

Figure 12. Gravity gradient tensor on the profile. These are estimated from the observed Bouguer anomalies on the profile shown in **Figure 10**.

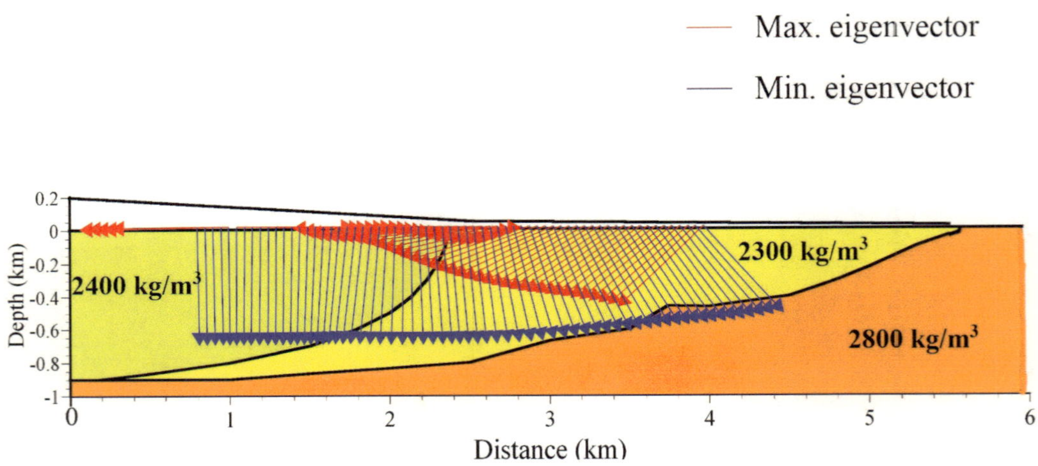

Figure 13. Eigenvectors of the gravity gradient tensor on the profile shown in **Figure 12**. The maximum eigenvector and minimum eigenvector are indicated by red and blue, respectively. The dips of eigenvectors are given clockwise from x-axis to z-axis.

the Quaternary layer areas. As shown in the previous subsection and in **Figure 10**, there is no conspicuous density contrast. Because the gravity gradient tensor is very sensitive to the subsurface structure near the surface, if a low density layer existed near the surface, the minimum eigenvector of the tensor would point to the low density layer, namely vertically.

5. Surface deformations due to strike-slip fault

The gravity anomaly and gravity gradient tensor cannot recognize or extract the fault location and shape if a conspicuous density contrast does not exist. Dip-slip faults in the thick sedimentary layers and very low-dip faults and lateral faults in the isotropic medium would be examples.

The MTL has both low-dip and lateral fault components. Thus, the gravity anomaly and gravity gradient tensor analyses could not extract the fault trace of the MTL nor estimate the fault dip. The remaining problem is how a low-dip reverse fault could move as a lateral fault: the contradiction between its fault dip and its fault motion.

Thus far, the fault dip and the fault shape have been discussed only as a material boundary because they have been estimated from geophysical explorations such as seismic surveys and gravity surveys. We think that this is the main reason for the contradiction. Using numerical simulations, we thus examined the fault dip of the MTL from the viewpoint of the fault motion, namely, by restoration of the tectonic topography caused by the fault motion.

In the numerical simulations, we employed dislocation modeling, which is often used for quantitative interpretation of crustal deformation caused by earthquakes and/or volcanic activities (e.g., [48–52]). Surface or interior displacements or strains can be calculated by considering dislocations on a plane embedded in an elastic isotropic half-space (**Figure 14**). The dislocation modeling employed here was Okada's dislocation solutions [53] and is defined within a range of the linear elasticity. Consequently, the modeled structures could not be compared directly with the actual geological structures. However, if we simply assume that the geological structures are the accumulation of the deformations caused by the fault motions, dislocation modeling can be a very useful tool because it provides the essential deformation pattern from the minimum number of parameters (fault parameters and elastic constants of the medium).

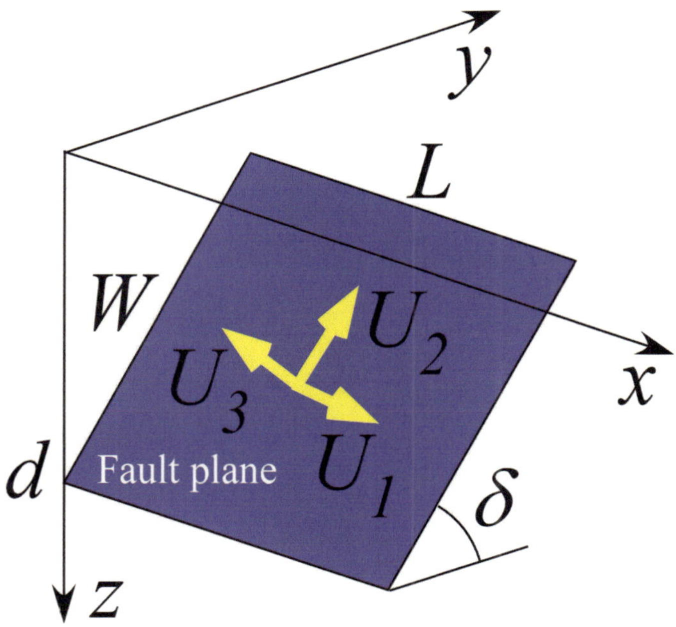

Figure 14. Dislocation plane. *W*, *L*, *d*, and δ are width, length, depth, and dip of the fault. U1, U2, and U3 are strike-slip, dip-slip, and tensile components of arbitrary dislocation. Positive U1 and U2 means left-lateral slip and thrust slip with $0 < \delta < \pi/2$.

In this study, we evaluated how the displacement fields at the terminus of the lateral fault would change if the fault dip was changed. A conspicuous vertical deformation field formed by lateral fault motion occurred at the terminus of the fault, and this did not occur at other areas (**Figure 15**). Geological structures formed by lateral faulting, such as the pull-apart basin, are the result of the accumulation of vertical deformation (subsidence) at the fault terminus area (e.g., [54–56]). In the numerical simulations for restoring these structures, if the geological evidence suggested that the active zone of a lateral fault shifted along its strike direction over time, we expressed its tectonic history by superimposing analytical solutions for different fault lengths (e.g., [57–59]). Hence, we paid attention to the vertical deformation pattern at the fault terminus.

In **Figure 16**, we show predicted vertical deformations at the terminus due to lateral faulting with dips of 45, 60, 70, 75, 80, and 90°. The length and width of the assumed fault model were 30 and 15 km, respectively, and the top of the model was the surface. We set a right-lateral motion of 1 m on the dislocation plane in each model, and a Poisson's ratio of 0.25 was assumed. **Figure 16** shows deformation fields within a distance of 5 km around the fault, and the deformation is the total of both the vertical and the horizontal deformations caused by the right-lateral fault motion. The light-purple thin line in this figure indicates vertical deformation of zero.

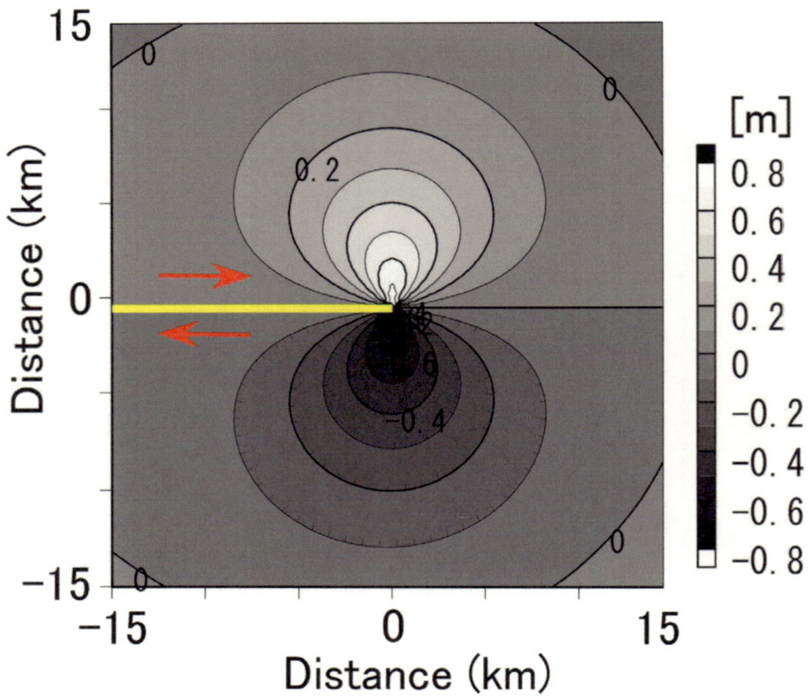

Figure 15. Deformation pattern at the fault terminus caused by right-lateral motion of 1 m. The fault dip is 90°. Yellow line indicates the location of the vertical fault, and red arrows indicate directions of fault motion.

http://dx.doi.org/10.5772/67964

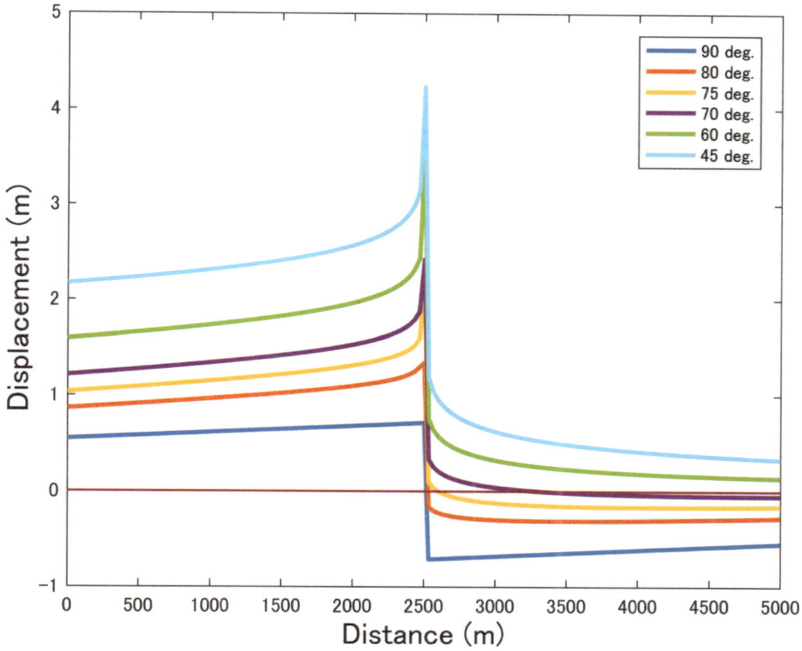

Figure 16. Predicted vertical deformations at the terminus due to lateral faulting with dips of 45, 60, 70, 75, 80, and 90°. The light-purple thin line indicates vertical deformation of zero.

From **Figure 16**, it was found that the basement on the footwall side would not subside if the fault dip is lower than 70°. Specifically, if the fault dip is lower than 70°, the areas around the terminus of the lateral fault would become uplift areas and areas of subsidence would not form. This also indicates that tectonic basins, such as a pull-apart basin, caused by lateral fault motions could not form if the lateral fault has a dip lower than 70°. On the other hand, if the fault dip is higher than 70°, the basement of the footwall side of the lateral fault would begin to subside around the terminus. Additionally, in the case of a fault dip of 90°, uplift of the basement of the hanging wall side would be equal to the subsidence of the basement of the footwall side. This indicates that a fault dip higher than 70° would be required to form tectonic basins due to lateral fault motions. From **Figure 16**, it was found that even if fault scale and lateral motion are the same, the surface deformation patterns would be different as an effect of fault dip. Specifically, surface deformations are basically dependent on fault dip.

6. Discussion

Gravity anomaly analyses, gravity gradient tensor analyses, and seismic surveys are important and useful for finding subsurface structures, and these subsurface structures play an important role for understanding the tectonics and formation processes of the area of concern. In the numerical simulations for restoring the subsurface structures, tectonic topography, and geological structures, we referred often to the subsurface structure given by geophysical explorations or geological surveys and made basic or initial structure models for the simulations (e.g., [55, 56, 60]).

We paid attention to the contradiction of the MTL being a right-lateral fault in the Quaternary, despite the fact that several seismic surveys have estimated its fault dip to have low angle.

For confirmation, we attempted to extract the MTL trace from the Bouguer anomaly by using several excellent edge emphasis techniques, but high dip-angle traces were not found. In addition, a state-of-the-art dip estimation technique using eigenvector analyses of the gravity gradient tensor could not suggest a high fault dip of the MTL. These facts showed that the MTL does not have a conspicuous density contrast at its motion boundary and that it has low dip as a material boundary.

Considering these results and the actual fault motion of the MTL, we paid attention to the deformation field at the fault terminus having an important role in the formation of tectonic basins caused by lateral fault motions, and we attempted to evaluate the fault dip from the viewpoint of fault motion. As a result, it was found that even if fault scale and lateral motion are the same, the surface deformation patterns at the fault terminus are controlled by fault dip. Specifically, if fault dip is lower than 70°, subsidence areas would not form at the fault terminus, but if fault dip is higher than 70°, both a subsidence area and an uplift area would form at the terminus.

As the present tectonic topography is an accumulation of uplift and subsidence caused by fault motions, we can apply the results shown here to the field as a first approximation without viscoelasticity and plastic flow of the crust at the geological time scale. Consequently, we can roughly estimate the fault dip from the deformed basement and/or sedimentary layer shape.

Conspicuous late Quaternary uplift and subsidence have not been found at the footwall of the MTL because (1) Pliocene layers distribute conspicuously around the eastern portion, the so-called Gojo area, of the footwall of the MTL [61]; (2) there are no new Quaternary layers in boring data obtained around the Kokawa area [8]; and (3) only two marine sediment clay layers of the last interglacial stage (Ma 12: discrimination name in Japan) and alluvium (Ma 13: discrimination name in Japan) in the marine sediment clay layers (Ma 1–13) found in Osaka Bay and Osaka basin have been found in the Wakayama Plain [62]. In addition, trench excavation studies show that the MTL is a vertical or subvertical fault, as was pointed out by Tsutsumi et al. [3], Yamazaki et al. [63], and others. Itoh et al. [12] also found high-dip faults in the Quaternary layer in the seismic profile given by Yoshikawa et al. [5]. Considering these geological and geomorphological evidences and the numerical simulation results of this study, we believe that a fault dip of 70–75° is a reasonable dip angle for the MTL. With a dip of 70–75°, the fault would be able to move as a lateral fault and also have the possibility of reverse-fault motion.

Because the technique of restoring geological structures or topography by numerical simulations is a useful method for evaluating the relationship between fault motions and such structures (e.g., [54–58]), our approach, which is an extension of this technique, should contribute to research of active faults. Although this study employed the dislocation solution in

http://dx.doi.org/10.5772/67964

linear elastic medium as the procedure of numerical simulation, a more detailed and higher accuracy discussion of fault motion and fault shape will be possible by employing a more realistic method such as finite element modeling, discrete element modeling, or other higher techniques (e.g., [64–66]).

On the other hand, new questions that should be discussed and solved in the future were revealed: (1) Why does the motion boundary of the MTL not correspond to the material boundary? (2) How did this structure or contradiction occur? (3) Can the process be explained from the viewpoints of both structural geology and geophysics? (4) Is the process compatible with regional tectonics? (5) Do faults of this type exist in other areas? To discuss and solve these problems, the ideas, suggestions, and interpretations of the tectonics and structures shown by Sato et al. [15] may be useful and helpful, although it is difficult to accept directly lateral fault motion of a low-dip fault. Therefore, we need to continue to study the MTL in the future.

7. Conclusions

Because the MTL is a very important tectonic line for understanding the tectonics in and around southwest Japan, many geological, geophysical, and topographical surveys have been conducted on it. Several reflection surveys have indicated that the MTL has very low dips of less than 55°. As these low-dip faults would not generally move as a lateral fault, the MTL presents a contradiction between its fault dip and fault motion. In addition, because the MTL is proximal to several cities and because fault dip is an important factor for hazard map creation, it is important to estimate the fault dip of the MTL as a seismogenic fault correctly.

In this study, we first attempted to extract the MTL trace and estimate the fault dip from the Bouguer anomaly and gravity gradient tensor, as a confirmation of previous and conventional results. However, these methods could not reveal any high dip-angle faults, because the MTL does not have a conspicuous density contrast at its boundary and has low dip as a material boundary.

To estimate the fault dip of the MTL from the viewpoint of fault motion, we next attempted to evaluate deformation fields at the fault terminus by numerical simulations based on dislocation theory in an elastic medium. The result showed that the surface deformation patterns at the fault terminus are controlled by fault dip if fault scale and lateral motion are the same. If the fault dip was lower than 70°, subsidence areas did not form at the fault termination. However, if the fault dip was higher than 70°, both subsidence and uplift areas formed at the terminus. Considering the accumulated geological and geomorphological evidence for the motion of the MTL in the Quaternary, and our numerical simulation results, we suggest that a fault dip of 70–75°, at which the fault would be able to move as a lateral fault with a reverse-fault component, is a reasonable value for the fault dip of the MTL (as a seismogenic fault).

Our results and suggestions indicate that the motion boundary of the fault does not correspond with the material boundary. In the future, it will be important to research this formation mechanism, including whether or not this type of fault exists in other areas.

Acknowledgements

This study was executed as a part of the Comprehensive Research and Survey for the Median Tectonic Line Fault System (Eastern Margin of Kongo Mountains-Southern Margin of Izumi Mountains; FY 2013-2015) organized by the Research and Development Bureau, MEXT, and the DPRI, Kyoto University. This work was supported partially by Japan Society for the Promotion of Science (JSPS) KAKENHI Grant Number 15K14274. The authors are grateful to the agency. And, the authors are also most grateful to Yoichi Fukuda for very useful comments that have considerably improved the manuscript. In addition, the authors are most grateful to Ana Pantar for her editorial advices and cooperation.

Author details

Shigekazu Kusumoto[1]*, Keiji Takemura[2] and Yasuto Itoh[3]

*Address all correspondence to: kusu@sci.u-toyama.ac.jp

1 Graduate School of Science and Technology for Research, University of Toyama, Toyama, Japan

2 Kyoto University, Sakyo-ku, Kyoto, Japan

3 Osaka Prefecture University, Sakai, Osaka, Japan

References

[1] Fitch T J. Plate convergence, transcurrent faults, and internal deformation adjacent to southeast Asia and the western Pacific. Journal of Geophysical Research 1972; **101**: 4432-4460.

[2] McCaffrey R. Slip partitioning at convergent plate boundaries of SE Asia. Geological Society of London Special Publication 1996; **106**: 3-18.

[3] Tsutsumi H, Okada A, Nakata T, Ando M, Tsukuda T. Timing and displacement of Holocene faulting on the Median Tectonic Line in central Shikoku, SW Japan. Journal of Structural Geology 1991; **13**: 227-233.

[4] Tsutsumi H, Okada A. Segmentation and Holocene surface faulting on the Median Tectonic Line, southwest Japan. Journal of Geophysical Research 1996; **101**: 5855-5871.

[5] Yoshikawa S, Iwasaki Y, Ikawa T, Yokota H. Geological structure of the MTL in west Wakayama by reflection seismic survey. Memories of the Geological Society of Japan 1992; **40**: 177-186 (in Japanese with English abstract).

[6] Yusa Y, Takemura K, Kitaoka K, Kamiyama K, Horie S, Nakagawa I, Kobayashi Y, Kubotera A, Sudo Y, Ikawa T, Asada M. Subsurface structure of Beppu Bay (Kyushu,

Japan) by seismic reflection and gravity survey. Zisin 2nd series 1992; **45**: 199-212 (in Japanese with English abstract).

[7] Ito T, Ikawa T, Yamakita S, Maeda T. Gently north-dipping Median Tectonic Line (MTL) revealed by recent seismic studies, southwest Japan. Tectonophysics 1996; **264**: 51-63.

[8] Mizuno K, Tsukuda E, Takahashi M, Momohara A, Uchiyama T. Subsurface geology of the Wakayama Plain, southwestern Japan based on the deep boring survey. Journal of Geological Society of Japan 1999; **105**: 235-238 (in Japanese with English abstract).

[9] Kawamura T, Onishi M, Kurashino E, Ikawa T, Ito T. Deep seismic reflection experiment using a dense receiver and sparse shot technique for imaging the deep structure of the Median Tectonic Line (MTL) in east Shikoku, Japan. Earth Planets Space 2003; **55**: 549-557.

[10] Ito T, Kojima Y, Kodaira S, Sato H, Kaneda Y, Iwasaki T, Kurashimo E, Tsumura N, Fujiwara A, Miyauchi T, Hirata N, Harder S, Miller K, Murata A, Yamakita S, Onishi M, Abe S, Sato T, Ikawa T. Crustal structure of southwest Japan, revealed by the integrated seismic experiment Southwest Japan 2002. Tectonophysics 2009; **472**: 124-134.

[11] Shigematsu N, Fujimoto K, Tanaka N, Furuya N, Mori H, Wallis S. Internal structure of the Median Tectonic Line fault zone, SW Japan, revealed by borehole analysis. Tectonophysics 2012; **532-535**: 103-118.

[12] Itoh Y, Takemura K, Kusumoto S. Neotectonic intra-arc basins within southwest Japan — Conspicuous basin-forming process related to differential motion of crustal blocks. In: Itoh Y. (ed.) Mechanism of Sedimentary Basin Formation: Multidisciplinary Approach on Active Plate Margins. Rijeka: InTech; 2013, pp. 191-206.

[13] Anderson E M. The Dynamics of Faulting and Dyke Formation with Applications to Britain. 2nd edition. Edinburgh: Oliver and Boyd; 1951. 206 p.

[14] Turcotte D, Schubert G. Geodynamics. 2nd edition. Cambridge: Cambridge University Press; 2002. 456 p.

[15] Sato H, Kato N, Abe S, Horne A V, Takeda T. Reactivation of an old plate interface as a strike-slip fault in a slip-partitioned system: Median Tectonic Line, SW Japan. Tectonophysics 2015; **644-645**: 58-67.

[16] Abrahamson N A, Somerville P. Effects of the hanging wall and footwall on ground motions recorded during the Northridge earthquake. Bulletin of Seismological Society of America 1996; **86**: S93-S99.

[17] Takemura M, Moroi T, Yashiro K. Characteristics of strong ground motions as deduced from spatial distributions of damages due to the destructive inland earthquakes from 1891 to 1995 in Japan. Zisin 2 1998; **50**: 485-505 (in Japanese with English abstract).

[18] Irikura K, Miyake H. Recipe for predicting strong ground motion from crustal earthquake scenarios. Pure and Applied Geophysics 2011; **168**: 85-104. doi 10.1007/s00024-010-0150-9.

[19] Komazawa M. Gravity grid database of Japan. DVD Edition, Digital Geoscience Map P-2. Tsukuba: Geological Survey of Japan, AIST. 2013.

[20] Blakely R, Simpson R W. Approximating edges of source bodies from magnetic or gravity anomalies. Geophysics 1986; **51**: 1494-1498.

[21] ten Brink, U S, Ben-Avarham Z, Bell R E, Hassouneh M, Coleman D F, Andreasen F, Tibor G, Coakley B. Structure of the Dead Sea Pull-apart Basin from Gravity Analyses. Journal of Geophysical Research 1993; **98**: 21877-21894.

[22] Shichi R, Yamamoto A, Kimura A, Aoki H. Gravimetric evidences for active faults around Mt. Ontake, central Japan: specifically for the hidden faulting of the 1984 Western Nagano Prefecture Earthquake. Journal of Physics of the Earth 1992; **40**: 459-478.

[23] Kudo T, Kono Y. Relationship between distributions of shallow earthquakes and gradient of gravity anomaly field in southwest Japan. Zishin 2 1999; **52**: 341-350 (in Japanese with English abstract).

[24] Yamamoto A. Gravity anomaly atlas of the Ishikari Plain and its vicinity, Hokkaido, Japan. Geophysical Bulletin of Hokkaido 2003; **66**: 33-62 (in Japanese with English abstract).

[25] Li L, Huang D, Han L, Ma G. Optimized detection filters in the interpretation of potential field data. Exploration Geophysics 2014; **45**: 171-176. doi: 10.1071/EG13059.

[26] Zhang X, Yu P, Tang R, Xiang Y, Zhao C-J. Edge enhancement of potential field data using an enhanced tilt angle. Exploration Geophysics 2014; **46**: 276-283. doi:10.1071/EG13104.

[27] Miller H G, Singh V. Potential field tilt—a new concept for location of potential field sources. Journal of Applied Geophysics 1994; **32**: 213-217.

[28] Wijns C, Perez C, Kowalezyk P. Theta map: edge detection in magnetic data. Geophysics 2005; **70**: L39-L43.

[29] Cooper G R J, Cowan D R. Enhancing potential field data using filters based on the local phase. Computers and Geosciences 2006; **32**: 1585-1591.

[30] Ma G. Edge detection of potential field data using improved local phase filter. Exploration Geophysics 2013; **44**: 36-41.

[31] Ferreira F J F, de Souza J, de B e S Bongiolo A, de Castro L G. Enhancement of the total horizontal gradient of magnetic anomalies using the tilt angle. Geophysics 2013; **78**: J33-J41.

[32] MEXT (Ministry of Education, Culture, Sports, Science and Technology, Japan), DPRI (Disaster Prevention Research Institute, Kyoto University). Comprehensive Research and Survey for the Median Tectonic Line Fault System (Eastern Margin of Kongo Mountains—Southern Margin of Izumi Mountains), Heisei 25 Fiscal Year Report. Uji: DPRI; 2014.

[33] Kusumoto S, Fukuda Y, Takemoto S, Yusa Y. Three-dimensional subsurface structure in the eastern part of the Beppu-Shimabara Graben, Kyushu, Japan, as revealed by Gravimetric Data. Journal of the Geodetic Society of Japan 1996; **42**: 167-181.

[34] Talwani M, Lamar W J, Landisman M. Rapid gravity computations for two-dimensional bodies with application to the Mendocino submarine fracture zone. Journal of Geophysical Research 1959; **64**: 49-59.

[35] Torge W. Gravimetry. Berlin: Walter de Gruyter; 1989. 465 p.

[36] Hofmann-wellenhof B, Moritz H. Physical Geodesy. Berlin: Springer; 2005. 403 p.

[37] Zhang C, Mushayandebvu M F, Reid A B, Fairhead J D, Odegrad M E. Euler deconvolution of gravity tensor gradient data. Geophysics 2000; **65**: 512-520.

[38] Beiki M, Pedersen L B. Window constrained inversion of gravity gradient tensor data using dike and contact models. Geophysics 2011; **76**: I59-I70.

[39] Martinetz C, Li Y, Krahenbuhl R, Braga M A. 3D inversion of airborne gravity gradiometry data in mineral exploration: a case study in the Quadrilatero Ferrifero, Brazil. Geophysics 2013; **78**: B1-B11.

[40] Cevallos C. Automatic generation of 3D geophysical models using curvatures derived from airborne gravity gradient data. Geophysics 2014; **79**: G49-G58.

[41] Li X. Curvature of a geometric surface and curvature of gravity and magnetic anomalies. Geophysics 2015; **80**: G15-G26.

[42] Beiki M, Pedersen L B. Eigenvector analysis of gravity gradient tensor to locate geologic bodies. Geophysics 2010; **75**: I37-I49.

[43] Kusumoto S. Dip distribution of Oita-Kumamoto Tectonic Line located in central Kyushu, Japan, estimated by eigenvectors of gravity gradient tensor. Earth Plan Space 2016; **68**: 153. doi: 10.1186/s40623-016-0529-7.

[44] Beiki M. TSVD analysis of Euler deconvolution to improve estimating magnetic source parameters: an example from the Asele area, Sweden. Journal of Applied Geophysics 2013; **90**: 82-91.

[45] Kusumoto S. Estimation of dip angle of fault or structural boundary by eigenvectors of gravity gradient tensors. Butsuri-Tansa 2015; **68**: 277-287 (in Japanese with English abstract).

[46] Kusumoto S. Eigenvector of gravity gradient tensor for estimating fault dips considering fault type, Progress in Earth and Planetary Science. 2017.

[47] Mickus K L, Hinojosa J H. The complete gravity gradient tensor derived from the vertical component of gravity: a Fourier transform technique. Journal of Applied Geophysics 2001; **46**: 159-174. (Accepted with minor revision).

[48] Miura S, Ueki S, Sato T, Tachibana K, Hamaguchi H. Crustal deformation associated with the 1998 seismo-volcanic crisis of Iwate volcano, northeastern Japan, as observed by a dense GPS network. Earth Planets Space 2000; **52**: 1003-1008.

[49] Owen S, Segall P, Lisowski M, Murray M, Bevis M, Foster J. The January 30, 1997, eruptive event on Kilauea volcano, as monitored continuous GPS. Geophysical Research Letters 2000; **27**: 2757-2760.

[50] Fialko Y, Simons M, Agnew D C. The complete (3-D) surface displacement field in the epicentral area of the 1999 Mw 7.1 Hector mine earthquake, California, from space geodetic observations. Geophysical Research Letters 2001; **28**: 3063-3066.

[51] Jonsson K M, Hsu Y J, Segall P, Yu S B. Fault geometry and slip distribution of the 1999 Chi-Chi, Taiwan, earthquake imaged from inversion of GPS data. Geophysical Research Letters 2001; **28**: 2285-2288.

[52] Lasserre C, Peltzer G, Crampe F, Klinger Y, Van der Woerd J, Tapponnier P. Coseismic deformation of the 2001 Mw=7.8 Kokoxili earthquake in Tibet, measured by synthetic aperture radar interferometry. Journal of Geophysical Research 2005; **110**: doi: 10.1029/2004JB003500.

[53] Okada Y. Surface deformation due to shear and tensile faults in a half-space. Bulletin of Seismological Society of America 1985; **75**: 1135-1154.

[54] Rodgers D A. Analysis of pull-apart basin development produced by en echelon strike-slip faults. Special Publications, International Association of Sedimentologists 1980; **4**: 27-41.

[55] Katzman R, ten Brink U S, Lin J. Three-dimensional modeling of pull-apart basins: Implications for the tectonics of the Dead Sea Basin. Journal of Geophysical Research 1995; **100**: 6295-6312.

[56] Kusumoto S, Takemura K, Fukuda Y, Takemoto S. Restoration of the depression structure at the eastern part of central Kyushu, Japan, by means of the dislocation modeling. Tectonophysics 1999; **302**: 287-296.

[57] Itoh Y, Kusumoto S, Furubayashi T. Quantitative evaluation of the Quaternary crustal deformation around the Takayama basin, central Japan: A paleomagnetic and numerical modeling approach. Earth and Planetary Science Letters 2008; **267**: 517-532. doi: 10.1016/j.epsl.2007.11.062.

[58] Tamaki M, Kusumoto S, Itoh Y. Formation and deformation process of the late Paleogene sedimentary basins in the southern central Hokkaido, Japan: paleomagnetic and numerical modeling approach. Islands Arc 2010; **19**: 243-258.

[59] Kusumoto S, Itoh Y, Takano O, Tamaki M. Numerical Modeling of Sedimentary Basin Formation at the Termination of Lateral Faults in a Tectonic Region where Fault Propagation has Occurred. In: Itoh, Y. (ed.), Mechanism of sedimentary basin formation: multidisciplinary approach on active plate margins. Rijeka: InTech; 2013, pp. 273-303.

[60] Itoh Y, Takano O, Kusumoto S, Tamaki M. Mechanism of longstanding Cenozoic basin formation in central Hokkaido: an integrated basin study on an oblique convergent margin. Progress in Earth and Planetary Science 2014; **1**: 6. doi: 10.1186/2197-4284-1-6.

http://dx.doi.org/10.5772/67964

[61] Mizuno K, Momohara A. Shobudani Formation, terrace and alluvial In: Ichihara, M. (ed.), Osaka Group. Osaka: Sohgen-sya; 1993, pp. 145-157

[62] KG-net and Kansai Geo-informatics Research Committee, Shin-Kansai geo-informatics—Wakayama Plain. Yodogawa-kohgi, Osaka, 2011.

[63] Yamazaki H, Tsukuda E, Okumura K, Kinugasa Y, Okada A, Nakata T, Tsutsumi H, Hasegawa S. Trench excavation of the Okamura fault in the Median Tectonic Line fault system at Saijo, southwest Japan. Memories of the Geological Society of Japan 1992; **40**: 129-142 (in Japanese with English abstract).

[64] Cundall P A, Strack O D L. A discrete numerical model for granular assemblies, Géotechnique 1979; **29**: 47-65.

[65] Hardy S, McClayc K, Muñoz J A. Deformation and fault activity in space and time in high-resolution numerical models of doubly vergent thrust wedges. Marine and Petroleum Geology 2009; **26**: 232-248.

[66] Kusumoto S, Itoh Y, Takemura K, Iwata T. Displacement fields of sedimentary layers controlled by fault parameters: the discrete element method of controlling basement motions by dislocation solutions. Earth Sciences 2015; **4**: 89-94.

[67] Hayama Y, Yamada T. Median Tectonic Line at the stage of its origin in relation to Plutonism and Mylonitization in the Ryoke belt. Memories of the Geological Society of Japan 1980; **18**: 5-26.

[68] Geological Survey of Japan, AIST (ed.). Seamless digital geological map of Japan 1: 200,000. May 29, 2015 version. Geological Survey of Japan, National Institute of Advanced Industrial Science and Technology, Tsukuba.

Chapter 5

Fission Track Thermochronology of Late Cretaceous Sandstones of the Izumi Group Adjacent to the Median Tectonic Line Active Fault System in Southwest Japan

Yasuto Itoh, Paul F. Green, Keiji Takemura and Tomotaka Iwata

Additional information is available at the end of the chapter

http://dx.doi.org/10.5772/67962

Abstract

Fission track (FT) thermochronology was applied to the Late Cretaceous turbidite sandstones of the Izumi Group adjacent to the Median Tectonic Line active fault system in southwest Japan. Apatite FT analyses revealed the following three stages of cooling (uplift) events: 95–78 Ma (Cenomanian–Campanian) from >130°C, 74–46 Ma (Campanian–middle Eocene) from approximately 100°C, and 27–7 Ma (late Oligocene–late Miocene) from approximately 70°C. By contrast, zircon FT analysis indicated cooling from >300°C at ca. 70 Ma. Apparent discrepancies between the cooling initiation times obtained using the two analytical methods indicate the distinct provenances of tuffaceous sandstones of the Izumi Group. The second episode is likely related to regional exhumation events on the eastern Eurasian margin. The latest event, which terminated by the end of the Miocene, appears to have been manifested in the strong deformation of the arc under a compressive stress provoked by the resumed subduction of the Philippine Sea Plate.

Keywords: fission track, thermochronology, Cretaceous, Izumi Group, southwest Japan

1. Introduction

Reflecting transient shifts in the convergence modes of oceanic plates, island arcs are sites of complicated deformation events, which provide insight into the timeline of plate motions. Although such tectonic records can be obtained as temporal changes in the sedimentary environments in basin areas, the deformation history of exhumed terranes is difficult to assess because of a lack of direct evidence. This paper presents a case study of fission track thermochronology of the Late Cretaceous sandstones of the Izumi Group that constitute a mountain range along a longstanding crustal break called the Median Tectonic Line (MTL) on the eastern Eurasian convergent margin. The present study provides the first-ever quantitative information on recurrent uplifts of the southwestern Japan arc, which was a part of the continental rim before the Neogene backarc rifting.

2. Geological background

During the Cretaceous, the MTL acted as a sinistral transcurrent fault along the eastern Eurasian margin [1]. As a result of repeated slips on the fault and the propagation of its termination, an enormous narrow pull-apart basin was generated with the eastward migration of the depocenter. The present study was focused on the easternmost part of the basin, which was buried with the Maastrichtian marine sediments of the Izumi Group [2].

A sedimentological study [3] has shown that the Izumi Group basically consists of stacks of mega-units. Facies associations indicate that the constituents of the units are classified into the depositional systems including main channels with overspilled deposits, distributary channels, and sheet flows.

Outcrop sandstone samples of the Izumi Group were obtained from two sites on a reservoir (**Figure 1**). Sp. 15040401 was taken on the surface of a reverse fault (within 30 cm) called the Negoro–Minami Fault, whose Quaternary movement has been confirmed [2]. Sp. 15040402 was taken from a site at least 60 m from the same fault.

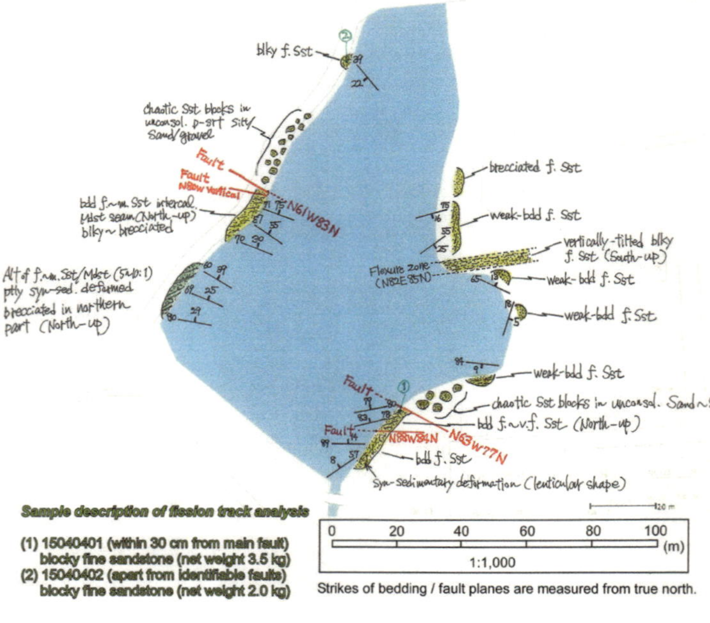

Figure 1. Sampling locality map. See the prologue of this section for the regional index.

3. Analytical methods and results

In this chapter, the authors provide new Apatite Fission Track Analysis (AFTA®) and Zircon Fission Track Analysis (ZFTA™) data and the associated thermal history interpretations for two samples of outcropping Maastrichtian sandstones from southwest Japan. AFTA and ZFTA results have been used to identify, characterize, and quantify the main heating and cooling episodes that have affected the section from which the samples were taken. This information was then synthesized to provide a regionally consistent thermal history framework, from which the history of the structural and tectonic development at the sample location can be understood.

3.1. Sample preparation

Outcrop samples were crushed in a jaw crusher and then ground to sand grade in a rotary disc mill. The ground samples were rinsed to remove dust, then dried, and processed by the conventional mineral separation techniques, namely heavy liquid and magnetic separator in order to obtain heavy minerals. Acquired apatite grains were mounted in epoxy resin sheet set on hot glass slides. It was polished to reveal internal crystal surfaces and etched for 20 s in 5 M HNO_3 in room temperature (20°C) for visualization of spontaneous fission tracks. Zircon grains were embedded in FEP Teflon sheet between heated microscope slides on a hot plate, polished to expose internal surfaces, and then etched in a molten KOH:NaOH eutectic etchant at ~220°C, for an appropriate time (usually 10–100 h) to reveal properly etched fission tracks contained in a high proportion of the grains.

After the etching procedure, all mounts were cut to 1.5×1.0 cm, and soaked in detergent, alcohol, and distilled water to clean them up. They were sealed in close contact with low-uranium detectors made of muscovite wrapped within heat-shrink membrane. Each batch of mounted samples was stacked between two pieces of uranium standard glass prepared in a similar fashion. It was then settled in an aluminum can for irradiation.

After neutron irradiation, mica detectors were removed from the mounted grains and standard glasses, and then etched in hydrofluoric acid to visualize the induced fission tracks produced by ^{235}U inherently contained in the apatite/zircon and standard glass.

3.2. Data quality

Excellent apatite yields were obtained from the two samples (**Table 1**), with fission track ages in 20 or more grains and 100 track lengths or more measured (the usual "target" numbers considered necessary for an analysis of the highest quality) in both samples. The apatite grains that were analyzed in both samples were of high quality. Thus, high-quality data were obtained for both samples, and the resulting thermal history constraints are regarded as highly reliable.

Both samples provided excellent yields of detrital zircon, with 20 single grain ages measured in both samples, resulting in very high-quality data. As with the apatites, the quality of the polished and etched surfaces in all analyzed zircon grains was high, and the resulting data are of very high quality.

Sample no. (ID in paper)	Loc. (lat./long.)	Lithology (remarks)	Stratigraphic division	Strat. age (Ma)	Raw wt. (g)	Washed wt. (g)	Apatite yield*
GC1181-3 (15040401)	34.2800/135.3100	Mass. f. sandstone (adjacent to fault)	Izumi Group (Maastrichtian)	71–65	2420	310	Excellent
GC1181-4 (15040402)	34.2800/135.3100	Mass. f. sandstone (apart from fault)	Izumi Group (Maastrichtian)	71–65	2040	360	Excellent

* Yield is based on the quantity of mineral separates which suit age determination. Excellent: >20 grains; good: 15–19 grains; fair: 10–14 grains; poor: 5–9 grains; very poor: <5 grains.

Table 1. Details of fission track samples and apatite yields.

3.3. Fission track ages

Fission track ages were calculated using the standard fission track age equation with the zeta calibration method (Eq. (5) of [4]). We determined ages by means of the external detector method (EDM; [5]). The EDM has a merit that fission track ages can be determined for single grains. In case of apatite, tracks are counted for each mount in 20 grains. If the desired number is not present in those samples, all available grains are counted. The actual number depends on the availability of suitably etched and oriented grains. We exclusively analyzed grains oriented with surfaces parallel to the crystallographic c-axis. Such grains are found based on etching characteristics, together with morphological evidence within euhedral grains. The grain mount is then sequentially scanned, and the first-identified 20 suitably oriented grains are treated by following procedures.

We count tracks within an eyepiece graticule that is divided into 100 grid squares. The number of spontaneous tracks (N_s) within a certain number of grid squares (N_a) was recorded in each grain. Then, we count the number of induced tracks (N_i) in the corresponding location of the external detector made of mica. Densities of spontaneous and induced tracks (ρ_s and ρ_i) are calculated by dividing the track counts by the total area counted, which is determined as the product of N_a and the area or each grid square. Eventually, fission track ages are calculated by substituting track counts (N_s and N_i) for track densities (ρ_s and ρ_i).

We carried out neutron irradiations in a well-thermalized flux (X-7 facility; Cd ratio for Au ~98) in the research reactor HIFAR of Australian Atomic Energy Commission. Track counts in mica external detectors attached to two pieces of Corning Glass Works standard glass CN5 (containing ~11 ppm Uranium) in the irradiation canister at each end of the sample stack provided us with total neutron fluence. Twenty-five fields were normally counted in each detector to determine track densities within external detectors irradiated adjacent to uranium standard glasses. The track density (ρ_D) was determined when we divided the total track count (N_D) by the total area counted. Covering the whole area of the detector, positions of the counted fields were arranged in a 5 × 5 grid. This is convenient to sample across the detector while gathering sufficient counts for the achievement of a precision of ~±2% within a reasonable time for typical track densities between approximately 5×10^5 and 5×10^6.

In the irradiation facility described above, small flux gradient is often observed over the length of the sample package. In case a detectable gradient is present, we convert the track count in the external detector adjacent to each standard glass to a track density (ρ_D), and a value for each mount in the stack is estimated based on linear interpolation. Or else, when no significant gradient is detected, we pool the track counts in the two external detectors to give a single value of ρ_D, based on which fission track ages for each sample are calculated. Detailed information of the apatite dating is given in **Tables 2** and **3** for Sp. 15040401 and 15040402, respectively.

GC1181-3 Apatite											**15040401**
Counted by: MEM											
Slide ref	**Current grain no**	N_s	N_i	N_a	ρ_s	ρ_i	**RATIO**	**U (ppm)**	**Cl (wt%)**	**F.T. AGE (Ma)**	
G1264-2	3	3	10	36	1.324E+05	4.414E+05	0.300	3.8	0.35	78.3 ±	51.6
G1264-2	5	12	55	100	1.907E+05	8.740E+05	0.218	7.5	0.86	57.1 ±	18.3
G1264-2	6	8	27	56	2.270E+05	7.662E+05	0.296	6.5	0.94	77.4 ±	31.2
G1264-2	7	25	127	60	6.621E+05	3.364E+06	0.197	28.7	0.23	51.5 ±	11.4
G1264-2	8	5	34	60	1.324E+05	9.005E+05	0.147	7.7	0.22	38.5 ±	18.5
G1264-2	9	5	20	40	1.986E+05	7.945E+05	0.250	6.8	0.10	65.3 ±	32.7
G1264-2	11	3	9	32	1.490E+05	4.469E+05	0.333	3.8	0.97	87.0 ±	58.0
G1264-2	12	8	41	100	1.271E+05	6.515E+05	0.195	5.6	0.83	51.0 ±	19.8
G1264-2	14	9	9	32	4.469E+05	4.469E+05	1.000	3.8	0.29	257.5 ±	121.6
G1264-2	17	2	10	36	8.828E+04	4.414E+05	0.200	3.8	0.22	52.3 ±	40.6
G1264-2	18	7	17	64	1.738E+05	4.221E+05	0.412	3.6	0.97	107.3 ±	48.3
G1264-2	19	23	123	30	1.218E+06	6.515E+06	0.187	55.5	0.21	48.9 ±	11.2
G1264-2	22	8	16	70	1.816E+05	3.632E+05	0.500	3.1	0.44	130.0 ±	56.4
G1264-2	23	9	30	70	2.043E+05	6.810E+05	0.300	5.8	0.90	78.3 ±	29.9
G1264-2	25	16	38	49	5.189E+05	1.232E+06	0.421	10.5	0.09	109.7 ±	32.8
G1264-2	26	22	79	49	7.135E+05	2.562E+06	0.278	21.8	0.09	72.7 ±	17.7
G1264-2	27	13	61	100	2.066E+05	9.693 E+05	0.213	8.3	0.90	55.7 ±	17.1
G1264-2	29	7	29	49	2.270E+05	9.405E+05	0.241	8.0	1.00	63.1 ±	26.6
G1264-2	30	7	21	30	3.708E+05	1.112E+06	0.333	9.5	0.92	87.0 ±	38.0
G1264-2	33	0	5	36	0.000E+05	2.207E+05	0.000	1.9	0.30	0.0 ±	106.0
		192	761		2.776E+05	1.100E+06		9.4			
Area of basic unit = 6.293E-07 cm^{-2}						Ages calculated using a zeta of 392.9 ± 7.4 for CN5 glass					
χ^2 = 24.656 with 19 degrees of freedom						ρ = 1.337E+06cm^{-2} ND =2212					
$P(\chi^2)$ = 17.2%						ρ_D interpolated between the top of can; ρ = 1.328E+06cm^{-2} ND = 1045					
Age Dispersion = 11.181% (did not converge}						bottom of can; ρ = 1.484E+06cm^{-2} ND = 1167					
N_s/Ni =0.252 ± 0.020						**POOLED AGE = 65.9 ± 5.6 Ma**					
Mean Ratio= 0.301 ± 0.044						CENTRAL AGE = 67.0 ±6.1 Ma					

Table 2. Details of apatite fission track analysis for Sp. 15040401.

GC1181-4 Apatite											**15040402**	
Counted by: MEM												
Slide ref	**Current grain no**	N_s	N_i	N_a	ρ_s	ρ_i	**RATIO**	**U (ppm)**	**Cl (wt%)**	**F.T. AGE (Ma)**		
G1264-3	3	6	12	30	3.178E+05	6.356E+05	0.500	5.4	0.87	130.8 ±	65.5	
G1264-3	5	2	13	42	7.567E+04	4.919E+05	0.154	4.2	0.95	40.5 ±	30.8	
G1264-3	6	8	10	40	3.178E+05	3.973E+05	0.800	3.4	0.91	208.1 ±	98.9	
G1264-3	7	6	29	50	1.907E+05	9.217E+05	0.207	7.8	0.77	54.5 ±	24.5	
G1264-3	8	7	19	42	2.648E+05	7.189E+05	0.368	6.1	0.85	96.7 ±	42.8	
G1264-3	10	12	50	70	2.724E+05	1.135E+06	0.240	9.6	0.89	63.1 ±	20.4	
G1264-3	11	7	21	80	1.390E+05	4.171E+05	0.333	3.5	0.28	87.5 ±	38.3	
G1264-3	12	2	12	60	5.297E+04	3.178E+05	0.167	2.7	0.47	43.9 ±	33.6	
G1264-3	13	1	11	24	6.621E+04	7.283E+05	0.091	6.2	0.61	24.0 ±	25.1	
G1264-3	14	10	21	42	3.783E+05	7.945E+05	0.476	6.7	1.07	124.7 ±	48.0	
G1264-3	15	14	43	70	3.178E+05	9.761 E+05	0.326	8.3	0.36	85.5 ±	26.4	
G1264-3	16	1	14	60	2.648E+04	3.708E+05	0.071	3.1	0.71	18.9 ±	19.5	
G1264-3	20	1	6	30	5.297E+04	3.178E+05	0.167	2.7	0.26	43.9 ±	47.4	
G1264-3	21	7	25	35	3.178E+05	1.135E+06	0.280	9.6	1.02	73.6 ±	31.5	
G1264-3	22	9	19	49	2.919E+05	6.162E+05	0.474	5.2	1.04	124.0 ±	50.3	
G1264-3	26	10	33	80	1.986E+05	6.555E+05	0.303	5.6	0.90	79.6 ±	28.8	
G1264-3	27	7	19	36	3.090E+05	8.387E+05	0.368	7.1	0.88	96.7 ±	42.8	
G1264-3	28	10	25	100	1.589E+05	3.973E+05	0.400	3.4	0.91	104.9 ±	39.4	
G1264-3	33	8	39	64	1.986E+05	9.683E+05	0.205	8.2	0.87	54.0 ±	21.0	
G1264-3	34	14	27	25	8.899E+05	1.716E+06	0.519	14.5	0.59	135.6 ±	44.8	
G1264-3	79	9	21	36	3.973E+05	9.270E+05	0.429	7.9	0.98	112.3 ±	44.9	
		151	469		2.253 E+05	6.998E+05		5.9				

Area of basic unit = 6.293E-07 cm^{-2}	Ages calculated using a zeta of 392.9 + 7.4 for CN5 glass
χ^2 = 19.588 with 20 degrees of freedom	ρ =1.346E+06cm^{-2} ND=2212
$P(\chi^2)$=48.4%	ρ_D interpolated between the top of can; ρ = 1.328E+06cm^{-2} ND = 1045
Age dispersion =2.735% (did not converge)	bottom of can; ρ = 1.484E+06^{-2} ND = 1167
N_s/N_i= 0.322 ± 0.030	**POOLED AGE = 84.6 ± 8.3 Ma**
Mean ratio = 0.327 ± 0.038	CENTRAL AGE = 84.6 ± 8.3 Ma

Table 3. Details of apatite fission track analysis for Sp. 15040402.

3.4. Track length measurements

Applications of fission track dating to accessory apatites from crystalline basement rocks revealed that the technique was extremely thermally sensitive, suggesting that fission track ages could be reset at relatively low temperatures around 100°C over geological timescales (e.g. [6]). This was confirmed by the direct measurement of fission track ages in subsurface samples [7]. Integration of fission track ages with confined track length measurements [8] led to deeper understanding of the method. Early measurements showed that even in volcanic rocks, which have experienced only very low temperatures after initial post-eruption cooling, the mean confined track lengths (around 14–15 μm) were shorter than the induced tracks (~16 μm) in the same apatites. Green [9] showed that this can be understood in terms of the thermal annealing of these tracks at low temperatures (<50°C) over geological timescales, highlighting the sensitivity of the technique.

Full lengths of "confined" fission tracks were utilized for track length studies in apatite. They are defined as tracks that do not intersect the polished crystal surface but have been etched via other tracks or fractures. In this way, the whole length of track is surely etched, and measured using a digitizing tablet. The device is connected to a microcomputer, superimposed on the microscope field of view via a projection tube. This system, which is calibrated against a stage graticule ruled in 2-μm divisions, enables individual tracks to be measured in a precision of ±0.2 μm. Because of anisotropic annealing efficiency of fission tracks residing in apatite [10], we measured tracks only in prismatic grains, which show sharp polishing scratches with well-etched tracks having omnidirectionally narrow cone angle. According to the recommendations of Laslett et al. [11], only horizontal tracks were measured. We measured 100 tracks whenever possible. As for samples with low-track density or those in which only a limited number of apatite grains were obtained, fewer confined tracks may inevitably be available. Hence, the whole mount was scanned in order to find as many confined tracks as possible. A summary of the length measurement of fission tracks is presented in **Table 4**.

Sample number	Mean track length (μm)	Standard deviation (μm)	Number of tracks (*N*)	Number of tracks in length intervals (μm)																			
				1	2	3	4	5	6	7	8	9	10	11	12	13	14	15	16	17	18	19	20
Outcrop samples																							
GC1181-3	12.77 ± 0.20	2.00	103	–	–	–	–	–	–	3	–	–	4	7	19	17	22	22	8	–	–	1	–
GC1181-4	12.87 ±0.19	1.95	103								2	3	1	5	18	29	16	18	6	3	1	1	–

Track length measurements by: M. Moore for samples 3–4.

Table 4. Summary of length distribution of fission tracks.

3.5. Thermal history interpretation of AFTA data

A compilation of confined track length data in a large number of apatite samples showed that the form of the track length distribution was a sensitive indicator of the style of thermal history [12]. Laboratory studies [10, 13, 14], together with detailed mathematical analysis [11, 15, 16], have established that the reduction in track length causes a reduction in the fission track age, by reducing the proportion of tracks that can intersect a polished grain surface. They demonstrated that as a result of heating, the mean track length is progressively reduced and the tracks effectively "shrink" from each end, until some individual tracks may break up into several segments in the final stages. This realization underpins all subsequent studies involving the quantitative prediction of apatite fission track parameters and the extraction of thermal history information from such data. Laslett et al. [17] showed that the variation of the mean track length with temperature and time can be well described by a "fanning Arrhenius plot" model, in which contours of equal track length reduction form straight lines in a plot of time against inverse absolute temperature, with the slope of these lines (reflecting an "activation energy") increasing as the degree of annealing increases. Their improved definition of the kinetics of fission track annealing obtained by using the mean confined track length as the fundamental parameter, combined with a detailed understanding of the way in which the reduction in track length is manifested in the fission track age [14], provided the basis for making realistic predictions of apatite fission track parameters within usual geological situations.

A key step in this process is the transition from isothermal annealing models to the variable temperature behavior encountered in geological settings [18]. This step provides a way forward by adopting the principle of "equivalent time." It rules that the annealing rate of a track at any given time depends only on its length to which the track has already been reduced and the prevailing temperature, and not upon the history of how the track reached the present length. On these advances, Green et al. [19] developed methods of quantitatively modeling the response of fission tracks in apatite to various styles of thermal history. The basis of the approach is the recognition that track length reduction is the controlling process in determining the AFTA parameters that result from any particular history. Another important finding is that chlorine content exerts a systematic influence on annealing rates [10, 13, 20]. In practical applications of AFTA, variation in fission track age and track length with Cl weight content allows the identification of any anomalous grains that might represent unusual annealing properties (e.g., [21]). Such anomalies can be eliminated from the dataset prior to the extraction of thermal history solutions. In usual geological settings, differential annealing effects for individual samples are maximized in rocks that have been heated to the critical temperatures (typically 90–120°C) where the most sensitive (i.e., low-Cl content) apatite grains are totally annealed, whereas more resistant (i.e., high-Cl content) apatites remain unaffected. Actually, since most common apatites tend to contain less than 0.1 wt% Cl, it should be expected that the apatites analyzed in most published studies might be more sensitive than the Durango apatite that was used in the Laslett et al. [17] model. Thus, it is possible that much of the so-called anomalous Late Cenozoic cooling reported in many studies could be due to such compositional effects, rather than any innate deficiencies in the model [22]. A summary of the AFTA data is shown in **Table 5**.

Sample no. (ID in paper)	Present temp. (°C)	Stratigraphic age (Ma)	Measured mean FT length (μm)	Predicted* mean FT length (μm)	Measured FT age (Ma)	Predicted* FT age (Ma)
GC1181-3 (15040401)	10	71–65	12.77±0.20	14.99	65.9±5.6	70
GC1181-4 (15040402)	10	71–65	12.87±0.19	15.02	84.6±8.3	70

* Values predicted from the default thermal history, that is, assuming that each sample is now at its maximum temperature since deposition. The values refer only to tracks formed after deposition. Samples may contain tracks inherited from sediment provenance areas. Calculations refer to apatites within the measured range of composition for each sample.

Table 5. Summary of AFTA data in two outcrop samples.

3.6. Results of fission track analyses

Basic analytical data for apatite grains separated from Sp. 15040401 and 15040402 are presented in **Figures 2** and **3**, respectively. Both samples analyzed in this study showed unusual and quite distinctive Cl content histograms (**Figures 2B** and **3B**). Detrital apatites from common quartzo-feldspathic sandstones around the world show a typical histogram of Cl contents in which the majority of grains have Cl contents between 0 and 0.1 wt%, while a smaller number of grains give values up to ~0.5 wt% and occasional grains contain 0.5–1.0 wt% Cl.

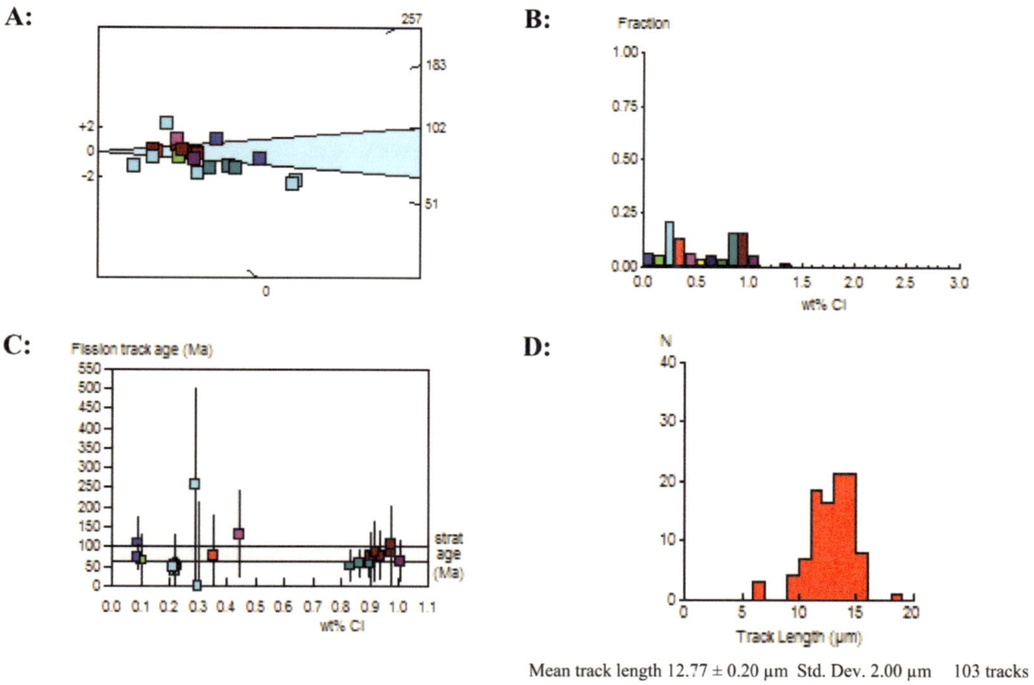

Figure 2. Fission track age data for Sp. 15040401. A: Radial plot of single grain ages. B: Distribution of Cl contents in apatite grains. C: Single grain age versus weight % Cl for individual apatite grains. D: Distribution of confined track lengths.

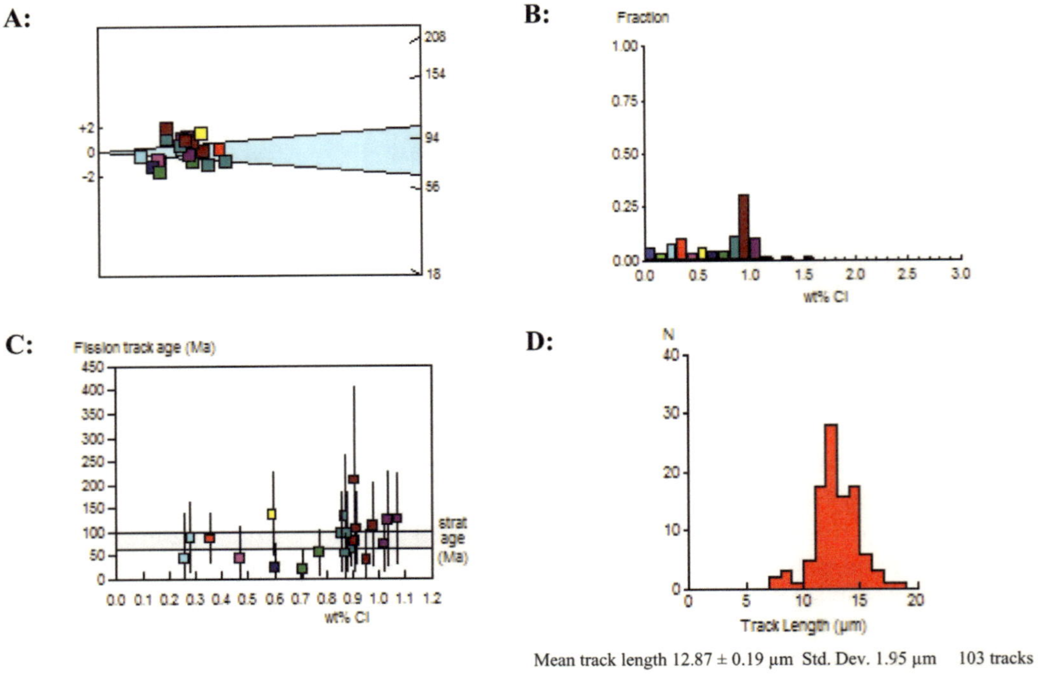

Figure 3. Fission track age data for Sp. 15040402. A: Radial plot of single grain ages. B: Distribution of Cl contents in apatite grains. C: Single grain age versus weight % Cl for individual apatite grains. D: Distribution of confined track lengths.

By contrast, both samples analyzed here showed a broad, bimodal distribution of Cl contents, with prominent peaks at approximately 0.2–0.4 wt% and between 0.8 and 1.0 wt%, with some grains containing as much as 1.5 wt% Cl. Such broad distributions of Cl content are commonly diagnostic of a derivation of apatites from tuffaceous volcanic units. In each of the samples analyzed for this paper, the measured distribution of the compositions has been employed in interpreting the AFTA data, using the methods outlined in the previous section.

In these figures, plots of fission track ages of single apatite grains against Cl content (**Figures 2C** and **3C**) are used to assess the degree of annealing. As an example, grains with a range of Cl contents from zero to some upper limit must have been totally annealed if those with the composition range all yield similar fission track ages that are significantly younger than the stratigraphic age. Or else, the sample displays a high degree of partial annealing in case that the fission track age falls rapidly with decreasing Cl content. As mentioned above, both samples analyzed in the present study showed a broad, bimodal distribution of Cl contents, and the unusual distribution of compositions has been employed in interpreting the AFTA data.

Figure 4 schematically presents the basic concept of the AFTA analysis, which was explained in detail in Section 3.5. Because only a limited number of outcrop samples have been analyzed in the present study, models were constructed for variable paleogeothermal gradients and on the assumption of a long-term constant surface temperature of 10°C. After all the necessary verifications, the thermal history interpretation was obtained for Sp. 15040401 and 15040402, as depicted in **Figures 5** and **6**, respectively.

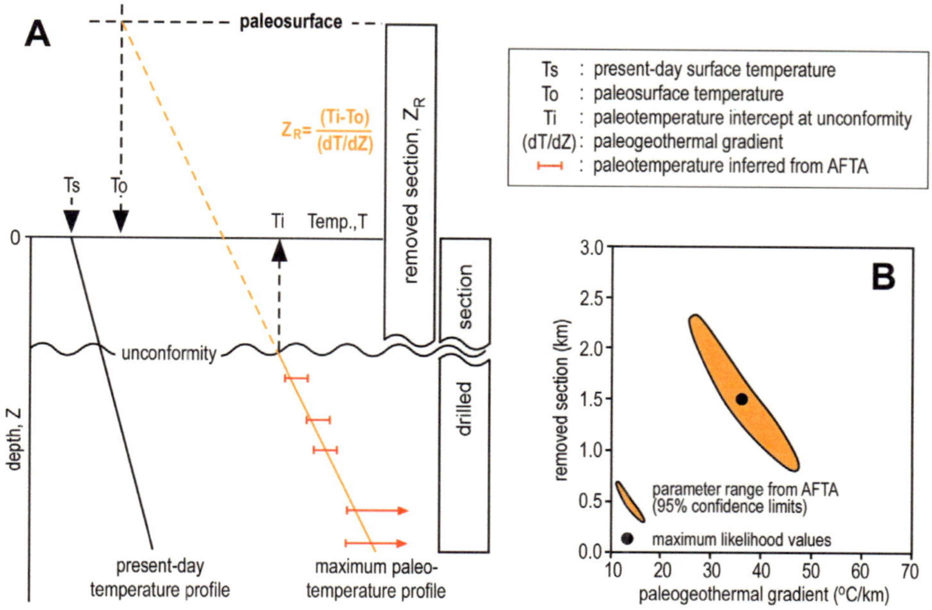

Figure 4. Schematic diagrams showing the AFTA methodology. Heating due to deeper burial (possibly often combined with elevated heat flow) and amounts of exhumation (or deeper burial) can be estimated by fitting a linear paleogeothermal gradient to a series of paleotemperature constraints from subsurface samples. It is then extrapolated to an assumed paleosurface temperature, as shown in a generalized section (A). Note that higher gradients extrapolate to lower values of removed section, while lower gradients correspond to larger amounts of removed section. Statistical analysis allows the definition of the range of values of paleogeothermal gradient and removed section within 95% confidence limits, represented by a hyperbolic ellipsoid in (B).

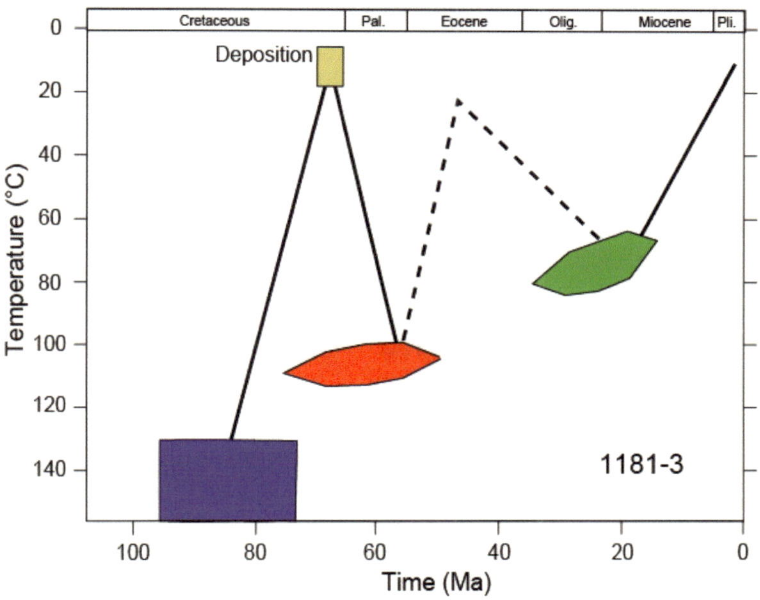

Figure 5. Thermal history interpretation of AFTA data for Sp. 15040401.

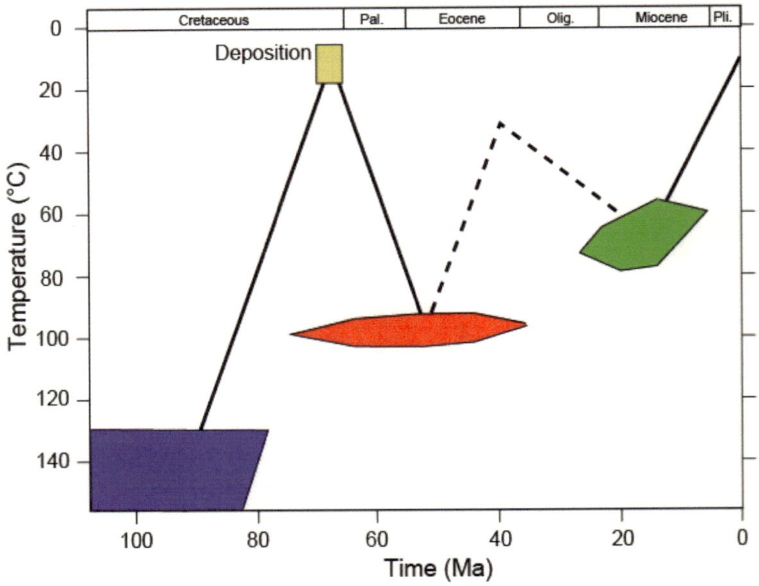

Figure 6. Thermal history interpretation of AFTA data for Sp. 15040402.

In **Figures 5** and **6**, we ran simulations of the data using Geotrack's in-house numerical model of the AFTA system and defined the range of conditions that give predictions consistent with the data with 95% confidence limits as depicted by Green and Duddy [23]. Our analyses have indicated that the evidence of higher temperatures in the past is not obtained from FT ages but instead from track length distribution. Since the effect of Cl contents on the mean track length was quantitatively given by Green and Duddy [24], we employed their result (**Figure 8** of [24]) to construct a self-consistent model.

The ZFTA results for Sp. 15040401 and 15040402 are presented in **Tables 6** and **7**, respectively, and integrated with AFTA results for the thermal history reconstruction detailed in the next section.

GC1181-3 Zircon									**15040401**	
Counted by: PFG										
Slide ref	**Current grain no**	N_s	N_i	N_a	ρ_s	ρ_i	**RATIO**	**U (ppm)**	**FT. AGE (Ma)**	
G1269-1	1	53	51	15	5.615E+06	5.403E+06	1.039	215.5	60.6 ±	12.0
G1269-1	2	95	71	27	5.591E+06	4.179E+06	1.338	166.7	77.9 ±	12.4
G1269-1	3	58	44	40	2.304E+06	1.748E+06	1.318	69.7	76.8 ±	15.5
G1269-1	8	189	141	50	6.007E+06	4.481E+06	1.340	178.8	78.1 ±	8.9
G1269-1	9	105	74	20	8.343E+06	5.880E+06	1.419	234.6	82.6 ±	12.7
G1269-1	11	132	147	40	5.244E+06	5.840E+06	0.898	233.0	52.4 ±	6.4
G1269-1	12	101	84	20	8.025E+06	6.674E+06	1.202	266.2	70.1 ±	10.5

GC1181-3 Zircon									**15040401**	
Counted by: PFG										
Slide ref	**Current grain no**	N_s	N_i	N_a	ρ_s	ρ_i	**RATIO**	**U (ppm)**	**FT. AGE (Ma)**	
G1269-1	13	116	94	25	7.373E+06	5.975E+06	1.234	238.4	71.9 ±	10.1
G1269-1	21	150	151	36	6.621E+06	6.665E+06	0.993	265.9	57.9 ±	6.8
G1269-1	24	93	65	25	5.911E+06	4.132E+06	1.431	164.8	83.3 ±	13.6
G1269-1	25	122	98	40	4.847E+06	3.893E+06	1.245	155.3	72.5 ±	10.0
G1269-1	26	119	95	50	3.782E+06	3.019E+06	1.253	120.4	73.0 ±	10.2
G1269-1	27	119	83	25	7.564E+06	5.276E+06	1.434	210.5	83.5 ±	12.1
G1269-1	33	139	90	25	8.835E+06	5.721E+06	1.544	228.2	89.9 ±	12.3
G1269-1	34	94	84	15	9.958E+06	8.899E+06	1.119	355.0	65.2 ±	9.9
G1269-1	35	111	90	40	4.410E+06	3.575E+06	1.233	142.6	71.9 ±	10.3
G1269-1	37	152	112	40	6.038E+06	4.449E+06	1.357	177.5	79.0 ±	10.0
G1269-1	38	91	68	40	3.615E+06	2.701E+06	1.338	107.8	77.9 ±	12.6
G1269-1	39	97	97	30	5.138E+06	5.138E+06	1.000	205.0	58.3 ±	8.5
G1269-1	41	99	80	30	5.244E+06	4.238E+06	1.237	169.0	72.1 ±	11.0
		2235	1819		5.611E+06	4.566E+06		182.2		

Area of basic unit = 6.293E-07 cm^{-2}

χ^2 = 21.327 with 19 degrees of freedom

$P(\chi^2)$ = 31.9%

Age Dispersion = 6.268% (did not converge)

N_s/N_i = 1.229 ± 0.039

Mean Ratio = 1.249 ± 0.038

Ages calculated using a zeta of 87.7 ± 0.8 for U3 glass

ρ = 1.336E+06cm^{-2} ND=2102

ρ_D interpolated between the top of can;
ρ =1.336E+06cm^{-2} ND=1051

bottom of can; ρ =1.336E+06cm^{-2} ND=1051

POOLED AGE = 71.6±2.8 Ma

CENTRAL AGE= 71.7 ± 3.0 Ma

Table 6. Details of zircon fission track analysis for Sp. 15040401.

GC1181-4 Zircon									**15040402**	
Counted by: PFG										
Slide ref	**Current grain no**	N_s	N_i	N_a	ρ_s	ρ_i	**RATIO**	**U (ppm)**	**F.T. AGE (Ma)**	
G1269-2	2	68	56	20	5.403E+06	4.449E+06	1.214	177.5	70.8 ±	12.9
G1260-2	3	80	57	24	5.297E+06	3.774E+06	1.404	150.6	81.7 ±	14.3
G1269-2	5	69	70	20	5.482E+06	5.562E+06	0.986	221.9	57.5 ±	9.8
G1269-2	9	159	55	25	1.011E+07	3.496E+06	2.891	139.5	167.2 ±	26.5

GC1181-4 Zircon									**15040402**	
Counted by: PFG										
Slide ref	**Current grain no**	N_s	N_i	N_a	ρ_s	ρ_i	**RATIO**	**U (ppm)**	**F.T. AGE (Ma)**	
G1269-2	10	75	57	24	4.966E+06	3.774E+06	1.316	150.6	76.6 ±	13.6
G1269-2	12	132	104	20	1.049E+07	8.263E+06	1.269	329.6	73.9 ±	9.9
G1269-2	13	56	79	12	7.416E+06	1.046E+07	0.709	417.3	41.4 ±	7.3
G1269-2	18	122	105	50	3.877E+06	3.337E+06	1.162	133.1	67.7 ±	9.2
G1269-2	19	58	47	16	5.760E+06	4.668E+06	1.234	186.2	71.9 ±	14.2
G1269-2	21	40	38	18	3.531E+06	3.355E+06	1.053	133.8	61.4 ±	14.0
G1269-2	22	80	46	24	5.297E+06	3.046E+06	1.739	121.5	101.1 ±	18.9
G1269-2	24	97	90	40	3.853E+06	3.575E+06	1.078	142.6	62.8 ±	9.3
G1269-2	30	126	102	50	4.004E+06	3.242E+06	1.235	129.3	72.0 ±	9.7
G1269-2	32	48	54	24	3.178E+06	3.575E+06	0.889	142.6	51.9 ±	10.4
G1269-2	33	82	80	16	8.144E+06	7.945E+06	1.025	317.0	59.8 ±	9.5
G1269-2	35	79	54	50	2.511E+06	1.716E+06	1.463	68.5	85.1 ±	15.2
G1269-2	36	54	36	18	4.767E+06	3.178E+06	1.500	126.8	87.3 ±	18.9
G1269-2	40	67	63	16	6.654E+06	6.257E+06	1.063	249.6	62.0 ±	11.0
G1269-2	46	84	65	30	4.449E+06	3.443E+06	1.292	137.3	75.3 ±	12.6
G1269-2	48	55	58	50	1.748E+06	1.843E+06	0.948	73.5	55.3 ±	10.5
		1631	1316		4.738E+06	3.823E+06		152.5		

Area of basic unit = 6.293E-07 cm^{-2}	Ages calculated using a zeta of 87.7 ± 0.8 for U3 glass
χ^2 = 58.002 with 19 degrees of freedom	ρ = 1.336E+06cm^{-2} ND = 2102
$P(\chi^2)$=0.0%	ρ_D interpolated between the top of can; ρ = 1.336E+06cm^{-2} ND=1051
Age dispersion =22.242%	Bottom of can; ρ = 1.336E+06cm^{-2} ND = 1051
N_s/N_i= 1.239 ± 0.046	POOLED AGE = 72.2 ± 3.2 Ma
Mean ratio = 1.274 ± 0.100	**CENTRAL AGE = 71.1 ± 4.8 Ma**

Table 7. Details of zircon fission track analysis for Sp. 15040402.

4. Discussion

The thermal history interpretation of AFTA data for two outcrop samples suggested three cooling (i.e., uplift) events around the study area. Among them, the Cenomanian–Campanian cooling from >130°C in the earliest of three episodes predates the deposition of the host sand-

http://dx.doi.org/10.5772/67962

stones, and appears to reflect a different provenance for the apatites compared to zircons. The Campanian–middle Eocene and late Oligocene–late Miocene cooling episodes clearly postdate the deposition of host rocks. Paleotemperatures in these episodes are interpreted as being due to deeper burial prior to cooling as a result of exhumation (uplift and erosion). Assuming a paleogeothermal gradient of 30°C/km, paleotemperatures of approximately 100°C prior to Campanian–middle Eocene cooling correspond to burial by ~3 km of section. Similarly, paleotemperatures around 70°C prior to late Oligocene–late Miocene cooling correspond to burial by ~2 km of section.

ZFTA data define cooling from >300°C at ca. 70 Ma, whereas AFTA data define an initial phase of cooling from >130°C that began in the interval of 95–78 Ma. This difference between the timing determined using the two techniques appears to indicate the different provenances of the two minerals, as indicated above.

Radiometric dates reported from the hinterlands seem to support the above-mentioned hypothesis of distinct provenances. **Figure 7** presents reliable radiometric ages obtained from the Cretaceous igneous rocks distributed in the central part of southwest Japan [25]. The ages show an explicit bimodal distribution, which agrees with the results of the present study. Although the previous dataset includes varied methods with different closing temperatures,

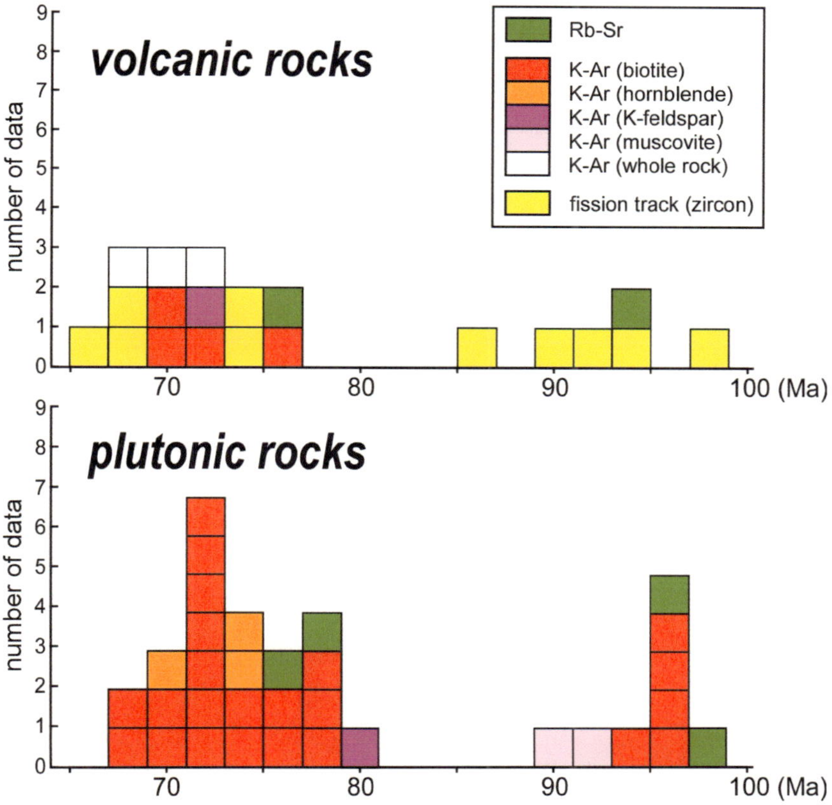

Figure 7. Histogram of radiometric ages reported from the Cretaceous igneous rocks distributed in the central part of southwest Japan [25].

ages obtained using the same technique and mineral (e.g., K–Ar ages on biotite) show clearly two peaks. Thus, the multiple igneous events on the Cretaceous Eurasian margin have been confirmed through the present AFTA and ZFTA analyses.

Figure 8 presents a complete reconstruction of the thermal history derived from AFTA and ZFTA analyses. As mentioned above, the present study has indicated distinct Cretaceous provenances for apatite and zircon grains yielded from outcrop sandstones of the Izumi Group. The second episode from the Campanian to the middle Eocene appears to correspond to a stage of regional uplift identified on the forearc of the northeastern Japan arc [26] although its driving force and/or effective area should be assessed through further study. The tectonic context of the youngest cooling (uplift) event terminated around the end of Miocene is intriguing. Based on their seismic interpretation on the backarc shelf of southwest Japan, Itoh and Nagasaki [27] argued that an arc contraction event at the timing had a regional impact under prevailing north–south tectonic stress probably because of the resumed convergence of the Philippine Sea Plate. In other words, younger deformation phases or shear heating linked to recent dextral slips on the MTL active fault system had no significant effect on the annealing of fission tracks in rock-forming minerals.

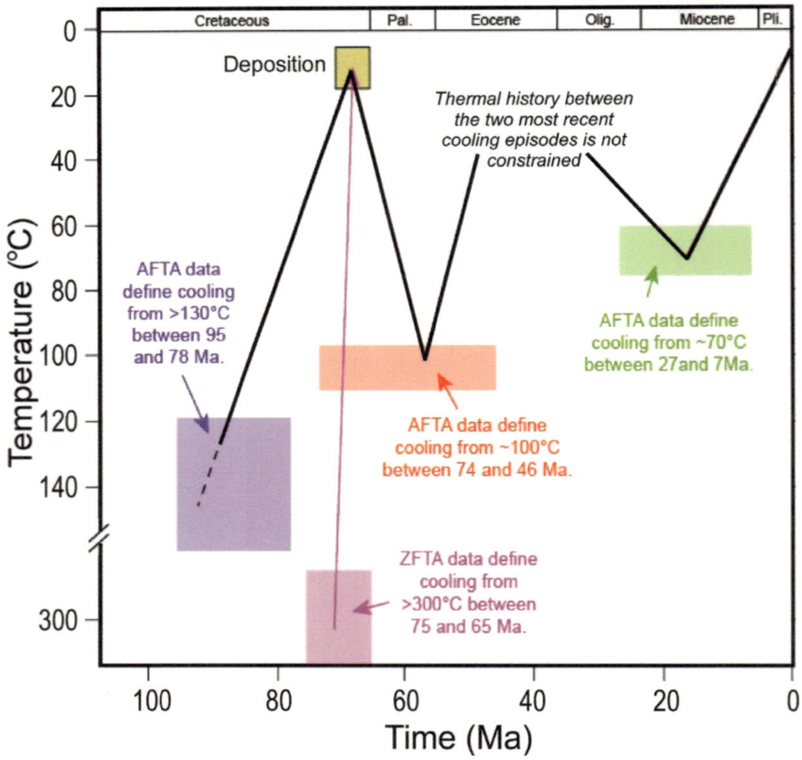

Figure 8. Preferred thermal history reconstruction of AFTA and ZFTA data in two outcrop sandstones of the Izumi Group in southwest Japan. As for the apatite analysis, thermal events were not supposed by FT ages but by track length distribution. As for the zircon analysis, the range of the cooling interval (75–65 Ma) was given as two sigma uncertainties on the weighted mean age.

http://dx.doi.org/10.5772/67962

Our research revealed a timeline of thermal episodes on the MTL. Because the fault has a quite long active trace, it is plausible that different tectonic history is applicable to remote segments of the MTL. Based on a systematic FT thermochronology, Tagami et al. [28] stated that thermal overprinting events adjacent to the eastern part of the MTL may be likely due to shear heating upon the fault during the Oligocene and Miocene. As presented in the geochemistry paper in this book (Chapter 6 by Kaneko et al.), the eastern portion of the longstanding fault may have evolved through different path from its central portion.

5. Conclusions

(1) Excellent yields of apatite and zircon were obtained from two sandstone samples of the Izumi Group along the MTL active fault system, resulting in very high-quality AFTA and ZFTA data, and the interpretations made in this study are regarded as highly reliable.

(2) AFTA data in both samples define three episodes of cooling from elevated paleotemperatures. If the results in both samples represent the effects of the same events, then the AFTA data define cooling that began in the following intervals:

95–78 Ma: Cenomanian–Campanian, from >130°C.

74–46 Ma: Campanian–middle Eocene, from ~100°C.

27–7 Ma: late Oligocene–late Miocene, from ~70°C.

(3) ZFTA data in the two samples are highly consistent, as are the AFTA data. These results suggest that no significant fault offsets exist between the two samples.

(4) ZFTA data suggest that the zircon grains cooled below ~250–300°C between 75 and 65 Ma. This interval overlaps the Maastrichtian depositional age range (71–65 Ma) for the sandstones, suggesting a rapidly cooled provenance contemporaneous with deposition.

(5) Cenomanian–Campanian cooling from >130°C in the earliest of the three episodes predates the deposition of the host sandstones, and appears to reflect a different provenance for the apatites than for the zircons. The Campanian–middle Eocene and late Oligocene–late Miocene-cooling episodes clearly postdate the deposition of the host sandstones. The paleotemperatures in these episodes were interpreted as being due to deeper burial prior to cooling as a result of exhumation (uplift and erosion). Assuming a paleogeothermal gradient of 30°C/km, paleotemperatures around 100°C prior to Campanian–middle Eocene cooling correspond to burial by ~3 km of section. Similarly, paleotemperatures of approximately 70°C prior to late Oligocene–late Miocene cooling correspond to burial by ~2 km of section.

(6) ZFTA data define cooling from >300°C at around 70 Ma, whereas AFTA data define an initial phase of cooling from >130°C that began in the interval of 95–78 Ma. This difference in the timing determined using the two techniques appears to indicate the different provenances of the two minerals, as indicated above.

(7) Radiometric dates reported from the hinterlands appear to support the hypothesis of distinct provenances. Cretaceous igneous rocks distributed within the central part of southwest Japan yielded ages showing an explicit bimodal distribution, which agrees with the results of the present AFTA and ZFTA analyses. Ages obtained using the same technique and mineral (e.g., K–Ar ages on biotite) show clearly two peaks and support the occurrence of multiple igneous events on the Cretaceous Eurasian margin.

(8) The second and third thermal episodes from the Campanian to the middle Eocene and from the late Oligocene to the late Miocene correspond, respectively, to stages of regional uplift identified around the forearc of northeast Japan and a strong arc contraction event under a prevalent north–south tectonic stress caused by resumed northerly convergence of the Philippine Sea Plate. The latest cooling (i.e., uplift) event also indicates that the Quaternary deformation phases or shear heating linked to vigorous dextral slips on the MTL active fault system had no significant effect on the annealing of fission tracks in rock-forming minerals in the fault-adjoining localities.

Acknowledgements

We are grateful to Mr. Masayoshi Shimizu of the Negoro–Ichijo Kiln for his support during our field survey and sampling. This study was executed as a part of the Comprehensive Research and Survey for the Median Tectonic Line Fault System (Eastern Margin of Kongo Mountains–Southern Margin of Izumi Mountains; FY2013-2015) organized by the Research and Development Bureau; the Ministry of Education, Culture, Sports, Science and Technology, and the Disaster Prevention Research Institute; Kyoto University. Constructive comments by Dr. Naoto Ishikawa greatly helped to improve an early version of the manuscript.

Author details

Yasuto Itoh[1]*, Paul F. Green[2], Keiji Takemura[3] and Tomotaka Iwata[4]

*Address all correspondence to: yasutokov@yahoo.co.jp

1 Graduate School of Science, Osaka Prefecture University, Osaka, Japan

2 Geotrack International, Brunswick West, VIC, Australia

3 Institute for Geothermal Sciences, Kyoto University, Beppu, Japan

4 Disaster Prevention Research Institute, Kyoto University, Uji, Japan

References

[1] Itoh Y, Uno K, Arato H. Seismic evidence of divergent rifting and subsequent deformation in the southern Japan Sea, and a Cenozoic tectonic synthesis of the eastern Eurasian margin. Journal of Asian Earth Sciences. 2006; 27: 933-942.

[2] Makimoto H, Miyata T, Mizuno K, Sangawa A. Geology of the Kokawa District, with Geological Sheet Map at 1:50,000. Tsukuba: Geological Survey of Japan, AIST; 2004. 89 p.

[3] Tanaka J. Sedimentary facies of the Cretaceous Izumi turbidite system, southwest Japan - an example of turbidite sedimentation in an elongated strike-slip tectonic basin. Journal of the Geological Society of Japan. 1989; 95: 119-128.

[4] Hurford AJ, Green PF. The zeta age calibration of fission track dating. Isotope Geoscience. 1983; 1: 285-317.

[5] Gleadow AJW. Fission track dating methods: what are the real alternatives? Nuclear Tracks. 1981; 5: 3-14.

[6] Wagner GA, Reimer GM. Fission track tectonics: the tectonic interpretation of fission track ages. Earth and Planetary Science Letters. 1972; 14: 263-268.

[7] Naeser CW, Forbes RB. Variation of fission track ages with depth in two deep drill holes. EOS (abstract). 1976; 57: 363.

[8] Bhandari N, Bhat SG, Lal D, Rajagopalan G, Tamhane AS, Venkatavaradan VS. Fission fragment tracks in apatite: recordable track lengths. Earth and Planetary Science Letters. 1971; 13: 191-199.

[9] Green PF. On the cause of the shortening of spontaneous fission tracks in certain minerals. Nuclear Tracks. 1980; 4: 91-100.

[10] Green PF, Duddy IR, Gleadow AJW, Tingate PR, Laslett GM. Thermal annealing of fission tracks in apatite 1. A qualitative description. Chemical Geology (Isotope Geoscience Section). 1986; 59: 237-253.

[11] Laslett GM, Kendall WS, Gleadow AJW, Duddy IR. Bias in measurement of fission track length distributions. Nuclear Tracks. 1982; 6: 79-85.

[12] Gleadow AJW, Duddy IR, Green PF, Lovering JF. Confined fission track lengths in apatite—a diagnostic tool for thermal history analysis. Contributions to Mineralogy and Petrology. 1986; 94: 405-415.

[13] Green PF, Duddy IR, Gleadow AJW, Tingate PR, Laslett GM. Fission-track annealing in apatite: track length measurements and the form of the Arrhenius plot. Nuclear Tracks. 1985; 10: 323-328.

[14] Green PF. The relationship between track shortening and fission track age reduction in apatite: combined influences of inherent instability, annealing anisotropy, length bias and system calibration. Earth and Planetary Science Letters. 1988; 89: 335-352.

[15] Galbraith RF, Laslett GM. Some calculations relevant to the thermal annealing of fission tracks in apatite. Proceedings of the Royal Society, London. 1988; A419: 305-321.

[16] Galbraith RF, Laslett GM. Apatite fission track analysis: geological thermal history analysis based on a three-dimensional random process of linear radiation damage. Philosophical Transcripts of the Royal Society of London. 1990; 332: 419-438.

[17] Laslett GM, Green PF, Duddy IR, Gleadow AJW. Thermal annealing of fission tracks in apatite 2. A quantitative analysis. Chemical Geology (Isotope Geoscience Section). 1987; 65: 1-13.

[18] Duddy IR, Green PF, Laslett GM. Thermal annealing of fission tracks in apatite 3. Variable temperature behaviour. Chemical Geology (Isotope Geoscience Section). 1988; 73: 25-38.

[19] Green PF, Duddy IR, Laslett GM, Hegarty KA, Gleadow AJW, Lovering JF. Thermal annealing of fission tracks in apatite 4. Quantitative modelling techniques and extension to geological timescales. Chemical Geology (Isotope Geoscience Section). 1989; 79: 155-182.

[20] Gleadow AJW, Duddy IR. Fission track analysis: a new tool for the evaluation of thermal histories and hydrocarbon potential. APEA Journal. 1981; 23: 93-102.

[21] Crowhurst PV, Green PF, Kamp PJJ. Appraisal of (U-Th)/He apatite thermochronology as a thermal history tool for hydrocarbon exploration: an example from the Taranaki Basin, New Zealand. AAPG Bulletin. 2002; 86: 1801-1819.

[22] Green P. The importance of validating annealing models (abstract). In: Proceedings of 10th International Conference on Fission Track Dating and Thermochronology; Amsterdam. 2004. p. 53.

[23] Green PF, Duddy IR. Synchronous exhumation events around the Arctic including examples from Barents Sea and Alaska North Slope. In: Vining BA, Pickering SC, editors. Petroleum Geology, From Mature Basins to New Frontiers. Proceedings of the 7th Petroleum Geology Conference. London: The Geological Society; 2010. pp. 633-644. DOI: 10.1144/0070633

[24] Green PF, Duddy IR. Thermal history reconstruction in sedimentary basins using apatite fission-track analysis and related techniques. In: Analyzing the Thermal History of Sedimentary Basins - Methods and Case Studies. SEPM (Society for Sedimentary Geology); 2012. pp. 65-104.

[25] Matsuura H, Kurimoto C, Sangawa A, Bunno M. Geology of the Hirone District, with Geological Sheet Map at 1:50,000. Tsukuba: Geological Survey of Japan, AIST; 1995. 110 p.

[26] Takano O. Variation in forearc basin configuration and basin-filling depositional systems as a function of trench slope break development and strike-slip movement: examples from the Cenozoic Ishikari-Sanriku-oki and Tokai-oki-Kumano-nada forearc basins, Japan. In: Itoh Y, editor. Mechanism of Sedimentary Basin Formation-Multidisciplinary Approach on Active Plate Margins. Rijeka: InTech; 2013. http://dx.doi.org/10.5772/56751

[27] Itoh Y, Nagasaki Y. Crustal shortening of Southwest Japan in the Late Miocene. The Island Arc. 1996; 5: 337-353.

[28] Tagami T, Lal N, Sorkhabi RB, Nishimura S. Fission track thermochronologic analysis of the Ryoke belt and the Median Tectonic Line, southwest Japan. Journal of Geophysical Research. 1988; 93: 13705-13715.

Chapter 6

Alteration Reaction and Mass Transfer via Fluids with Progress of Fracturing along the Median Tectonic Line, Mie Prefecture, Southwest Japan

Yumi Kaneko, Toru Takeshita, Yuto Watanabe,

Norio Shigematsu and Ko-Ichiro Fujimoto

Additional information is available at the end of the chapter

http://dx.doi.org/10.5772/68112

Abstract

We have analyzed mass transfer in the cataclasite samples collected from the Median Tectonic Line, southwest Japan, in which the degree of fracturing is well correlated with the bulk rock chemical compositions determined by the X-ray fluorescence (XRF) analysis. The results of "isocon" analysis indicate not only a large volume increase up to 110% but also the two-stage mass transfer during cataclasis. At the first stage from the very weakly to weakly fractured rocks, the weight percents of SiO_2, Na_2O, and K_2O increase, while those of TiO_2, FeO, MnO, MgO, and CaO decrease. At the second stage from the weakly to moderately and strongly fractured rocks, the trend of mass transfer is reversed. The principal component analysis reveals that the variation of chemical compositions in the cataclasite samples can be mostly interpreted by the mass transfer via fluids and by the difference in chemical composition in the protolith rocks to lesser degree. Finally, the changes in the modal composition of minerals with increasing cataclasis analyzed by the X-ray diffraction (XRD) with the aid of "RockJock" software clearly elucidate that the mass transfer of chemical elements was caused by dissolution and precipitation of minerals via fluids in the cataclasite samples.

Keywords: Median Tectonic Line, cataclasite, mass transfer, isocon analysis, dissolution and precipitation of minerals

1. Introduction

In the Japanese islands, a various scale of faults in different orientations develop as shown by the distribution of active faults (e.g., see [1]), and part of a significant amount strain rate as measured with a global positioning system (GPS) [2] could be explained by a various amount of displacement rate along these faults. Since a lower part of crust underneath the Japanese islands arc is fairly ductile due to a high geothermal gradient, c. 30°C/km, it is only sustained by a thin rigid upper crust of c. 10 km thick (i.e., thin-skinned plate [3]). Therefore, it is very important to elucidate how high the strength (i.e., constitutive equations) along faults is, and how fast the displacement occurs along them in the upper crust, which enables us to discuss about the stain rate and differential stress in the crust, and hence dynamic processes in it.

It has been interpreted that the fault zones, which are originally formed as relatively short narrow segments, are linked to become long faults, and then widen to become mature and weak fault zones [4, 5]. During the fault development, the fault zones formed at the wide depth range above the brittle-ductile transition depth become weakened by frictional sliding along mica or clay minerals with lower internal coefficients of friction accommodated by pressure solution creep (e.g., see [6, 7]), the amount of which increases with increasing deformation aided by alteration of fault rocks [8, 9]. As shown in the present study, new minerals which have different chemical compositions from those of the dissolved minerals and stable under new temperature and pressure conditions precipitate at the dilatant site caused by microfracturing from the solution in cataclasite constituting a fault core (e.g., filling minerals between amphibole microboudine, see [10]). These series of processes could be called dissolution-precipitation creep in a broad sense, in contrast to that in single-phase mineral aggregates studied by many researchers (e.g., see [11]). Hence, to analyze dissolution and precipitation of minerals based on element migration in cataclasite via fluids as conducted in the present study may be equivalent to analyze dissolution-precipitation creep in the broad sense.

In the present research, element migration via fluids in the Cretaceous granitoid cataclasite core samples from the borehole drilled through the Median Tectonic Line (MTL) in Mie Prefecture, southwest Japan [12], has been analyzed. We have first quantified the degree of cataclasis, based on the fracture density (number/cm) measured on the thin sections. Then, we have analyzed altered minerals in both plagioclase and fine-grained materials (ultracataclasite) in the cataclasite samples mostly using a scanning electron microscopy with energy dispersive X-ray spectrometry (SEM-EDS). We further analyzed the bulk rock compositions with an X-ray fluorescence (XRF) and modal composition of minerals constituting the cataclasite samples with an X-ray diffraction (XRD) with the aid of the software "RockJock" [13] as a function of degree of cataclasis. Further, the "isocon" method [14] is employed to analyze the mass transfer in the cataclasite, and further the origin of large variation in chemical composition has been analyzed with the method of principal component analysis (PCA, e.g., see [15]). Finally, we discussed the development of the cataclasite zone along the MTL based on chemical reaction and mass transfer which occurred in this zone.

2. Geological setting and sample description

The MTL (**Figure 1**) was originally formed as the boundary normal fault between the Cretaceous Ryoke granitic and low-P/T type metamorphic rocks, and Cretaceous high-P/T type Sambagawa metamorphic rocks during exhumation of the latter rocks in the earliest Paleogene (c. 60 Ma, e.g., see [16]). In the upper plate consisting of the Ryoke rocks, a shear zone consisting of the so-called the Kashio mylonite was formed in the latest Cretaceous, along which a large amount of sinistral displacement is inferred to have occurred [17–19]. As the time passed, the mylonite zone in the Ryoke rocks, which was developed into the MTL, was overprinted by cataclastic deformation during exhumation and cooling. The Ryoke granitoid cataclasite samples of drill core analyzed in this study are from a borehole drilled by the Geological Survey of Japan, AIST, at the Matsusaka–Iitaka observatory (ITA), Mie Prefecture, Japan. This borehole was drilled through the Median Tectonic Line in Mie Prefecture, southwest Japan, at the depth of 473.9 m and further drilled through the Sambagawa metamorphic rocks down to the depth of 600 m (**Figures 1** and **2**, see [12]). The positions of the MTL at both outcrops and borehole reveal a fault plane orientation of the MTL as N86°E56°N, which is consistent with the attitude of planar fabrics in the gouge zone in direct proximity to the MTL [12, 20]. The protolith of the Ryoke granitoids distributed in this area is mostly tonalite, called Hatai tonalite or Arataki granodiorite [21, 22].

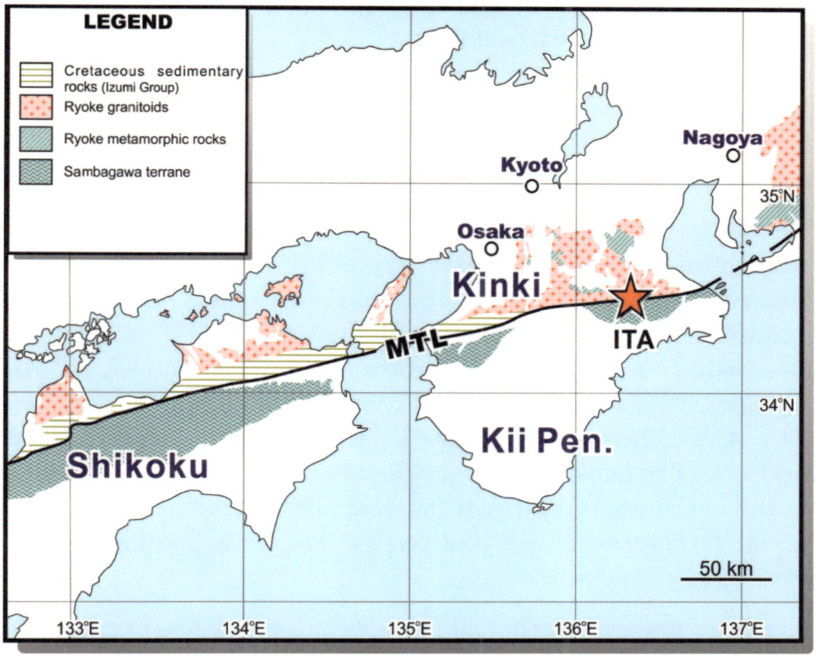

Figure 1. Index map of the Ryoke and Sambagawa belts with the MTL and locality of the Matsusaka-Iitaka (ITA) borehole.

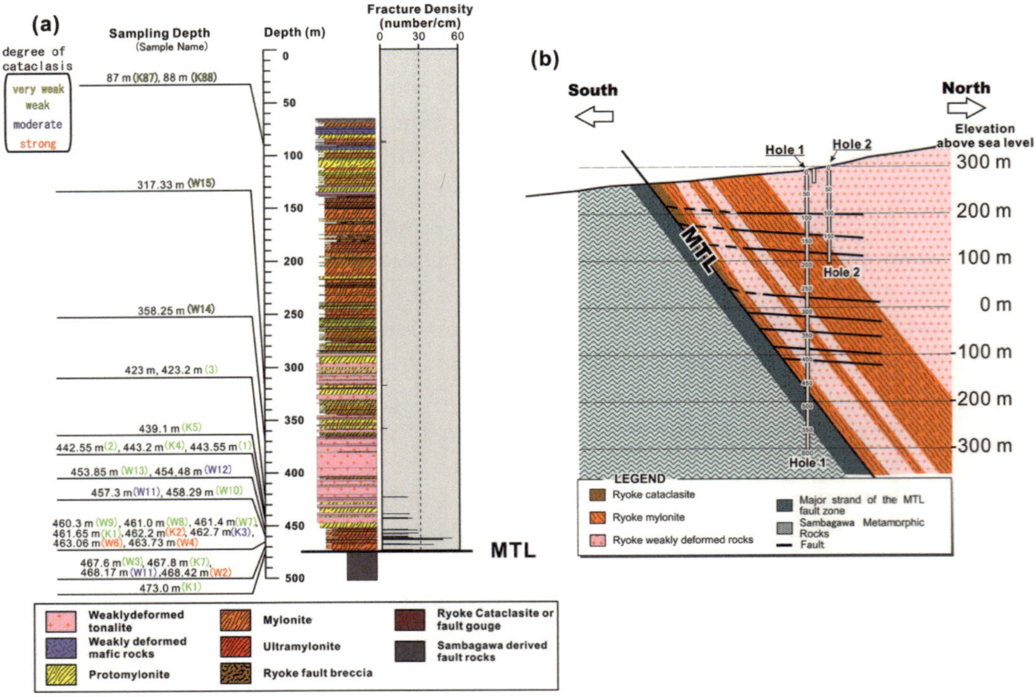

Figure 2. (a) Sample localities of the cataclasite samples in a lithology column of the Matsusaka-Iitaka (ITA) borehole. (b) Cross section of the ITA borehole. Modified after Ref. [12].

3. Results

3.1. Fracture density in the cataclasite

Based on observations of the cataclasite samples by naked eyes, they can be divided into four groups in terms of the degree of cataclasis: very weakly, weakly, moderately and strongly fractured cataclasites (**Figure 2**). Further, based on observations of some of the cataclasite samples under an optical microscope, the fracture density (FD) is determined to vary for one very weakly (W14; FD = 4.8/cm), seven weakly (K13, K14, K7, K5, K4, K11, W3; FD = 15.7–25.6/cm), three moderately (K6, K3, W12; FD = 30.1–37.7/cm) and two strongly (W2, K2; FD = 43.9–59.2/cm) fractured cataclasite (**Table 1**). It should be noted that non-fractured rocks cannot be found near the cataclasite zone constituting the upper plate of the MTL (**Figure 1**), and hence the very weakly fractured cataclasite samples collected at the depth of 87, 88, 317 and 358 m should be only considered as a relatively fracture-free samples among the analyzed cataclasite samples.

Microphotographs of these cataclasite samples with different degree of cataclasis are shown in **Figure 3**. As seen in this figure, in the very weakly fractured cataclasite sample (W14), a mylonite foliation consisting of elongate quartz aggregates is still well preserved, although the plagioclase grains are pervasively sericitized (**Figure 3a, b**). In the weakly fractured cataclasite sample (W10), although anastomosing fractures along which white mica is grown are pervasively

W14#	W3$	K13$	K14$	K7$	K5$	K4$	K1$	W12%	K6%	K3%	W2&	K2&
4.83	15.65	17.69	19.83	22.75	23.25	25.58	25.61	30.14	33.92	37.74	43.92	59.19

Table 1. Results for fracture densities in the cataclasite samples. Degree of cataclasis: #, very weak; $, weak; %, moderate; &, strong.

developed, the original clasts consisting of stretched quartz aggregates are still preserved (**Figure 3c, d**). In the strongly fractured sample (W4), plagioclase and quartz clasts become very fine grained by strong comminution, and not only white mica, but also chlorite are grown in the fine-grained part (ultracataclasite, **Figure 3e, f**).

3.2. Mineral phases grown in altered plagioclase and fine-grained aggregates (ultracataclasite)

From the observations of thin sections of the cataclasite samples under an optical microscope, it is quite clear that the increase of fracture density (i.e., the degree of cataclasis) is accompanied by the increase of the degree of alteration (i.e., growth of new minerals by metamorphic reaction via fluids, **Figure 3**). Since the comminuted minerals and newly grown minerals are very fine grained, it is very difficult to identify these mineral phase under an optical microscope. Therefore, analysis of microstructures and identification of these fine-grained mineral phases are mostly conducted with a SEM (JSM-5310) and EDS (Oxford-7068). As mentioned below, in weakly to strongly fractured rocks, white mica (phengite, here referred to as muscovite) is grown either replacing original plagioclase altered to albite (weakly fractured Sample K14, **Figure 4a**), or as fracture-filling minerals (i.e., veins, weakly fractured samples W10, **Figure 3c, d**). The existence of white mica is shown by a large amount of both K and Al (**Figure 4a**). In the former sample, prehnite is also grown replacing albite, shown by a large amount of both Ca and Al (**Figure 4a**).

In a strongly fractured cataclasite samples (K2), the fragments of albite and quartz are identified by a large amount of Na, and that of Si alone, respectively (**Figure 4b**). Chlorite shown by a large amount of both Mg and Fe makes an anastomosing network surrounding the fractured plagioclase and quartz. Further, the presence of fine-grained sphene is shown by a large amount of both Ti and Ca. In this sample, both long axis of fractured clasts and grown minerals are aligned in the top-down direction, forming a foliated cataclasite [23].

3.3. Bulk rock composition in the cataclasite samples

Bulk rock composition in the cataclasite samples determined with an XRF analysis is shown in **Table 2**, and also the stratigraphic changes (**Figure 5**) are shown with different colors defined by the different degree of fracturing in the cataclasite samples. Although the compositions in the cataclasite samples with different degree of cataclasis greatly vary, there is a general tendency of variation in the composition with increasing cataclasis, as mentioned below. It is noted, however, that the composition of the protolith tonalite mylonite is not well constrained, because it was not possible to obtain samples free from fractures.

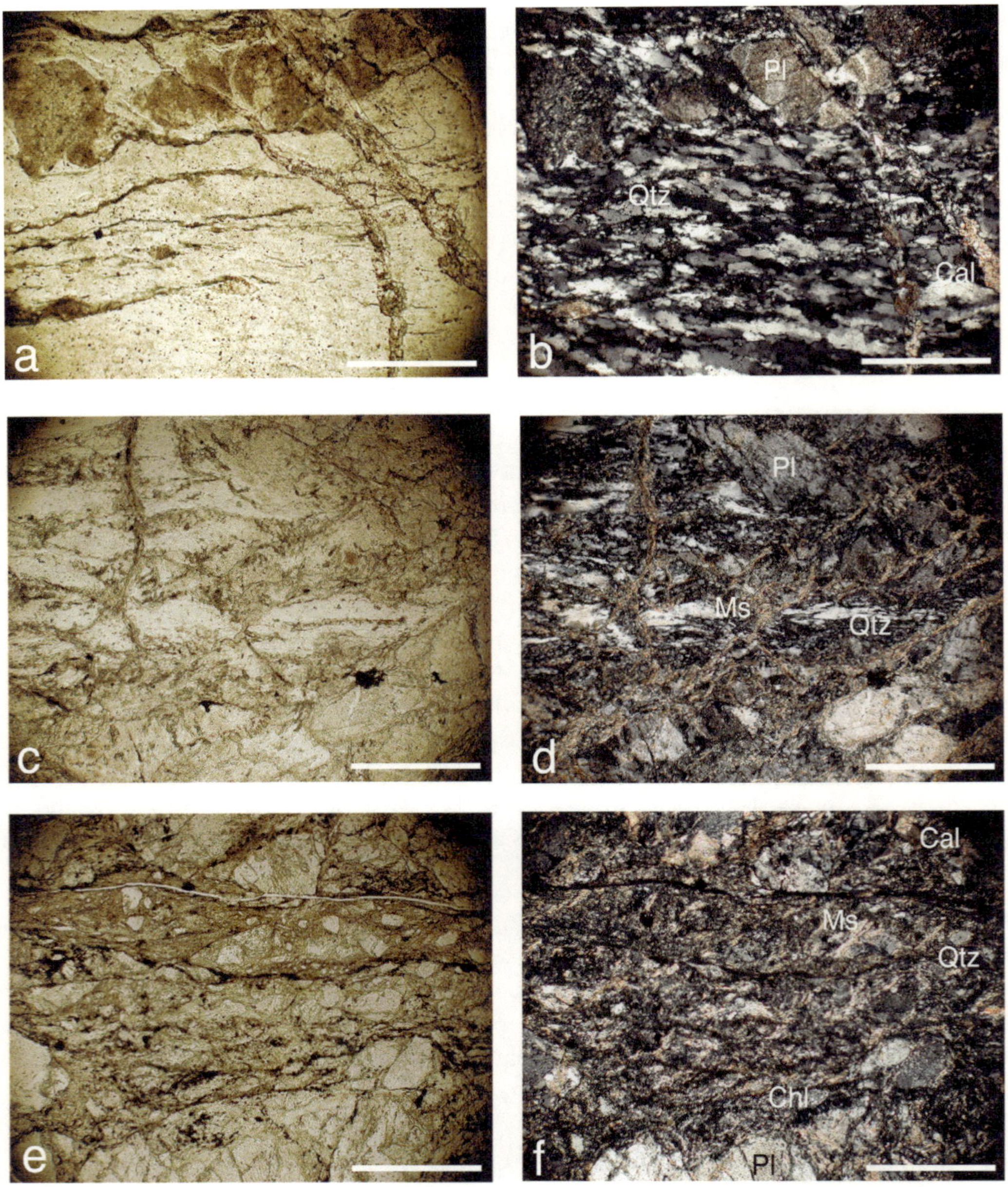

Figure 3. Microphotographs of the cataclasite samples. (a, b) Very weakly fractured sample (W14). (c, d) Weakly fractured sample (W10). (e, f) Strongly fractured sample (W4). (a), (c) and (e) are taken with plane polarized light, while (b), (d) and (f) with crossed polarized light. Scale bars are all 0.4 mm. Abbreviation: Qtz, quartz; Pl, plagioclase; Cal, calcite; Ms, muscovite; Chl, chlorite.

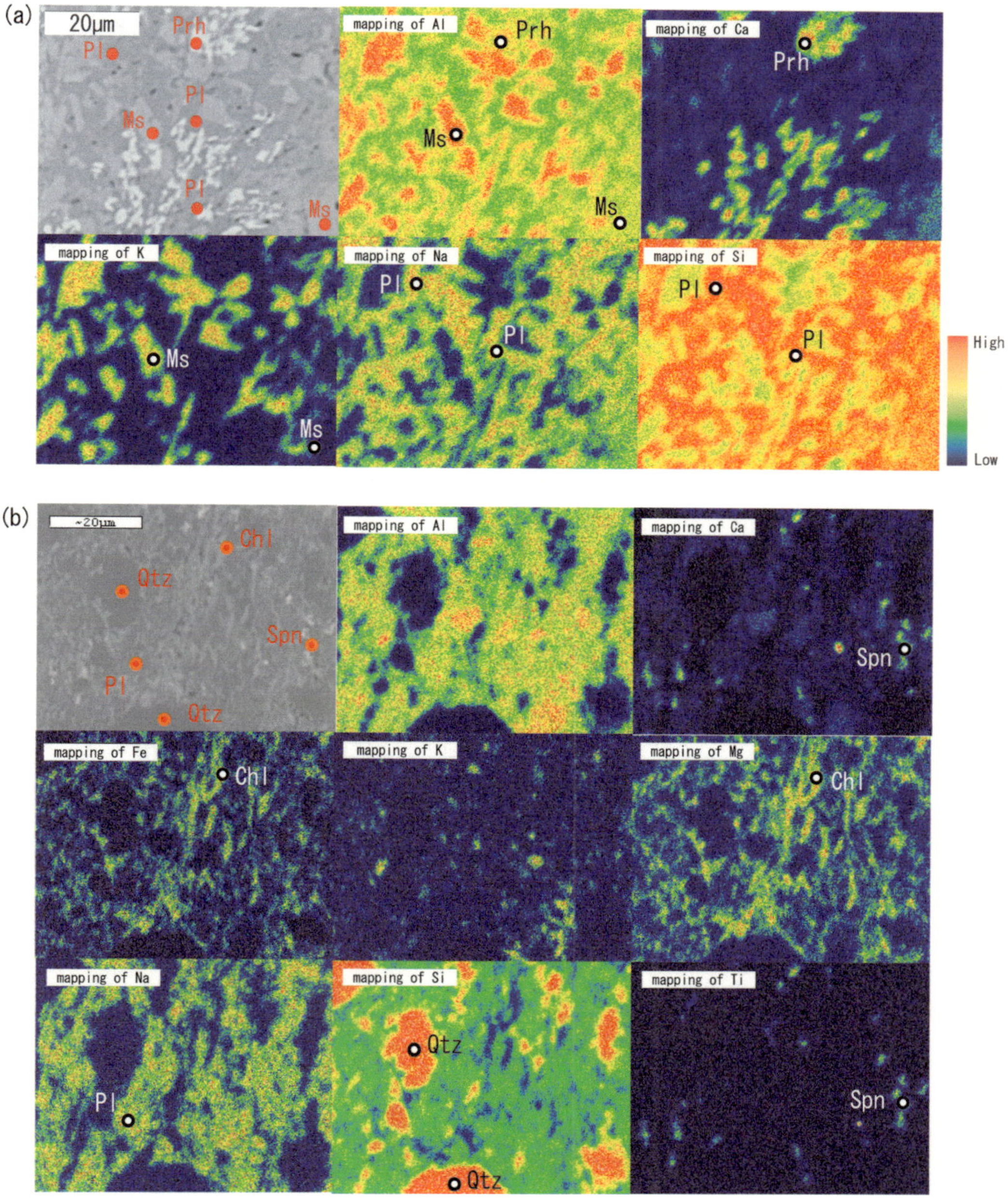

Figure 4. SEM-BSE image and EDS mapping for the cataclasite samples of different degree. See text for explanations. (a) Sample K14 (weakly fractured). (b) Sample K2 (strongly fractured). Abbreviation: Qtz, quartz; Pl, plagioclase; Cal, calcite; Ms, muscovite; Chl, chlorite; Prh, prehnite; Rt, rutile; Spn, sphene.

Oxide (wt%)	W15[#]	W14[#]	W13[$]	W12[%]	W10[$]	W9[$]	W8[$]	W7[$]	W6[&]	W5[&]	W4[&]	W2[&]	W1[$]
SiO_2	63.81	70.88	72.83	61.82	76.38	69.98	72.90	75.28	50.76	50.16	50.89	55.65	70.85
TiO_2	0.68	0.30	0.02	0.99	0.11	0.21	0.20	0.09	0.65	0.61	0.63	0.37	0.28
Al_2O_3	17.15	14.70	12.16	15.70	12.53	13.62	13.75	13.23	17.04	17.72	17.90	16.19	14.07
FeO*	4.57	2.40	0.74	4.75	1.15	2.20	2.15	1.13	6.45	5.79	5.95	3.88	2.51
MnO	0.09	0.05	0.02	0.10	0.02	0.05	0.05	0.03	0.14	0.12	0.12	0.10	0.04
MgO	1.86	0.85	0.13	2.38	0.29	0.56	0.46	0.23	6.73	5.39	5.39	4.12	0.84
CaO	4.73	2.52	2.43	5.03	1.46	3.59	1.93	1.43	7.25	7.07	7.29	8.49	2.47
Na_2O	3.98	4.19	3.90	3.40	4.09	3.37	3.96	4.07	2.39	2.76	2.79	2.36	4.26
K_2O	0.89	2.47	2.81	1.76	2.50	3.15	2.53	3.11	1.61	2.18	2.23	2.03	2.09
P_2O_5	0.17	0.08	0.02	0.17	0.03	0.06	0.06	0.02	0.13	0.12	0.13	0.09	0.08
L.O.I	2.07	1.55	4.95	3.89	1.44	3.20	2.01	1.39	6.84	8.08	6.68	6.72	2.51
Zr	134.545	117.519	77.11	114.205	92.154	121.429	141.781	95.541	66.156	75.547	76.008	67.056	177.51
Total	100.95	100.47	100.85	100.81	99.19	101.01	101.14	101.99	100.70	101.49	100.07	100.48	100.48

Oxide (wt%)	K1[$]	K2[&]	K3[%]	K4[$]	K5[$]	K6[%]	K7[$]	K88[#]	K87[#]	4-6[$]	4-4[$]	4-1[$]	3B-4[$]
SiO_2	74.98	59.06	49.42	62.12	63.16	49.44	70.49	58.61	72.12	61.65	62.57	54.00	61.76
TiO_2	0.11	0.62	0.60	0.60	0.64	0.62	0.15	0.77	0.34	0.63	0.68	0.88	0.62
Al_2O_3	12.81	15.73	15.88	17.00	16.73	17.66	13.41	16.74	14.11	16.49	16.51	17.97	15.56
FeO*	1.36	5.03	6.11	4.18	4.22	5.82	1.33	6.43	2.77	4.58	4.48	6.48	4.45
MnO	0.03	0.11	0.14	0.08	0.08	0.13	0.04	0.15	0.05	0.10	0.09	0.15	0.08
MgO	0.30	3.88	6.29	1.79	1.78	5.69	0.70	3.54	0.75	2.20	2.06	3.71	2.11
CaO	1.93	5.78	8.51	4.29	4.56	7.33	3.99	6.06	2.56	5.28	5.25	7.11	5.08
Na_2O	4.05	2.86	1.98	3.53	3.90	1.74	3.56	3.07	3.23	3.19	3.50	3.38	3.13
K_2O	2.12	1.41	1.80	2.23	1.91	2.68	2.13	1.42	2.81	1.54	1.16	0.92	2.07
P_2O_5	0.04	0.12	0.12	0.16	0.16	0.09	0.05	0.17	0.09	0.15	0.17	0.17	0.15
L.O.I	2.27	5.40	9.15	4.02	2.85	8.79	4.14	3.05	1.17	4.19	3.53	5.24	5.00
Zr	100.969	94.352	75.156	122.799	123.296	71.647	106.766	133.586	175.7	111.26	135.92	104.55	115.46
Total	102.42	101.41	101.02	100.51	100.97	101.40	101.56	100.01	99.65	101.46	101.25	101.28	101.63

Oxide (wt%)	3B-2[$]	2C-5[$]	2C-3[$]	2C-1[$]	2B-1[$]	2A-3[$]	2A-1[$]	1B-2[$]	1A-3[$]	1A-1[$]
SiO_2	60.30	58.43	61.63	58.92	56.23	58.51	59.05	64.58	63.18	64.05
TiO_2	0.68	0.79	0.59	0.70	0.79	0.75	0.74	0.57	0.54	0.57
Al_2O_3	15.73	17.72	16.58	16.10	16.81	15.71	16.50	15.61	15.99	16.47
FeO*	5.23	5.40	4.15	5.14	6.34	5.74	5.71	3.90	3.74	3.81
MnO	0.09	0.11	0.08	0.11	0.14	0.14	0.13	0.08	0.08	0.08

Oxide (wt%)	3B-2$	2C-5$	2C-3$	2C-1$	2B-1$	2A-3$	2A-1$	1B-2$	1A-3$	1A-1$
MgO	2.49	2.40	1.90	2.33	2.91	2.71	2.76	1.67	1.61	1.64
CaO	4.79	3.91	4.09	5.15	5.31	5.07	4.53	4.59	5.22	4.29
Na_2O	3.06	3.61	3.39	3.93	3.11	3.10	3.37	3.79	3.87	3.77
K_2O	2.27	2.68	2.68	1.93	2.29	2.22	2.01	1.45	1.50	1.62
P_2O_5	0.16	0.19	0.14	0.17	0.17	0.17	0.16	0.14	0.14	0.14
L.O.I	5.20	4.76	4.77	5.51	5.91	5.90	5.04	3.61	4.13	3.56
Zr	121.53	138.35	123.87	118.34	117.95	117.17	105.65	114.16	116.39	114.16
Total	100.72	101.45	101.60	101.08	101.31	101.32	101.81	101.59	100.97	101.62

Table 2. Results for bulk rock compositions determined with the X-ray fluorescence analysis in the cataclasite samples. Degree of cataclasis: #, very weak; $, weak; %, moderate; &, strong.

SiO_2 wt.% greatly varies with increasing degree of cataclasis. First, it increases from 58.6 to 72.1 wt.% in the very weakly fractured rocks to 54.0–76.4 wt.% in the weakly fractured rocks, and then decreases to 49.4–62.8 wt.% and 50.2–59.1 wt.% in the moderately and strongly fractured rocks, respectively (**Table 2**). Associated with this change in SiO_2 wt.%, there is a general tendency that the wt.% of FeO, MgO, and CaO first decreases from the very weakly to weakly fractured rocks, and then increases from the weakly to the moderately and strongly fractured rocks. On the other hand, the wt.% of Na_2O and K_2O first increases and then decreases with increasing degree of cataclasis.

As shown in the compositional changes of major elements in the stratigraphic column (**Figure 5**), the compositional changes are quite abrupt at some horizons, in particular, at two horizons where the depth from the surface is 463.0 and 468.5 m, respectively. Perhaps, these sudden changes in the composition of cataclasite, which are also shown for cataclasites along the San Andreas Fault by [24], indicate the fault zone defined by very strong cataclasis and chemical alteration in the whole cataclasite zone.

3.4. Modal composition of minerals in the cataclasite measured by an XRD

Modal composition of minerals in the cataclasite samples is measured by an XRD (**Table 3**). It is generally clear that the modes of minerals constituting the protolith tonalite mylonite such as albite, quartz, K-feldspar, amphibole, and rutile decrease, while those of newly grown minerals, mostly clay and mica minerals (i.e., chlorite and white mica), and calcite, prehnite, and sphene increase with increasing cataclasis in the analyzed cataclasite samples. However, it should be noted that chlorite, which is normally abundant only in the moderately and strongly fractured cataclasite, is already abundant in some of the very weakly fractured samples (W15 and K88), where the mode of quartz is lower and that of plagioclase is higher than the respective modes in the other very weakly fractured samples (W14 and K87).

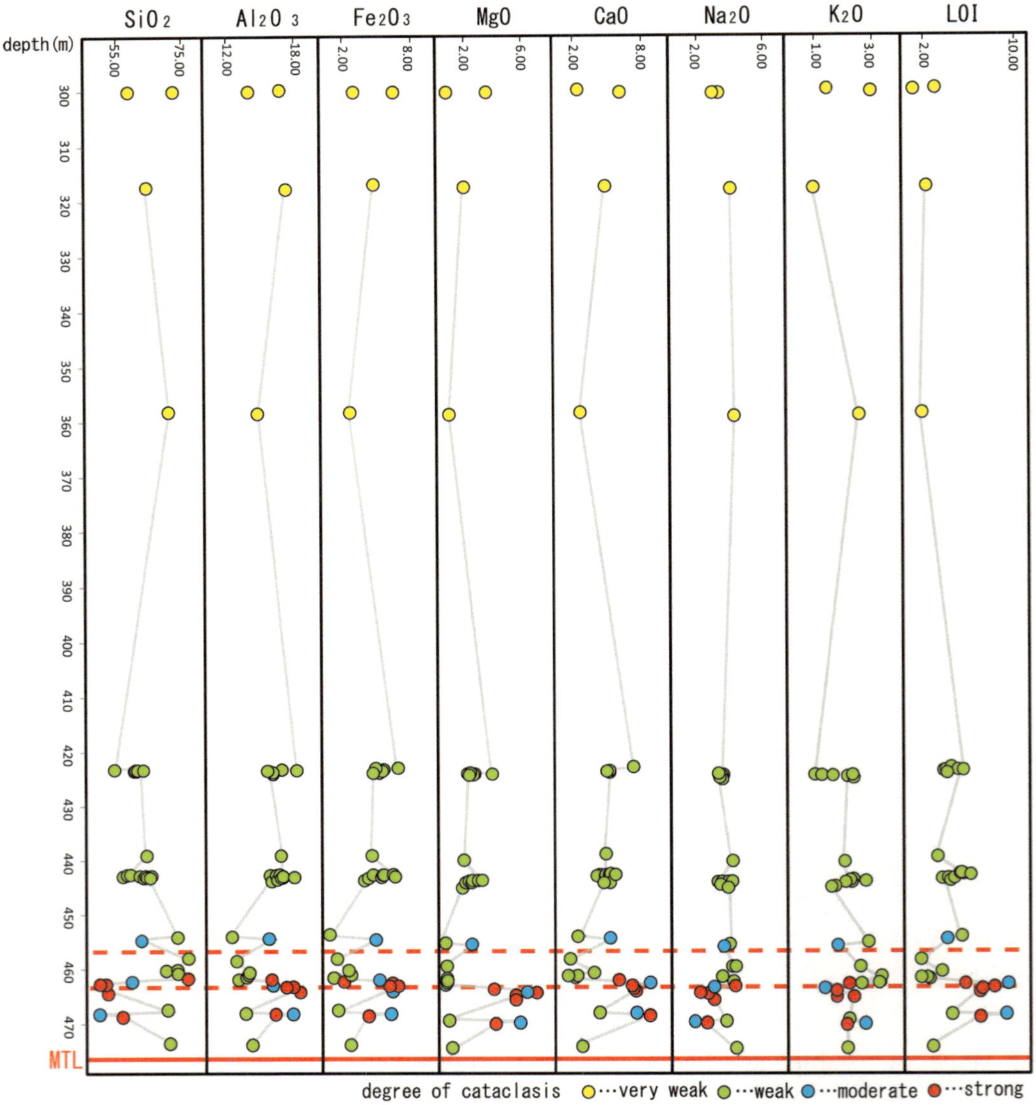

Figure 5. Spatial variation of the bulk rock compositions in the direction of the borehole in the cataclasite samples. Red dashed lines indicate inferred faults.

Therefore, the facts suggest that there are already large variations in the composition of the protolith tonalite mylonite, which could follow their own changes in the modal composition of minerals during cataclasis and alteration. This possibility will be explored in the following section using the principal component analysis (PCA), following which the more refined results of modal composition of minerals in the cataclasite samples will be presented.

Mineral (weight%)	W14	W15	K1	K2	K3	K4	K5	K6	K7	K87	K88
Quartz	34.23	26.75	39.05	27.10	18.86	25.85	24.71	18.29	37.36	36.48	19.70
Kspar	0.00	0.00	0.00	0.00	0.00	0.00	0.00	0.00	0.00	15.97	0.27
Plagioclase	46.85	54.15	37.60	32.93	20.64	43.93	44.01	17.08	32.59	35.74	46.15
Calcite	1.80	1.24	3.24	6.52	12.87	4.32	1.40	12.03	6.23	0.43	1.64
Amphibole	0.00	0.00	0.00	0.00	0.45	0.00	0.00	0.00	0.00	0.00	3.06
Pyrite	0.00	0.00	0.00	0.11	0.07	0.00	0.16	0.00	0.00	0.06	0.34
Sphene	1.81	0.86	0.29	0.16	0.56	0.00	2.09	0.00	0.00	1.07	1.61
Anatase	0.21	0.41	0.00	0.00	0.00	0.00	0.00	0.00	0.00	0.08	0.00
Rutile	0.24	0.00	0.13	0.00	0.00	0.00	0.16	0.00	0.00	0.44	0.00
Epidote	0.17	0.36	0.34	2.67	1.18	0.00	2.27	0.39	0.00	0.62	1.34
Zircon	0.00	0.00	0.00	0.00	0.00	0.00	0.09	0.00	0.00	0.00	0.00
Chabazite	0.00	0.15	0.09	0.00	0.00	0.00	0.21	0.00	0.00	0.00	0.71
Prehnite	0.69	0.00	0.00	0.69	1.02	0.00	3.21	1.01	0.26	0.81	2.16
Total non-clay and non-mica	86.00	83.92	80.74	70.18	55.65	74.10	78.31	48.80	76.44	91.70	76.98
Kaolinite	0.00	0.44	0.22	0.93	2.19	0.00	0.00	1.76	0.00	0.00	1.48
Muscovite	4.83	1.69	12.12	9.58	13.16	14.28	5.89	19.64	13.73	1.07	0.69
Glauconite	0.08	1.83	1.47	0.00	0.00	0.00	0.00	0.00	1.49	0.00	0.75
Talc	0.00	0.00	0.08	1.76	3.76	0.37	0.00	1.54	0.00	0.00	2.00
Chlorite	5.64	13.02	3.22	18.20	24.06	11.83	13.32	23.61	2.65	6.27	18.64
Smectite	0.00	0.49	0.00	0.00	0.03	0.00	0.00	0.00	0.00	0.00	0.00
Total clay and mica	10.55	17.47	17.11	30.47	43.20	26.48	19.21	46.55	17.87	7.34	23.56
Total	96.55	101.39	97.85	100.65	98.85	100.58	97.52	95.35	94.31	99.04	100.54

Table 3. Results of analyses of modal compositions of minerals measured with an XRD in the cataclasite samples.

4. Interpretation

4.1. Principal component analysis (PCA)

In order to investigate why the amount of these chemical elements varies in the cataclasite samples, we have conducted the principal component analysis (PCA) using the free software Easy PCA [15] for the bulk chemical compositions determined for the samples W1, 2, 4, 5, 6, 7, 8, 9, 10, 12, 13, 14, and 15 and K1, 2, 3, 4, 5, 6, 7, 87, and 88. Based on the results of PCA, it has been found that the contribution rates of the first and second principal components are 76 and 15%, respectively, the sum of which exceeds 90%. Hence, only the first and second principal components are considered to interpret the net results. The calculation of factor loadings shows that for the first principal component all the chemical compositions have a similar value, and SiO_2, Na_2O, and K_2O, and the other chemical compositions show opposite sign for the contribution to the first principal component (**Figure 6a**). On the other hand, for the second principal component, Na_2O, K_2O, P_2O_5 and LOI (loss of ignition) have larger factor loadings than the other chemical compositions, and hence greatly contribute to it (**Figure 6b**). It should be noted that the second principal component increases with the increasing amount of K_2O, while it decreases with the increasing amount of Na_2O. Finally, the values of the first and second principal components of the analyzed bulk chemical compositions from the cataclasite samples with the different fracture densities are plotted in the X-Y coordinates, where the horizontal and vertical axes show the values of first and second principal components, respectively (**Figure 7**). According to this diagram, it has been found that for those samples where the values of the second principal component are mostly positive and greater than −0.1, the values of the first principal component decrease with increasing fracture density, although the very weakly and weakly fractured, and moderately and strongly fractured samples cannot be distinguished in terms of the value of the first principal component, respectively. This group

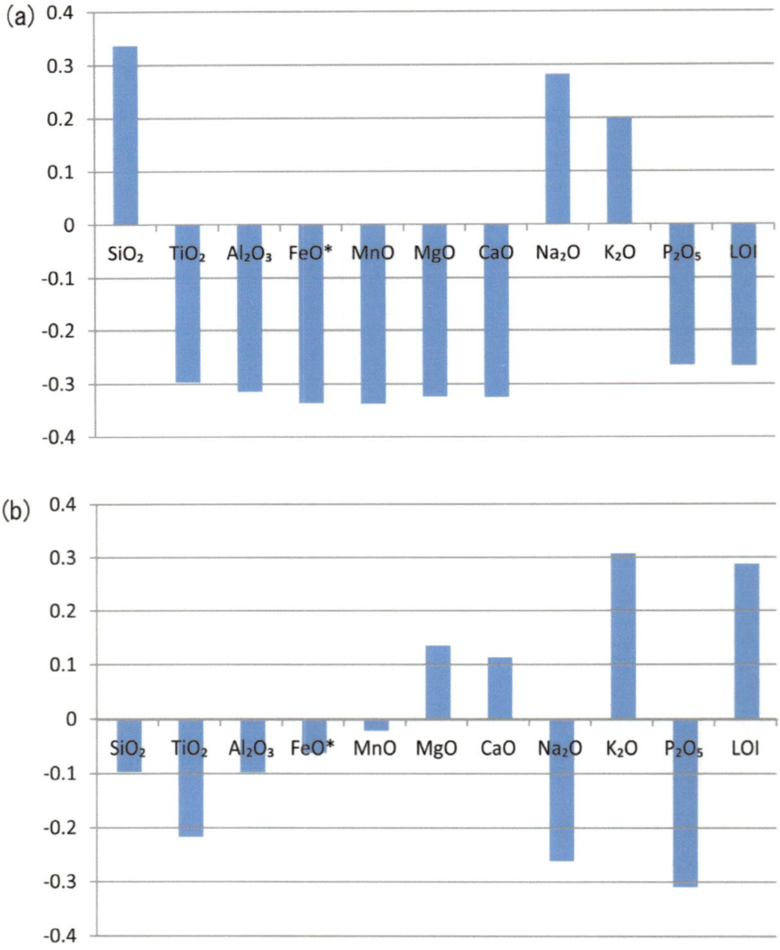

Figure 6. Principal component analysis. (a) Factor loading for the first principal component and (b) factor loading for the second principal component.

http://dx.doi.org/10.5772/68112

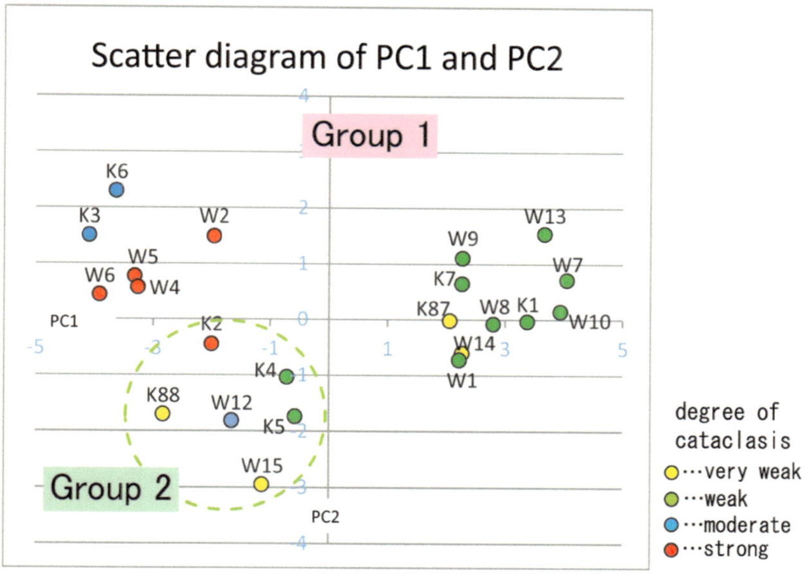

Figure 7. Plots of the first and second principal component values in the X-Y coordinates.

of samples is called Group 1. On the other hands, the plots of those samples where the values of the second principal component are mostly negative and smaller than −1.0 except for K2 are clustered, and they are grouped into Group 2 here.

We have now plotted the modal composition of minerals determined with an XRD analysis in the cataclasite samples K1, K2, K3, K4, K5, K6, K7, K87, K88, and W14 and W15 as a function of fracture densities separately for Group 1 and 2 to interpret the principal components (**Figure 8**). From this figure, it should be first noted that K-feldspar is only contained in one of the very weakly fractured samples of Group 1 (K87). It should be also noted that the amount of quartz in the very weakly fractured samples is higher in Group 1 than Group 2 sample, while the amount of plagioclase in them is higher in Group 2 than Group 1 sample. These results are consistent with the results of principal component analysis that the second principal component increases with increasing amount of K_2O (major component of K-feldspar) toward Group 1, while it decreases with increasing amount of Na_2O (major component of plagioclase) toward Group 2. The amount of muscovite increases with increasing degree of fracturing for both Groups of samples. The amount of chlorite increases with increasing degree of fracturing in Group 1 samples, while it shows a constant and high value in the very weakly to strongly fractured samples in Group 2 samples. Finally, the amount of calcite increases with increasing degree of fracturing in the cataclasite samples for both Groups.

It has now become clear that the difference between Groups 1 and 2 can be attributed to the difference in modal composition of minerals constituting the protolith. Group 1 samples are derived from more felsic, while Group 2 samples are derived from more mafic protoliths. The interpretation is consistent with the fact that these cataclasite samples from Group 2 contain a large amount of chlorite even at the initial stage of fracturing (in the very weakly and weakly

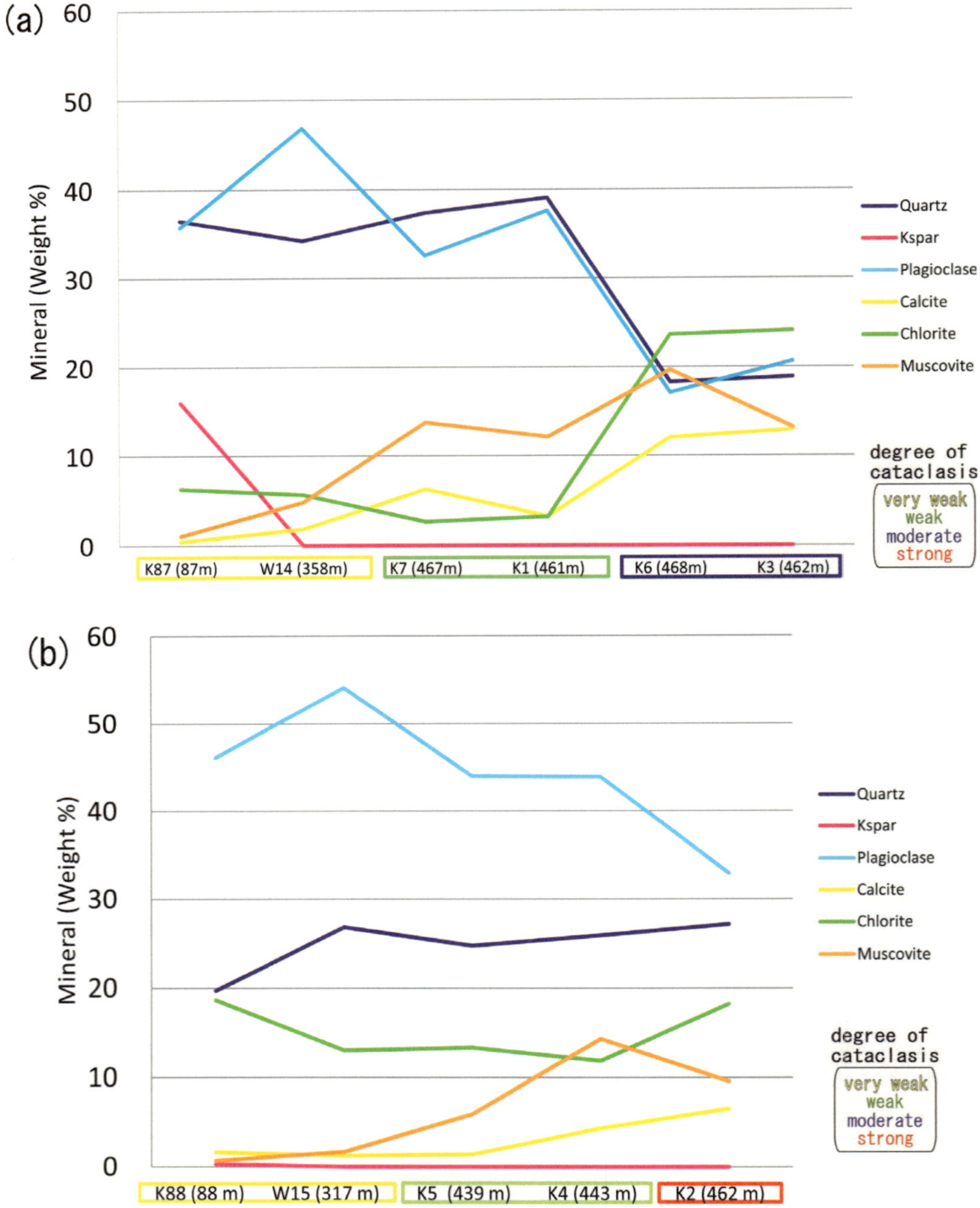

Figure 8. Plots of modal compositions of minerals with increasing fracture densities measured with an XRD in the cataclasite samples. (a) Group 1 and (2) group 2.

fractured samples), which is in accord with the fact that a large amount of biotite and amphibole altered to chlorite is observed in the cataclasite samples W15, K4, K5, and K88 belonging to Group 2 under an optical microscope. Therefore, the second principal component represents the difference in protolith (i.e., mafic vs. felsic). On the other hand, the value of the first principal component for Group 1 samples changes with changing fracture densities. Therefore, the first principal component perhaps represents the degree of element migration via fluid (i.e., dissolution and precipitation), which is essentially controlled by the fracture densities in the cataclasite samples. The inference is consistent with the facts that while the modal ratios of plagioclase and quartz clasts gradually decrease due to dissolution, those of muscovite and chlorite increase due to precipitation with the increasing density of fractures for Group 1 samples.

4.2. Isocon diagram

In order to analyze the mass changes with increasing cataclasis, we applied the isocon method by [14]. In the isocon diagram, the mass of each major chemical element in the protolith (X) and cataclasite (Y) is plotted as a point in the X-Y coordinate. If there is no mass change in the cataclasite and transfer for any particular chemical element during cataclasis, the slope of the straight line connecting the point (X, Y) and origin is 1. However, the slope is not 1 for most of the chemical elements, which is changed by both mass change in the cataclasite and mass transfer for any particular element during cataclasis. First, we will analyze the mass change during cataclasis based on the slope of the straight line for an immobile element (i.e., immobile isocon). If the slope is less than 1, mass is inferred to increase (i.e., the concentration is decreased). On the other hand, if the slope is more than 1, mass is inferred to decrease (i.e., the concentration is increased). Further, if there is no change in density of rocks, this mass increase or decrease can be treated as volume increase or decrease. In this paper, the mass increase or decrease is rephrased as volume increase or decrease assuming no density change in the cataclasite samples. These mass increase and decrease can be expressed using the slope of immobile isocon (S) by the following equation.

$$\text{Mass increase} = \left[\left(\frac{1}{S}\right) - 1\right] \tag{1}$$

In this analysis, we use Zr as an immobile element (e.g., see [25]), because at the temperature conditions (<300°C) where the cataclasite samples experienced chemical reactions via fluids, it is considered that zircon cannot be dissolved into fluids [26]. In this study, although we measure the weight percents of other trace elements, we only plot the ones of ZrO_2 and TiO_2 determined by the XRF analysis. Further, the mass transfer of any particular chemical element can be inferred by comparing the slope of straight line for any particular element with that of the immobile isocon. If the slope for any particular element is larger than that of the immobile isocon, the element flows into the system and its mass increases. On the other hand, if the slope for any particular element is smaller than that of the immobile isocon, the element flows out the system and its mass decreases.

In the present study, we have analyzed the mass change and transfer for the following three cases: the very weakly versus weakly fractured samples, the weakly versus moderately fractured samples, the weakly versus strongly fractured samples (**Figure 9**). Since both moderately and strongly fractured samples show the same trend in terms of the mass change and transfer relative to the weakly fractured samples as mentioned below, both chemical compositions are compared with those of the weakly fractured samples. From these analyses, it has been

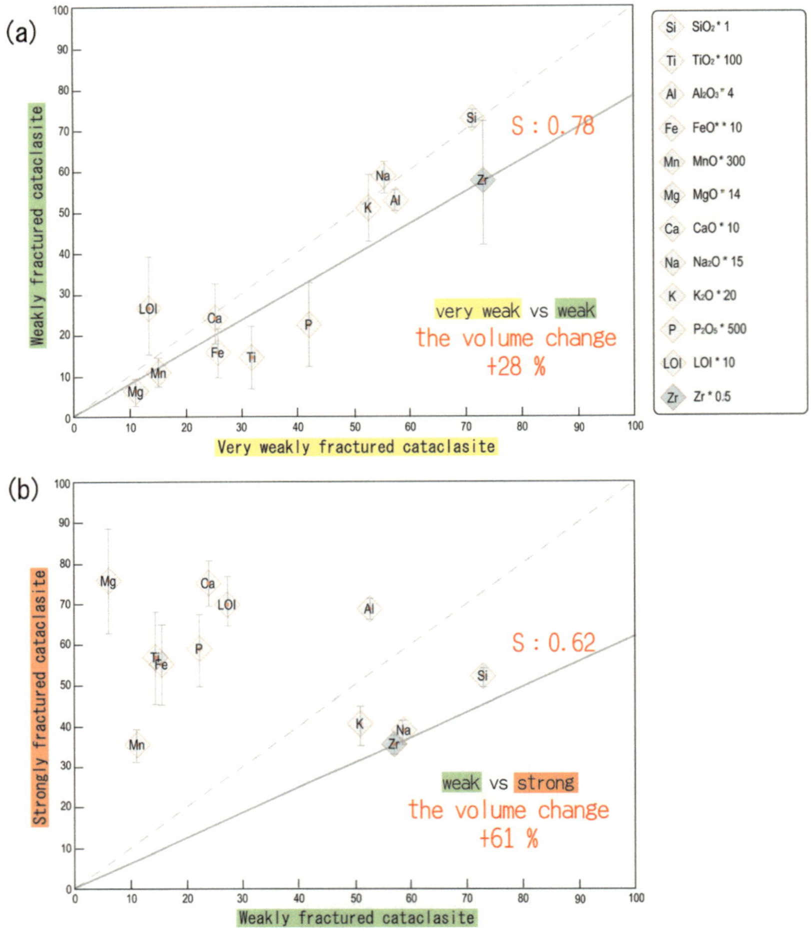

Figure 9. Isocon diagram for the Group 1 cataclasite samples with different degree of cataclasis. (a) Very weakly versus weakly fractured, and (b) weakly versus strongly fractured samples. The isocon diagram for weakly versus moderately fractured samples is not shown.

inferred that there is 28% mass increase from the very weakly to weakly fractured samples during cataclasis. Further, there could be 56 and 61% mass increase from the weakly to moderately and strongly fractured samples, respectively.

For the element transfer, from the very weakly to weakly fractured rocks the weight percents of SiO_2, Na_2O, K_2O, Al_2O_3, and LOI increase, while those of TiO_2, FeO*, MnO, MgO, and CaO decrease (**Figure 9a**). The increase of K_2O can be interpreted by alteration of plagioclase to white mica, which is evidenced by both microstructural observations with a SEM-EDS (**Figure 4**) and modal analysis of constitutive minerals by an XRD (**Figure 8**) in the differently fractured cataclasite samples. The increase of white mica is accompanied by the increase of LOI. The increase of SiO_2 is caused by the formation of quartz veins, which are observed in both core samples and thin sections. The increase of Na_2O is perhaps caused by the formation of albite from oligoclase constituting the protolith tonalite partly mylonitized, which is not completely albitized.

http://dx.doi.org/10.5772/68112

From the weakly to moderately and strongly fractured rocks, the weight percents of FeO*, MgO, CaO, TiO_2, MnO, Al_2O_3, and LOI increase, while those of SiO_2, Na_2O and K_2O are nearly constant (**Figure 9b**). This is interpreted by both further alteration of albite and increase of volume fraction of the ultracataclasite, where the rocks are not only comminuted, but also new minerals precipitated from the solution in the open space formed by fracturing. The increase of CaO is interpreted by not only the precipitation of laumontite and sphene in the ultracataclasite part, but also the replacement of albite by prehnite. The increase of FeO* is interpreted by the precipitation of chlorite and iron sulfide (e.g., pyrite). The increase of MgO, MnO, and LOI is interpreted by the precipitation of chlorite, that of Al_2O_3 by the precipitation of chlorite and laumontite, and that of TiO_2 by the precipitation of sphene from rutile (**Figure 4**) in the ultracataclasite part.

4.3. Dissolution and precipitation reaction of minerals via fluids in the cataclasite

Based on the identification of minerals in plagioclase porphyroclast and ultracataclasite, the analyses of change of modal composition of minerals with an XRD and mass transfer using isocon diagrams with increasing cataclasis, we have inferred that the following dissolution and precipitation reactions occurred via fluids during cataclasis in the analyzed samples.

$$\begin{gathered}\text{Anorthite} \rightarrow \text{Albite} \\ CaAl_2Si_2O_8 + Na^+ + Si^{4+} \rightarrow NaAlSi_3O_8 + Ca^{2+} + Al^{3+}\end{gathered} \tag{2}$$

$$\begin{gathered}\text{Albite} \rightarrow \text{Muscovite [8]} \\ 3NaAlSi_3O_8 + K^+ + 2H^+ \rightarrow KAl_3Si_3O_{10}(OH)_2 + 6SiO_2 + 3Na^+\end{gathered} \tag{3}$$

$$\begin{gathered}\text{Anorthite} \rightarrow \text{Muscovite [8]} \\ 3CaAl_2Si_2O_8 + 2K^+ + 4H^+ \rightarrow 2KAl_3Si_3O_{10}(OH)_2 + 3Ca^{2+}\end{gathered} \tag{4}$$

$$\begin{gathered}\text{Orthoclase} \rightarrow \text{Muscovite [8]} \\ 3KAlSi_3O_8 + 2H^+ \rightarrow KAl_3Si_3O_{10}(OH)_2 + 6SiO_2 + 2K^+\end{gathered} \tag{5}$$

$$\begin{gathered}\text{Plagioclase} \rightarrow \text{Prehnite} \\ 2NaAlSi_3O_8 + 2CaO + H_2O \rightarrow Ca_2Al_2Si_3O_{10}(OH)_2 + Na_2O + 3SiO_2\end{gathered} \tag{6a}$$

$$CaAl_2Si_2O_8 + CaO + SiO_2 + H_2O \rightarrow Ca_2Al_2Si_3O_{10}(OH)_2 \tag{6b}$$

$$\begin{gathered}\text{Rutile} \rightarrow \text{Sphene} \\ TiO_2 + CaO + SiO_2 + H_2O \rightarrow CaTiSiO_4(OH)_2\end{gathered} \tag{7}$$

5. Discussion

5.1. Two-stage mass transfer during cataclasis

As mentioned earlier, two-stage mass transfer during cataclasis could be identified based on the isocon diagrams in the analyzed cataclasite samples. The elements analyzed can be divided into three groups from the viewpoint of mass transfer: SiO_2, Na_2O, and K_2O (the first group of elements), which increase at the first stage, and then do not change very much at the second stage; FeO*, MgO, CaO, TiO_2, and MnO (the second group of elements) which decrease at first stage, and then increase at the second stage; Al_2O_3 and LOI which increase at both stages. Here, the first stage denotes the cataclasis from the very weakly to weakly fractured cataclasites, and the second stage that from the weakly to moderately and strongly fractured cataclasites. This kind of two-stage mass transfer was also observed for the cataclasites derived from leucogranite, where K-feldspar is abundant [8], although the trend of mass transfer is different for some elements between the cataclasites originated from leucogranite and tonalite analyzed in the present study. In the case of the cataclasites derived from leucogranite, K_2O decreases during the cataclasis from leucogranite to ultracataclasite (first stage), and then increases from ultracataclasite to fault phyllonites (second stage), while CaO increases at the first stage, and then decrease at the second stage. For other elements, the trend of mass transfer at both stages is similar to that of the present case. The difference in the trend of mass transfer between the leucogranite and tonalite protoliths results from the facts that for the case of leucogranite protolith the clay mineral ultimately formed is muscovite, while for the case of tonalite the clay mineral ultimately formed is chlorite, and Ca-phases such as prehnite, laumontite, sphene, calcite are abundant in the moderately and strongly fractured rocks. However, the two-stage alteration and mass transfer for both cases could imply that mass transfer is internally balanced (cf. [27]), so that the first group of elements (CaO belongs to the first group, while K_2O belongs to the second group for the case of leucogranite protolith) migrated from the moderately and strongly fractured cataclasites to the weakly fractured cataclasites, while vice versa for the second group of elements. On the other hand, a large amount of mass increase is inferred in the cataclasite zone as a whole, and hence the mass transfer from the outside of the system is not precluded as mentioned in the subsequent text.

5.2. A large mass increase during cataclasis

In the present study, it has been inferred that there is 28% mass increase from the very weakly to weakly fractured samples during cataclasis. Further, there could be 56 and 61% mass increase from the weakly to moderately and strongly fractured samples, respectively. So, considering that the strongly fractured samples are derived from the weakly fractured samples, which themselves are derived from the very weakly fractured sample, there could be c. 110% mass increase in the strongly fractured cataclasite samples caused by mass transfer, which occurred during the entire period of fracturing. The origin of this large amount of mass increase in the cataclasite samples will be discussed in the subsequent text.

First of all, the inference of mass changes is greatly dependent on the selection of immobile elements. In the present study, although we select Zr as an immobile element, other studies used Ti [28], both Ti and Zr [24, 27] as immobile elements, or take the average slope of isocons of relatively immobile elements for the slope of immobile isocon [29]. However, we did not use Ti as an immobile element because a large amount of sphene is grown from rutile by chemical reaction aided by the presence of fluids. In fact, the slopes of isocons of TiO_2 and $ZrO_2$are greatly different. However, if we take the average slope of isocons of relatively immobile elements as the slope of immobile isocon, the present results could be greatly modified. We currently do not know how we should choose an immobile element or how we could infer the slope of isocon of immobile element. Hence, the mass change in the cataclasite samples during cataclasis remains to be investigated.

Nevertheless, we will discuss the origin of this large volume increase in the present cataclasite sample, referring to previous studies. Most of the studies on the volume changes caused by mass transfer during cataclasis reported a significant amount of volume loss for granite protolith (37 ± 10% volume loss, see, e.g., [28]). This is reasonable because a large amount of silica dissolved from quartz into fluids by pressure solution is expelled from cataclasites. In fact, microstructures such as solution seams indicating pressure solution are ubiquitous in the present cataclasites. However, a few studies indicate a significant amount of volume increase for granite protolith [27] and for protolith of mafic igneous rocks (gabbro) [29]. Ref. [27] argues that the volume change in phyllonite zones is so valuable ranging between −35 and +95% at a meter scale in the extensional stress field. Ref. [29] argues that extension fractures in hard rocks such as metabasite are sustained yielding volume increase, while these are easily compressed in soft rocks such as paragneiss yielding volume decrease caused by mass transfer during cataclasis. For the present case of tonalite protolith, since the strength could be between those of granite and gabbro, the volume loss could have been prohibited due to the relatively high strength. Also, the tectonic setting such as extensional stress field, which has not been fully known, may have favored the significant amount of volume increase in the cataclasites. However, the origin of the significant amount of volume increase inferred in the present study remains to be investigated, including the problem if this estimate of volume change is correct or not.

6. Summary

Based on microstructural, and bulk chemical and microchemical analyses of the cataclasite samples collected from the Median Tectonic Line, Mie Prefecture, southwest Japan, we have summarized the present work as mentioned below. In the present cataclasite samples, the progress of cataclasis, which is quantified by the fracture densities (FD, number/cm), is accompanied by the change in bulk chemical compositions determined by the XRF analysis. Both principal component and isocon analyses for the bulk rock compositions in the cataclasite samples reveal the origin of variation of chemical compositions, which mostly results from two-stage mass transfer during cataclasis. Further, a SEM-EDS analysis of altered plagioclase and ultracataclasite (i.e., fine-grained materials by comminution) and XRD analysis with the aid of "RockJock" software delineate that the mass transfer in these cataclasite samples is indeed caused by dissolution and precipitation of minerals, which occurred during cataclasis.

Acknowledgements

We thank Jun Kameda for assistance with XRD analysis with the aid of "RockJock." This research was supported by MEXT KAKENHI grant 26109004 to TT, NS, and KF. Also, this study was supported by the Joint Usage/Research Center Program of Earthquake Research Institute, The University of Tokyo.

Author details

Yumi Kaneko[1]*, Toru Takeshita[2], Yuto Watanabe[2], Norio Shigematsu[3] and Ko-Ichiro Fujimoto[4]

*Address all correspondence to: yumi.kaneko@japex.co.jp

1 Japan Petroleum Exploration Co., Ltd., Tokyo, Japan

2 Hokkaido University, Hokkaido, Japan

3 Advanced Industrial Science and Technology, Tsukuba, Japan

4 Tokyo Gakugei University, Tokyo, Japan

References

[1] Research Group on active fault. Active Faults in Japan (New Version): Distribution Map and Data. Tokyo: University of Tokyo Press; 1991. p. 440. ISBN:978-4-13-060700-1

[2] Sagiya T, Miyazaki S, Tada T. Continuous GPS Array and Present-day Crustal Deformation of Japan. Pure and Applied Geophysics. 2000;**157**:2303–2322. DOI: 10.1007/PL0002 2507

[3] Shimamoto T. Rheology of rocks and plate tectonics—From rigid plate to deformable plate. Kagaku. 1989;**59**:170–181 (in Japanese)

[4] Fusseis F, Handy MR, Schrank C. Networking of shear zones at the brittle-to-viscous transition (Cap de Creus, NE Spain). Journal of Structural Geology. 2006;**28**:1228–1243

[5] Schrank CE, Handy MR, Fusseis F. Multiscaling of shear zones and the evolution of the brittle-to-viscous transition in continental crust. Journal of Geophysical Research. 2008;**113**:B01407. DOI: 10.1029/2006JB004833

[6] Bos B, Spiers CJ. Frictional-viscous flow of phyllosilicate-bearing fault rocks: Microphysical model and implications for crustal strength profiles. Journal of Geophysical Research. 2002;**107**:1–13. DOI: org/10.1029/2001JB000301

[7] Holdsworth RE, van Diggelen EWE, Spiers CJ, de Bresser JHP, Walker RJ, Bowen L. Fault rocks from the SAFOD core samples: Implications for weakening at shallow depths along the San Andreas Fault, California. Journal of Structural Geology. 2011;**33**:132–144. DOI: 10.1016/j.jsg.2010.11.010

[8] Wibberley C. Are feldspar-to-mica reactions necessarily reaction-softening? Journal of Structural Geology. 1999;**21**:1219–1227

[9] Takeshita T, El-Fakharani AH. Coupled micro-faulting and pressure solution creep overprinted on quartz schist deformed by intracrystalline plasticity during exhumation of the Sambagawa metamorphic rocks, southwest Japan. Journal of Structural Geology. 2013;**46**:142–157. DOI: 10.1016/j.jsg.2012.09.014

[10] Misch P. Paracrystalline microboudinage of zoned grains and other criteria for synkinematic growth of metamorphic minerals. American Journal of Science. 1969;**267**:43–63

[11] Shimizu I. Kinetics of pressure solution creep in quartz: Theoretical considerations. Tectonophysics. 1995;**245**:121–134

[12] Shigematsu N, Fujimoto K, Tanaka N, Furuya N, Mori H, Wallis S. Internal structure of the Median Tectonic Line fault zone, SW Japan, revealed by borehole analysis. Tectonophysics. 2012;**532–535**:103–118. DOI: 10.1016/j.tecto.2012.01.024

[13] Eberl DD. User's guide to RockJock—A program for determining quantitative mineralogy from powder X-ray diffraction data. USA: Geological Survey Open-File Report; 2003. p. 47

[14] Grant JA. The isocon diagram-a simple solution to Gresens' equation for metasomatic alteration. Economic Geology. 1986;**81**:1976–1982

[15] Davis JC. Statistics and Data Analysis in Geology. New York: John Wiley and Sons Inc.; 2002. p. 638

[16] Kubota Y, Takeshita T. Paleocene large-scale normal faulting along the Median Tectonic Line, western Shikoku, Japan. Island Arc. 2008;**17**:129–151

[17] Hara I, Shyoji, K, Sakurai Y, Yokoyama S, Hide K. Origin of the Median Tectonic Line and its initial shape. In: Ichikawa K, editor. Memoirs of the Geological Society of Japan. Vol. 18. Tokyo: The Geological Society of Japan; 1980. pp. 27–49

[18] Sakakibara N. Qualitative estimation of deformation temperature and strain rate from microstructure and lattice preferred orientation in plastically deformed quartz aggregates. Journal of Geological Society of Japan. 1996;**102**:199–210

[19] Shimada K, Takagi H, Osawa H. Geotectonic evolution in transpression regime: Time and space relationships between mylonitization and folding in the southern Ryoke belt, eastern kii Peninsula, southwest Japan. Journal of Geological Society of Japan. 1998;**104**:825–844

[20] Shigematsu N, Otsubo M, Fujimoto K, Nobuaki Tanaka N. Orienting drill core using borehole-wall image correlation analysis. Journal of Structural Geology. 2014;**67**:293–299. DOI: org/10.1016/j.jsg.2014.01.016

[21] Hayama Y, Yamada T, Ito M, Kutsukake T, Masaoka K, Miyakawa K, Mochizuki Y, Nakai Y, Tainosho Y, Yoshida M, Kawarabayashi I. Tsumura Y. Geology of the Ryoke Belt in the eastern Kinki District, Japan—The phase-divisions and the mutual relations of the granitic rocks. Journal of Geological Society of Japan. 1982;**88**:451–466

[22] Hayama Y, Yamada T. Median Tectonic Line at the stage of its origin in relation to plutonism and mylonitization in the Ryoke belt. In: Ichikawa K, editor. Memoirs of the Geological Society of Japan. Vol. 18. 1980. pp. 5–26

[23] Jefferies SP, Holdsworth RE, Wibberley CAJ, Shimamoto T, Spiers CJ, Niemeijer AR, Lloyd GE. The nature and importance of phyllonite development in crustal-scale fault cores: An example from the Median Tectonic Line, Japan. Journal of Structural Geology. 2006;**28**:220–235

[24] Schleicher AM, Tourscher SN, van der Pluijm BA, Warr LN. Constraints on mineralization, fluid-rock interaction, and mass transfer during faulting at 2–3 km depth from the SAFOD drill hole. Journal of Geophysical Research. 2009;**114**:(B04202). DOI: 10.1029/2008JB006092

[25] Grant JA. Isocon analysis: A brief review of the method and applications. Physics and Chemistry of the Earth. 2005;**30**:997–1004. DOI: 10.1016/j.pce.2004.11.003

[26] Harley SL, Kelly NM, Moller A. Zircon behavior and the thermal histories of mountain chains. Element. 2007;**3**:25-30

[27] Hippertt, JF. Breakdown of feldspar, volume gain and lateral mass transfer during mylonitization of granitoid in a low metamorphic grade shear zone. Journal of Structural Geology. 1998;**20**:175–193

[28] Evans JP, Chester FM. Fluid-rock interaction in faults of the San Andreas system: Inferences from San Gabriel fault rock geochemistry and microstructures. Journal of Geophysical Research. 1995;**100**:13. DOI: 007-13020

[29] Zulauf GS. Palm S, Petschick R, Spies O. Element mobility and volumetric strain in brittle and brittle-viscous shear zones of the superdeep well KTB (Germany). Chemical Geology. 1999;**156**:135–149

Chapter 7

Structural Features Along the Median Tectonic Line in Southwest Japan: An Example of Multiphase Deformation on an Arc-Bisecting Fault

Yasuto Itoh and Tomotaka Iwata

Additional information is available at the end of the chapter

http://dx.doi.org/10.5772/67669

Abstract

A geological survey for the Late Cretaceous Izumi Group distributed on the Median Tectonic Line (MTL) active fault system in the central part of southwestern Japan has revealed varied deformation styles. Among the confined deformation zones found in the western and central parts of the study area, some are located far from the active trace of the MTL (Negoro Fault), at distances of up to 300–350 m. Such kink zones may have been generated during a contraction phase of the MTL from the end of the Pliocene to the early Pleistocene. We identified clear active foldings in a narrow zone sandwiched between a north dextral and a south reverse active fault. Western and eastern upheavals of the crustal sliver show ridge and domal active morphologies, respectively. Structural analysis was extended to the north of the MTL, where the Izumi Group has suffered multiphase deformation since the Cretaceous. The phase stripping method was introduced to extract the neotectonic trend, which successfully delineated complicated deformation zones related to the morphological divergence of the MTL active fault system.

Keywords: structural geology, active fault, active fold, Median Tectonic Line (MTL), southwest Japan

1. Introduction

The activity levels of neotectonic zones on convergent margins are usually assessed based on the geomorphological features or deformation of unconsolidated outcropping sedimentary layers by trench surveys. Such means achieve high resolution in describing superficial ruptures but provide little information on the mid-crustal architecture of faults, which is crucial for the evaluation of earthquake hazards. In contrast, geophysical researches exemplified by reflection seismic surveys can achieve great penetration into Earth's deep interiors but do not allow the deciphering of detailed time sequences of tectonic events. Geologic studies, in between, yield a reasonable definition, allowing the delineation of the upper crustal structure, which is useful in the understanding of the three-dimensional morphology of active faults. This has been, however, often misused for a one-sided interpretation, ignoring the cumulative processes of structures. In this chapter, the authors aim to extract a target deformation phase by stripping superimposed geologic features based on reliable tectonic models of structural build-up. The focus of the present field survey is the southwestern Japan arc, where the long-standing convergence of oceanic plates with variable obliqueness has provoked repeated slips upon an arc-bisecting fault and the complicated development of a damage zone.

2. Geological background

2.1. Median Tectonic Line (MTL)

The Median Tectonic Line (MTL) was formed during the Cretaceous, when a remarkably rapid northward movement of the Izanagi Plate resulted in the sinistral wrenching and eventual break-up of the eastern Eurasian margin [1]. Paleogeographic reconstruction suggested that the incipient MTL had been connected with the Central Sikhote Alin Fault to the north and constituted a regional transcurrent fault [2]. Propagating termination of the MTL was a site of pull-apart basin formation that was buried by an enormous amount of clastics, collectively called the Izumi Group [3].

After an essentially dormant period during the Paleogene, the MTL resumed activity under the strong influence of changes in the convergent mode of the Philippine Sea Plate. The Pliocene was marked by an intensive inversion related to the north-south compressive regime [4]. Although the inversion is most obviously observed on the backarc shelf, contemporaneous contraction features can also be identified on the forearc shelf [5]. Watershed mountain ranges emerged on the northern bank of the MTL, as was clarified by provenance studies of basin-filling clasts (e.g., [6]). The Philippine Sea Plate shifted its converging direction counterclockwise at ca. 2–1 Ma [7] and induced the production of vigorous dextral slips by the MTL.

2.2. Izumi Group

The present study area is located at the easternmost part of the basin, which was buried with the Maastrichtian marine sediments of the Izumi Group [8]. It is a series of marine sediments and is composed of the main facies (turbidite) and southern facies (non-turbidite). The main facies comprise voluminous turbidites associated with coarse clastics and intercalated by acidic tuff layers. It is divided into the Kada, Shindachi, Iwade and Kokawa Formations in ascending order. They form a large syncline with an east-plunging axis.

A previous sedimentological study [9] showed that the Izumi Group basically consists of stacks of mega-units. Facies associations indicate that the constituents of the units are classified into depositional systems, including main channels with overspilled deposits, distributary channels and sheet flows.

Zircon fission track ages obtained from acidic tuffs in the group range from 77 to 72 Ma, and molluscan fauna yielded from the unit is assigned to the latest Cretaceous [8].

3. Field observation

The results of our field survey are shown in **Figure 1** (upper) along with a previously developed geologic map of the same area [8] for comparison. In the following sections, we present a thorough description of structural features of the Izumi Group.

3.1. Confined deformation zones

The most remarkable fault damage zone was identified and described in the western part of the study area. **Figure 2** presents a plan view of a road construction site where deformed and brecciated sedimentary rocks of the Izumi Group are exposed. As shown in the figure, the azimuth of the damage zone inferred from the observation of surrounding exposures is coincident with the strike of the MTL, a fact which strongly indicates the development of the kink zone under the influence of shearing on the regional fault. It is noted, however, that the observatory is located far from the active trace of the MTL (Negoro Fault) with a distance of up to 300–350 m (see Figure 14 of Chapter 3). Hence, the kink zone may have been generated during a contraction phase of the MTL from the end of the Pliocene to the early Pleistocene. **Figure 3** shows typical photographs of the kink zone, with Photo 22 depicting mudstone and sandstone blocks in a chaotic matrix.

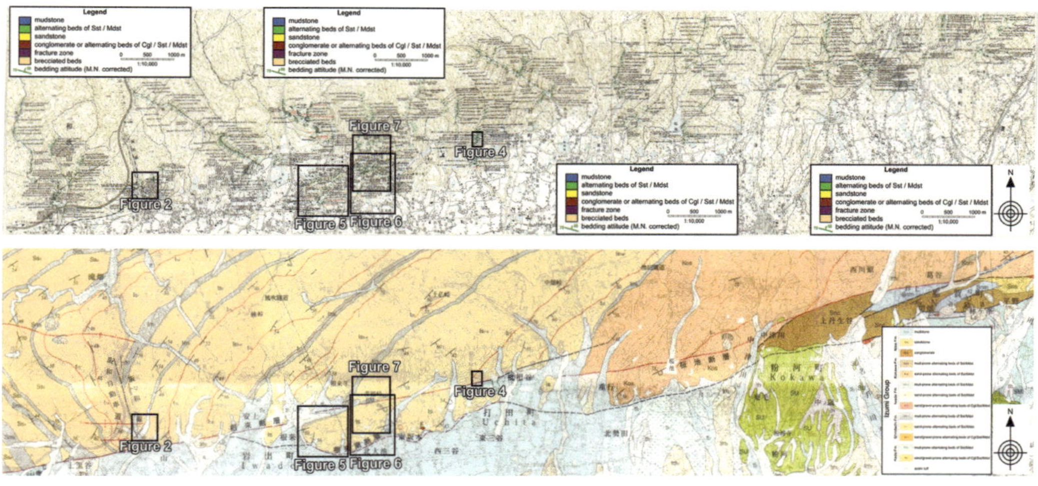

Figure 1. A field survey map of the present study (upper) together with a previous geologic map (lower; [8]). See prologue of this section for regional index.

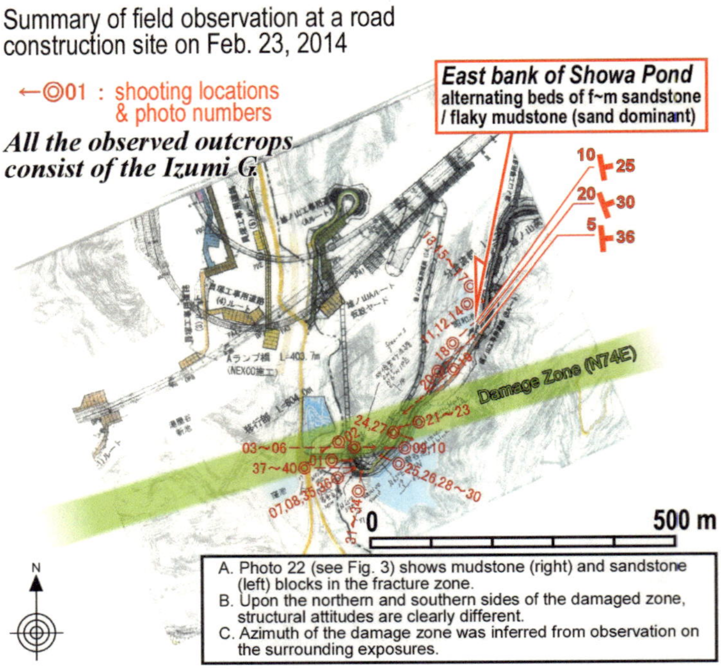

Figure 2. A summary of field observation on February 23, 2014, at a road construction site where a fault-related fracture zone was exposed. See **Figures 1** and **3** for the locality and example photographs, respectively.

Figure 3. Example photographs of the fracture zone described in **Figure 2**. Photo 22 shows mudstone (right; m) and sandstone (left; s) chaotic blocks in the fracture zone.

http://dx.doi.org/10.5772/67669

Another notable deformation adjacent to the MTL active fault system in the central part of the study area was described in detail. In **Figure 4**, the ENE-WSW-trending active dextral fault runs along a carriageway in the southernmost part [10]. Along a rivulet crossing the fault, a broad damage zone with a width of nearly 200 m was confirmed; in this zone, the bedding attitudes of the Izumi Group show complicated undulation discordant with the general trend of the plunged syncline. From what can be observed, the level of damage shows no clear tendency to decrease in proportion to the distance from the fault trace. Brecciated sedimentary rocks in

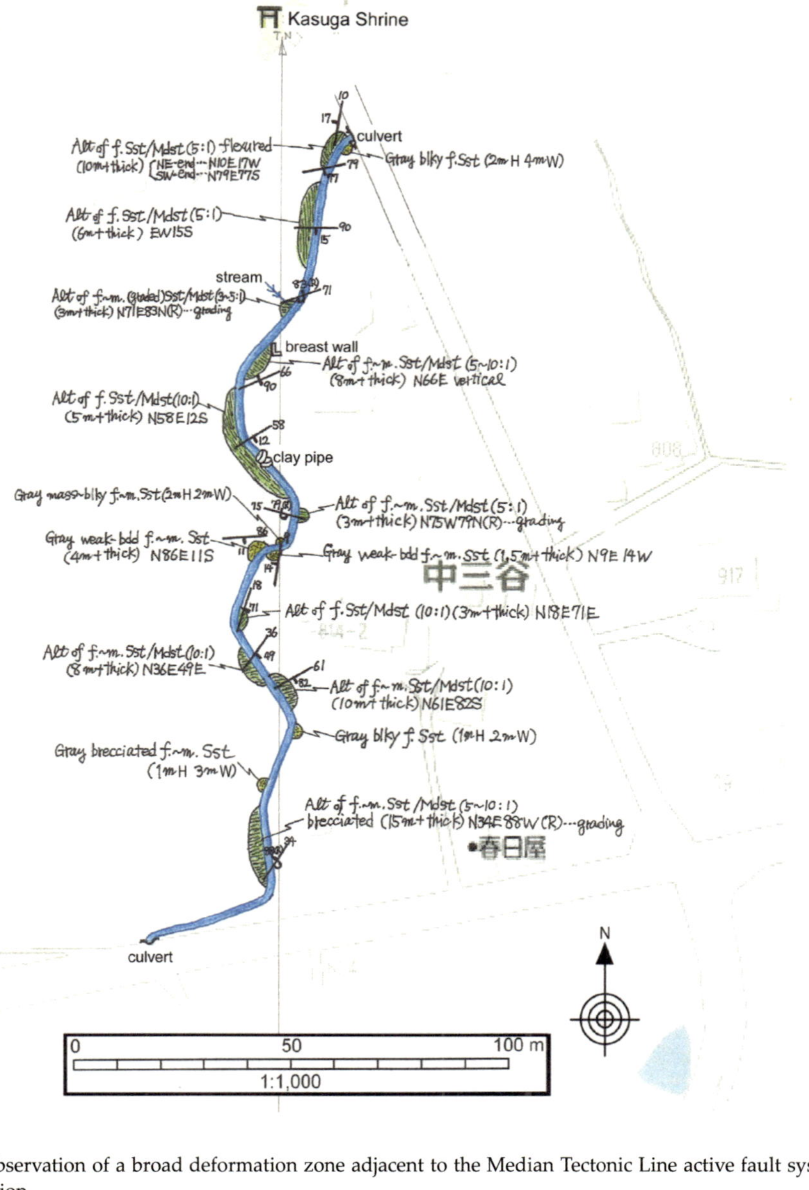

Figure 4. Field observation of a broad deformation zone adjacent to the Median Tectonic Line active fault system. See **Figure 1** for location.

the southern portion have steep structures parallel to the azimuth of the MTL, whereas the central and northern outcrops show often overturned bedding with flexure.

3.2. Active foldings

The MTL active fault system within the study area is specifically divided into two strands, the straightforward dextral Negoro Fault to the north and the sinuate reverse Negoro-Minami Fault to the south (see **Figure 1** (bottom)). An elongated sliver sandwiched between the faults, which mainly comprises sandy units of the Izumi Group, is expected to exhibit unique deformation acquired during recent periods. However, previous geologic [8] and geomorphological [10] studies have not remarked on its features. Hence, the authors attempted to describe the structural characteristics of the crust piece spooned up by the faults. In the following analysis, it was assumed that the initial structure of the Cretaceous system is horizontal since the block is situated outside of the pull-apart basin that experienced post-depositional contraction and harsh bending of the infill sediments.

Figure 5 shows the structural features of the Negoro-Maeyama Hill in the western part of the fault-bounded sliver. Although the previous geologic map (a; [8]) depicts tilting only on

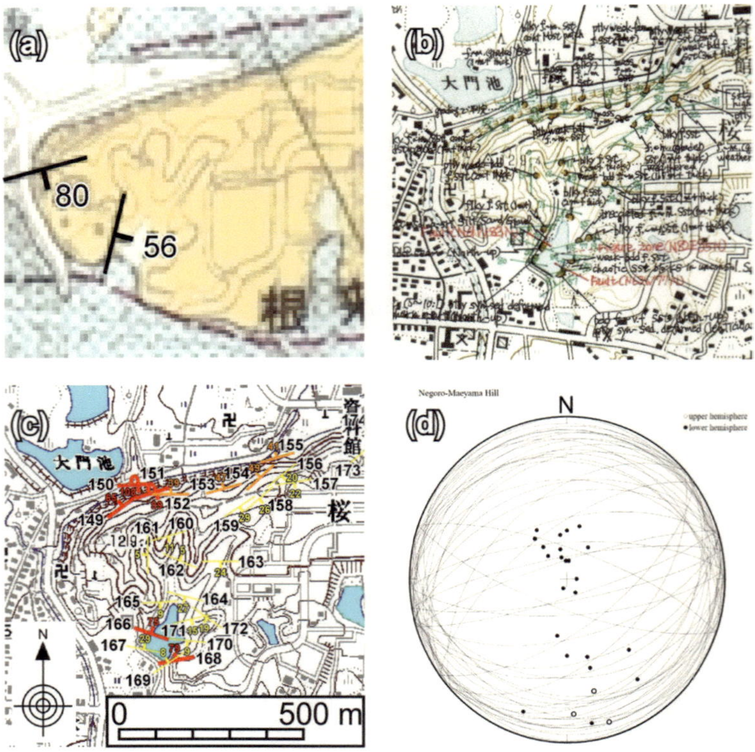

Figure 5. Structural analysis of the Negoro-Maeyama Hill (see **Figure 1** for the area of the maps a, b and c) between the Negoro and Negoro-Minami Faults. (a) Previous structural data based on a geologic map [8]. (b) A field survey map of the present study. (c) A compiled map showing the structural trends. Yellow, orange and red marks correspond to gentle (less than 30°), intermediate (between 30° and 60°) and steep (60° or more) dipping angles, respectively. (d) An equal-area projection of bedding girdles and their poles. Open dots represent overturned beds.

http://dx.doi.org/10.5772/67669

the hillside, the present detailed geologic investigation (b, c) clarified that the centre of the hill is characterized by gentle dipping, which is surrounded by steeply dipping (occasionally overturned) strata. A stereographic projection showing the poles and girdles of bedding planes (d) indicates that the initially flat-lying beds were folded around the ENE-WSW axis and formed a ridge parallel to the MTL.

Figure 6 shows the structural features of the Atago-yama Hill in the eastern part of the fault-bounded sliver. Similar to the analysis for the Negoro-Maeyama Hill, the present geologic investigation (b, c) clarified that the core of the hill is characterized by a gentle structure surrounded by steeply dipping (occasionally overturned) strata, although the previous geologic map (a; [8]) depicts tilting only on the hillside. A stereographic projection showing the

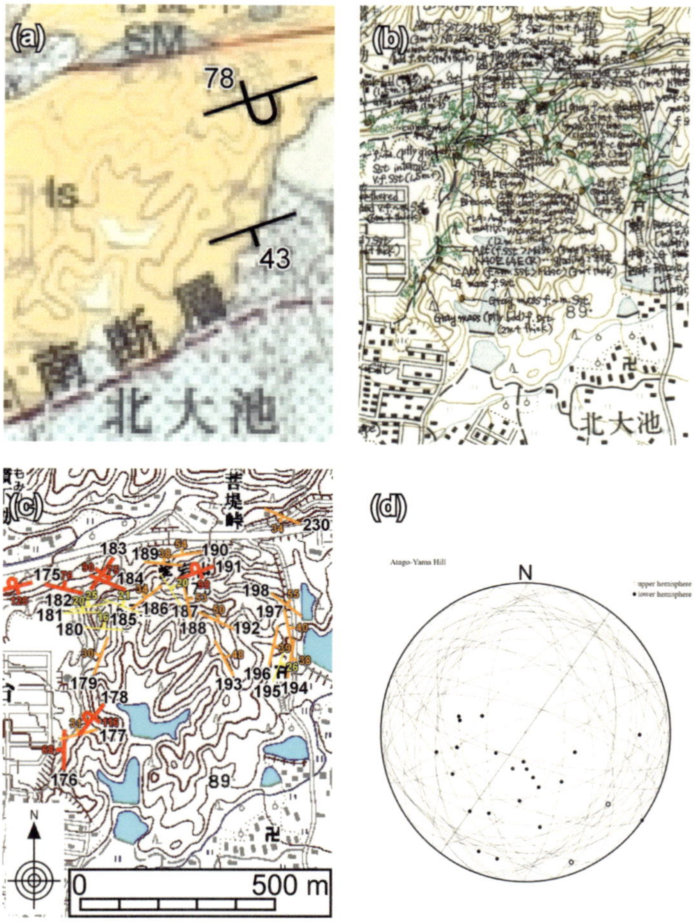

Figure 6. Structural analysis of the Atago-yama Hill (see **Figure 1** for the area of the maps a, b and c) between the Negoro and Negoro-Minami Faults. (a) Previous structural data based on a geologic map [8]. (b) A field survey map of the present study. (c) A compiled map showing the structural trends. Yellow, orange and red marks correspond to gentle (less than 30°), intermediate (between 30° and 60°) and steep (60° or more) dipping angles, respectively. (d) An equal-area projection of bedding girdles and their poles. Open dots represent overturned beds.

poles and girdles of bedding planes (d) indicates that the initially flat-lying beds were folded around the axes with varied azimuths and formed active domal morphologies.

As for the Atago-yama Hill, closer observation suggests the presence of monomictic unconsolidated gravel beds composed of angular sandstone pebbles and cobbles with finely ground matrices (see **Figure 7**). The conspicuous unit was interpreted as a reworked Quaternary sediment that originated from fault breccia.

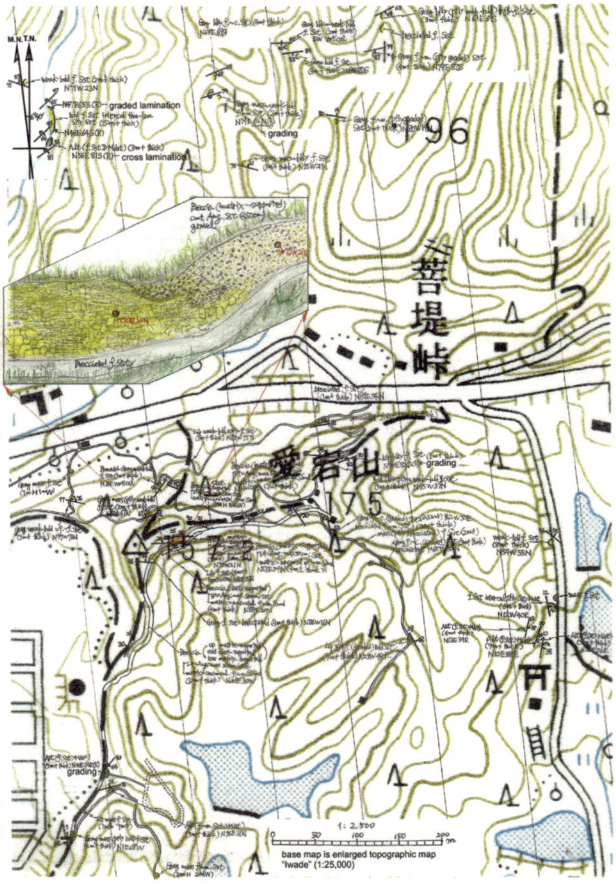

Figure 7. Detailed field observation of the Atago-yama Hill showing a sketch of the highly brecciated beds of the Izumi Group. See **Figure 1** for location.

4. Structural analysis

4.1. Recognition of the general trend

To extend our structural analysis north of the MTL, the general trend in the pull-apart basin buried by the Izumi Group had to be considered. As mentioned before, the primitive geologic framework was an east-plunging syncline. The previous geologic map [8] depicts the fold axis as being crosscut by the MTL, and only the north wing of the synclinal structure is preserved except in the westernmost part of the study area (**Figure 1** (bottom)). However, the present three-year geologic expedition obtained a considerably different structural model showing that the axis of syncline is traceable from the western to eastern ends of the study area as shown in **Figure 1** (top).

This discrepancy arises from different data densities. Although southeastward-dipping attitudes are common in the central and eastern parts of the map, as suggested in previous works, a careful inspection in the present study revealed kink zones with opposite tilting sense in many survey routes, which is conceived as overlapping neotectonic features developed through the waxing and waning of compressive stress linked with changes in the convergence modes of the Philippine Sea Plate since the Pliocene [11]. It is necessary, however, to separate multiphase deformation to comprehend tectonic episodes because the geologic structure described on the basis of the field survey is a product of intermittent structural developments. In the next section, a stripping method for deformation phases is introduced.

4.2. Phase stripping method

Figure 8 presents a map of the general structural trend of the Izumi Group. The initial plunged syncline appears to be well preserved throughout the study area, excluding the areas between the strike-slip and reverse fault strands of the MTL.

Based on the incipient tilting trend, A, the total deformation recorded during the field survey, C, is expressed as the following matrix product:

$$C = BA \tag{1}$$

where B is the desired tilting trend acquired since the Pliocene. In each field observation location, B can be estimated using a simple calculation as:

$$B = BA\,A^{-1} = C\,A^{-1} \tag{2}$$

In the present study, this phase stripping method was applied to structural data obtained from 472 sites.

Figure 8. A base map of the phase stripping analysis of the Izumi Group. Orange lines represent significant faults after a geologic map [8]. Peach lines with dipping angles define the initial structure (plunged syncline) of the Izumi Group, which buried the late Cretaceous pull-apart basin along the Median Tectonic Line.

4.3. Neotectonic structural trend

Figure 9 presents all the neotectonic structural attitudes plotted on an equal-area projection. It is obvious that the poles of bedding planes are aligned on a perpendicular plane with an ENE-WSW horizontal pole, which agrees with the azimuth of the MTL active fault system. Thus, the younger deformation likely developed during rising compressive episodes on the MTL since the Pliocene.

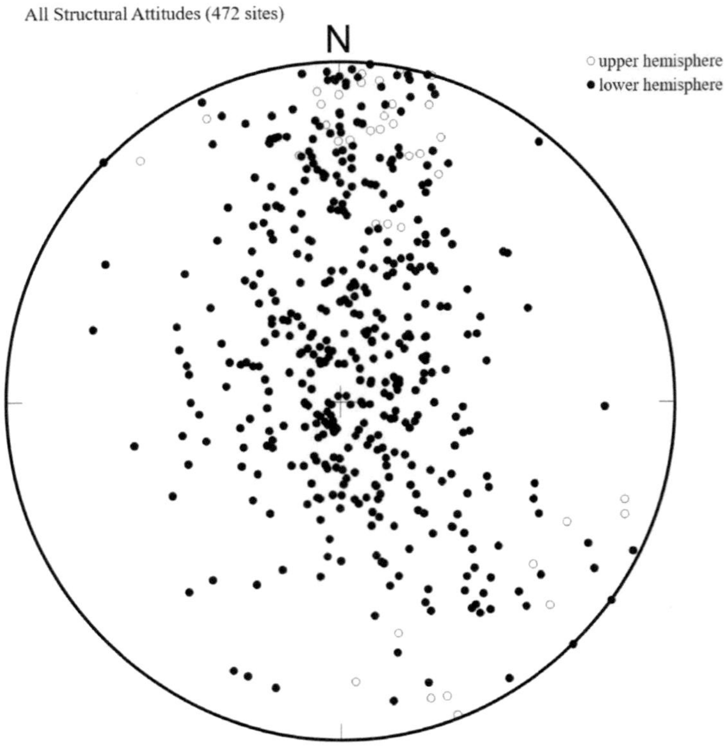

Figure 9. An equal-area projection of poles of bedding planes showing the extracted neotectonic deformation trend. Open dots represent overturned beds.

5. Discussion

The results of the structural analysis performed in the present study are summarized in **Figure 10**. The spatial variation in the extracted neotectonic feature clearly suggests the uneven deformation of the Izumi Group. Steeply dipping sites, which are highlighted by red symbols, are concentrated in the coexistent interval of the Negoro and Negoro-Minami Faults. It is noteworthy that the deformation is not confined between the two fault segments but extends north of the Negoro Fault, forming a broad damage zone. Another remarkable

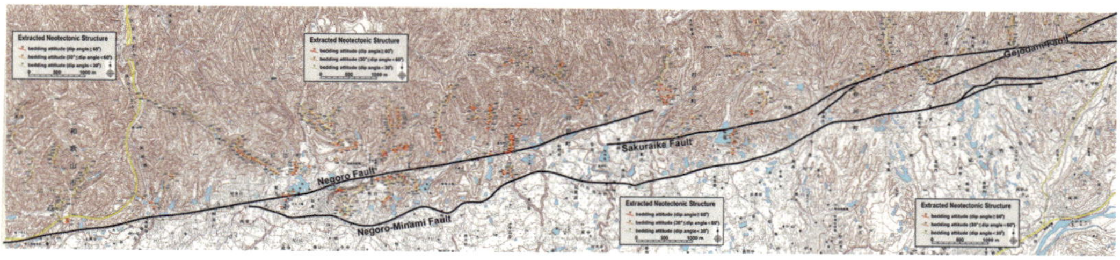

Figure 10. A map of bedding attitudes showing the extracted neotectonic deformation trend. Yellow, orange and red marks correspond to gentle (less than 30°), intermediate (between 30° and 60°) and steep (60° or more) dipping angles, respectively. The names of significant faults are as those given in the previous geologic map [8].

deformation is observed at the eastern termination of the Negoro Fault, where a complicated deformation is anticipated according to active dextral slips on the fault. In sharp contrast, the Cretaceous strata within the westernmost part of the study area, where the Negoro Fault has a straightforward morphology, are immune to serious deformation.

Multidisciplinary studies presented in several chapters of this publication are in close agreement with some of tectonic episodes around southwest Japan since the end of the Miocene. This section chronicles these episodes and outlines the empirical assessment of the sensitivities of the research methods used in these studies.

5.1. Regional contraction (ca. 6 Ma) and basin development related to lopsided convergence (5–3 Ma)

Around southwest Japan, one of the most noted tectonic events in the late Cenozoic era is remarkable north-south backarc contraction, which was originally described by Tai [12] based on the structural trend on the coastal area. Utilizing reflection seismic data, Itoh and Nagasaki [4] clarified that the deformation zone covers the whole of the backarc shelf. Although the activity level seems to diminish eastward, Itoh et al. [2, 13] demonstrated a coeval deformation on the eastern side of the southwestern Japan backarc. In the case that the regional event is linked with oceanic plate motion, the Philippine Sea Plate probably enforced northerly convergence at ca. 6 Ma. As reported in Chapter 5 of this publication, the fission track thermochronology of the Izumi Group has confirmed the impact of the regional exhumation event even on the forearc side.

In contrast to the preceding stage, the early Pliocene is characterized by emergence of an enormous area of extension and basin formation around the western part of the MTL [14]. North-south tensile stress was dominant around the western part of southwest Japan, whereas the eastern part of southwest Japan was a site of strong compression, which was accompanied by the collision of the Tanzawa landmass against the eastern forearc [15]. Between both disturbed ends of the island arc, the central part of southwest Japan was free from strong deformation, and basin formation was inactive throughout the period. It was assumed that such a tectonic trend is related to lopsided subsidence, namely the jump of the Euler pole of the Philippine Sea/Eurasian Plates.

5.2. Expansion of contractional domain (3–1 Ma)

The early Quaternary was an era of contractional deformation around southwest Japan. The eastern part of the arc was under a continuous compressive regime, which provoked the formation of regional unconformity in the eastern forearc at ca. 3.0–2.5 Ma (e.g., [16]). Watershed mountains simultaneously emerged along the MTL in the central part of southwest Japan (e.g., [6]); this was succeeded by the extensive deformation of the adjoining forearc basin [5]. A tension graben around the Beppu Bay in the western part of the island arc became inactive and changed the basin architecture. Such tectonic features may be related to the continued migration of the Euler pole of the Philippine Sea Plate. Within this publication, the reverse faulting and development of a half-graben along the MTL have been delineated in detail based on the reflection seismic survey results (Chapter 3).

5.3. Dominance of simple shear deformation (1 Ma–present)

The latest change in the convergence mode is thought to have occurred at ca. 1–2 Ma. Nakamura et al. [7] have described the submarine topography and shallow structure along the eastern forearc slope of southwest Japan and advocated a counterclockwise shift (NNW to WNW) in the convergent direction of the Philippine Sea Plate. The cessation of the uplift of the watershed mountain ranges [6] and the dominance of simple shear deformation [17] in the central part of southwest Japan is concordant with such a tectonic event.

The present chapter based on authentic geological survey has delineated the structural trend of the Izumi Group on the MTL active fault system that was provoked by uneven tectonic stress related to the morphological diversity of the active fault.

6. Conclusions

The present field survey for the Izumi Group distributed on the MTL active fault system in the central part of southwest Japan has revealed the following points.

(1) Confined deformation zones are present in the western and central parts of the study area. Some of them are located far from the active trace of the MTL (Negoro Fault), at distances of up to 300–350 m. Such kink zones may have been generated during a contraction phase of the MTL from the end of the Pliocene to the early Pleistocene.

(2) Active foldings were identified in the Negoro-Maeyama Hill and Atago-yama Hill, which are sandwiched between the Negoro and Negoro-Minami Faults. They show ridge and domal active morphologies, respectively.

(3) The phase stripping method was introduced to extract the neotectonic trend, which successfully delineated complicated deformation zones related to the morphological divergence of the MTL active fault system.

As for the detailed structural analysis of the study area, original large figures in this chapter are available at OPERA: Osaka Prefecture University Education and Research Archives (http://hdl.handle.net/10466/15058). References to color figures in this chapter are available in the public digital archive.

Acknowledgements

We would like to thank K. Takemura for his constructive comments, which greatly helped in the improvement of an early version of the manuscript. This study was executed as a part of the Comprehensive Research and Survey for the Median Tectonic Line Fault System (from the Eastern Margin of Kongo Mountains to the Southern Margin of Izumi Mountains; FY 2013–2015) organized by the Research and Development Bureau; the Ministry of Education, Culture, Sports, Science and Technology; and the Disaster Prevention Research Institute; Kyoto University.

Author details

Yasuto Itoh[1]* and Tomotaka Iwata[2]

*Address all correspondence to: yasutokov@yahoo.co.jp

1 Graduate School of Science, Osaka Prefecture University, Osaka, Japan

2 Disaster Prevention Research Institute, Kyoto University, Uji, Japan

References

[1] Fitch TJ. Plate convergence, transcurrent faults, and internal deformation adjacent to southeast Asia and the western Pacific. Journal of Geophysical Research. 1972; 77: 4432-4460

[2] Itoh Y, Uno K, Arato H. Seismic evidence of divergent rifting and subsequent deformation in the southern Japan Sea, and a Cenozoic tectonic synthesis of the eastern Eurasian margin. Journal of Asian Earth Sciences. 2006; 27: 933-942

[3] Noda A, Toshimitsu S. Backward stacking of submarine channel-fan successions controlled by strike-slip faulting: The Izumi Group (Cretaceous), southwest Japan. Lithosphere. 2009; 1: 41-59

[4] Itoh Y, Nagasaki Y. Crustal shortening of Southwest Japan in the Late Miocene. The Island Arc. 1996; 5: 337-353

[5] Takano O, Nishimura M, Fujii T, Saeki T. Sequence stratigraphic distribution analysis of methane-hydrate-bearing submarine-fan turbidite sandstones in the eastern Nankai Trough area: Relationship between turbidite facies distributions and BSR occurrence. Journal of Geography. 2009; 118: 776-792

[6] Oka Y. The formation of the Izumi range and the Osaka Group. The Quaternary Research. 1978; 16: 201-210

[7] Nakamura K, Renard V, Angelier J, Azema J, Bourgois J, Deplus C, Fujioka K, Hamano Y, Huchon P, Kinoshita H, Labaume P, Ogawa Y, Seno T, Takeuchi A, Tanahashi M, Uchiyama A, Vigneresse J-L. Oblique and near collision subduction, Sagami and Suruga Troughs – preliminary results of the French–Japanese 1984 Kaiko cruise, Leg 2. Earth and Planetary Science Letters. 1987; 83: 229-242

[8] Makimoto H, Miyata T, Mizuno K, Sangawa A. Geology of the Kokawa District, with geological sheet map at 1:50,000. Tsukuba: Geological Survey of Japan, AIST; 2004. 89 p

[9] Tanaka J. Sedimentary facies of the Cretaceous Izumi turbidite system, southwest Japan – an example of turbidite sedimentation in an elongated strike-slip tectonic basin. Journal of the Geological Society of Japan. 1989; 95: 119-128

[10] MEXT (Ministry of Education, Culture, Sports, Science and Technology, Japan), DPRI (Disaster Prevention Research Institute, Kyoto University). Comprehensive research and survey for the Median Tectonic Line fault system (Eastern Margin of Kongo Mountains – Southern Margin of Izumi Mountains), Heisei 27 fiscal year report. Uji: DPRI; 2016

[11] Itoh Y, Kusumoto S, Takemura K. Evolutionary process of Beppu Bay in central Kyushu, Japan: A quantitative study of the basin-forming process controlled by plate convergence modes. Earth, Planets and Space. 2014; 66: 74. doi:10.1186/1880-5981-66-74

[12] Tai Y. On the 'Shinji folded zone'. Memoir of the Geological Society of Japan. 1973; 9: 137-146

[13] Itoh Y, Nakajima T, Takemura A. Neogene deformation of the back-arc shelf of Southwest Japan and its impact on the palaeoenvironments of the Japan Sea. Tectonophysics. 1997; 281: 71-82

[14] Itoh Y, Kusumoto S, Takemura K. Characteristic basin formation at terminations of a large transcurrent fault – basin configuration based on gravity and geomagnetic data. In: Itoh Y, editor. Mechanism of sedimentary basin formation – Multidisciplinary approach on active plate margins. Rijeka: InTech; 2013. http://dx.doi.org/10.5772/56702

[15] Niitsuma N. Rupture and delamination of arc crust rupture and delamination of island arc crust due to the arc-arc collision in the South Fossa Magna, central Japan. The Island Arc. 1999; 8: 441-458

[16] Kameo K, Sekine T. Calcareous nannofossil biostratigraphy and the geologic age of the Anno Formation, the Awa Group, in the Boso Peninsula, central Japan. Journal of the Geological Society of Japan. 2013; 119: 410-420

[17] Itoh Y, Takemura K. Quaternary geomorphic trends within Southwest Japan: Extensive wrench deformation related to transcurrent motions of the Median Tectonic Line. Tectonophysics. 1993; 227: 95-104

Section 3

Research Frontiers of Diverse Tectonic Events: Backarc Rifting, Arc Volcanism, and Frontal Accretion

Chapter 8

Paleomagnetic Studies on Miocene Sequences of Hokutan and Tottori Groups in Southwest Japan: Implications for Middle Miocene Rotational Movement of Southwest Japan Block Associated with the Japan Sea Opening

Naoto Ishikawa, Takashi Suzuki and Shiro Ishida

Additional information is available at the end of the chapter

http://dx.doi.org/10.5772/67960

Abstract

Miocene sequences composed of volcanic rocks and overlying marine sediments distributing at the Japan Sea side of Southwest Japan have been considered to form related to the rifting and subsequent spreading of the Japan Sea back-arc basin in Miocene time. We performed paleomagnetic investigations on the sequences in the eastern San'in district, the Hokutan and Tottori Groups. Paleomagnetic analyses on samples from 33 sites indicated that characteristic magnetic components from five sites of volcanic rocks in the Hokutan Group and from four sites of marine sediments in the Tottori Group were regarded as primary components. An obtained paleomagnetic direction of the volcanic rocks has an easterly deflected declination (D = 23.9° ± 20.2°), while that of the marine sediments shows no significant deflection in declination (D = 17.8° ± 19.1°). Through the comparison with paleomagnetic data from the Miocene sequences in Southwest Japan, it is suggested that magnetic polarities of the volcanic and sedimentary sequences are assigned to C5Cn and C5Br-C5Bn, respectively, and that the eastern San'in district suffered a clockwise rotation of 24° at around 16 Ma after the early Miocene volcanic activity and before the middle Miocene marine transgression in the whole clockwise rotation process of Southwest Japan related to the Japan Sea opening.

Keywords: Southwest Japan, Japan Sea, paleomagnetism, clockwise rotation, Miocene, marine transgression

1. Introduction

In the San'in and Hokuriki districts on the Japan Sea side of Southwest (SW) Japan, Miocene sequences characterized by the lower volcanic units and the upper marine sedimentary ones are distributed mainly at three areas (**Figure 1**): the Hokuriku Group in the Hokuriku district, and the Hokutan/Tottori Groups, and the Iwami Group in the eastern and central parts of the San'in district, respectively. The lower volcanic units are composed of subaerial volcanic and pyroclastic rocks formed by intensive volcanic activities in early Miocene time, and the upper marine sedimentary sequences are characterized by an upward change in lithofacies from conglomerate and sandstone under shallow marine environment at its basal part to well-stratified mudstone under deep marine environment, indicating a marine transgression and subsequent rapid subsidence of sedimentary basins in middle Miocene time [1–6]. The Miocene sequences have been considered to form related to the rifting and subsequent spreading of the Japan Sea back-arc basin in Miocene time [3].

Paleomagnetic investigations in SW Japan, including researches on the Miocene sequences, have revealed a clockwise (CW) rotation of SW Japan in middle Miocene time associated with the formation of the Japan Sea [7–14]. Numerical analyses on compiled paleomagnetic

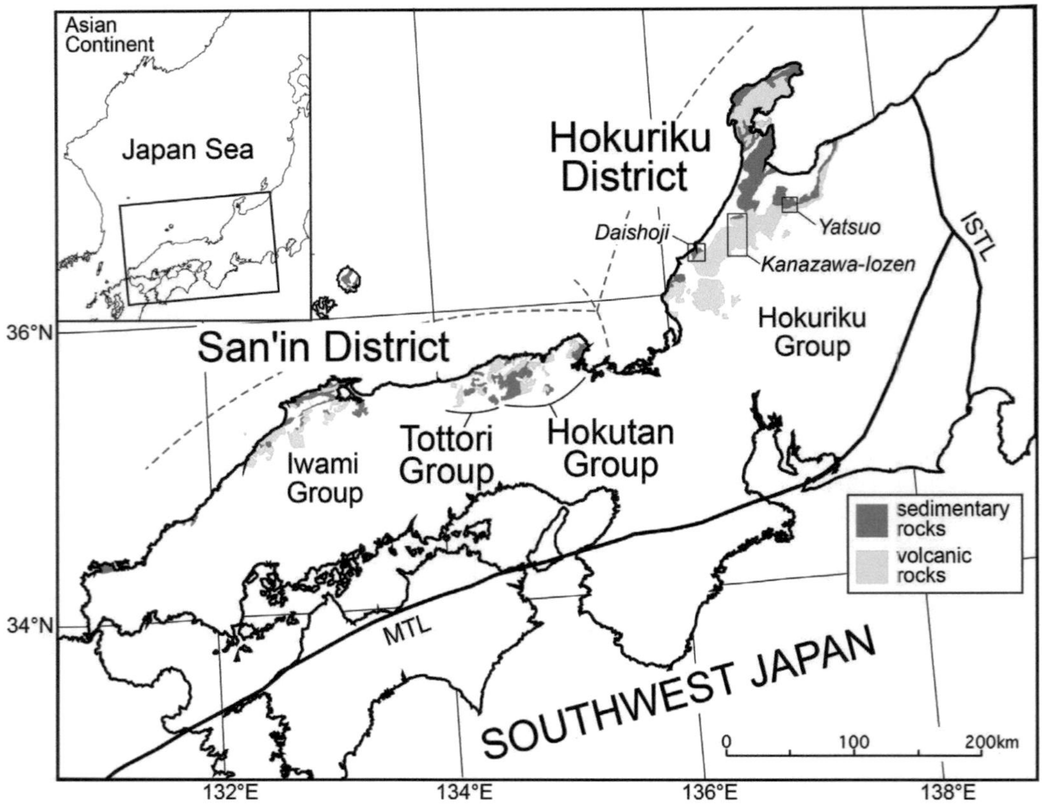

Figure 1. Map showing the distribution of Miocene volcanic and sedimentary sequences at the Japan Sea side in Southwest Japan modified from Ref. [21]. MTL: Median Tectonic Line. ISTL: Itoigawa-Shizuoka Tectonic Line.

data with geochronological age controls [7] indicated that SW Japan was rotated clockwise through about 47° at about 15 Ma for the duration as short as less than 1 m.y. A pivot of the rotation for SW Japan as a coherent block has been estimated between the Korean Peninsula and the western toe of the SW Japan block, and a rapid fan-shape opening of the Japan Sea has been proposed as an opening mode at the climax stage in the formation process of the Japan Sea [7, 15]. Shallow marine deposits at the basal part of the upper marine sedimentary units of the Miocene sequences provide molluscan fossils referred to the Kadonosawa Fauna [16] indicating subtropical to tropical environment [17]. Chinzei et al. [18, 19] suggested that the Kadonosawa Fauna are found in the strata assigned to the N8 and the lower N9 zones of planktonic foraminiferal biostratigraphy of Blow [20], and that the occurrence of the Kadonosawa Fauna in 16.5–15 Ma is attributed to a regional change in paleogeography around the Japan Sea related to the rapid CW rotation of SW Japan.

In order to clarify the temporal and regional relationship between the CW rotation of SW Japan and tectonic events related to the Japan Sea formation recorded in on-land geology in more detail, paleomagnetic and magnetostratigraphic investigations are still indispensable. We applied the investigations to the Miocene sequences of the Hokutan and Tottori Groups in the eastern San'in district (**Figure 1**). Paleomagnetic data from the two groups are rare, although strenuous paleomagnetic studies have been carried out for the Miocene sequences of the Hokuriku Group in the Hokuriku district mainly with magnetobiostratigraphic analyses and the Iwami Group in the central San'in district with radiometric age dating [8, 11, 13, 14] as mentioned later. We regarded the strata including the Kadonosawa Fauna as key beds, which are considered to represent the initiation of the marine transgression in middle Miocene time, and sampled volcanic and pyroclastic rocks, and sedimentary ones below and above the key bed along several sampling routes.

2. Geological setting and sampling

Paleomagnetic samplings were performed on the Hokutan Group distributed in Yakuno-Wadayama area around Mt. Kanatoko at the northern area of Kyoto and Hyogo Prefectures and the Tottori Group at the eastern part of Tottori Prefecture (**Figure 2**).

The Hokutan Group is divided into five formations, namely the Takayanagi, Yoka, Toyooka, Amino, and Tango Formations in ascending order [2, 22]. In the Yakuno-Wadayama area, the Yoka and Toyooka Formations are distributed, overlying the Takayanagi Formation of a basal conglomerate for the Hokutan Group [23, 24]. The Yoka Formation consists mainly of basaltic to andesitic lavas and pyroclastics. K-Ar ages of 19–20 Ma were reported from the Yoka Formation [25]. The Toyooka Formation is composed mainly of conglomerate and sandstone intercalating acidic to intermediate volcanics and is considered to have formed in sedimentary environment of lacustrine in lower and middle part and of shallow marine in the upper part [1]. Molluscan fossils of the Kadonosawa Fauna were found in the Toyooka Formation [24, 26]. The Amino Formation is composed mainly of alternations of sandstone and mudstone deposited in marine environment, which is accompanied with acidic to intermediate volcanics [1, 2, 22]. Planktonic foraminiferal fossils assigned to the lower part of Zone N9 of Blow [20] were found from mudstone layers in the lower part of the Amino

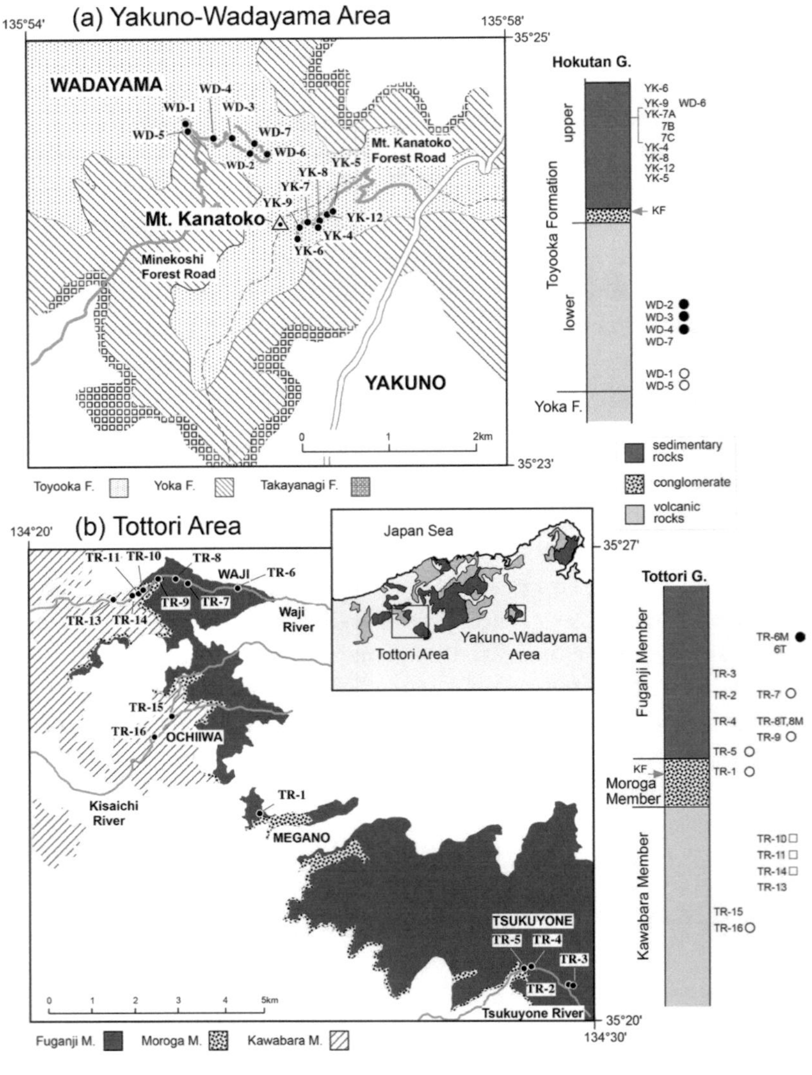

Figure 2. Simplified geological maps of (a) the Yakuno-Wadayama area around Mt. Kanatoko after [23] and (b) the Tottori area after [29]. Solid circles on the maps indicate locations of paleomagnetic sampling sites. Generalized columnar sections of (a) the Hokutan Group in the Yakuno-Wadayama area after [2] and (b) the Tottori Group after Ref. [29]. Relative stratigraphic positions for horizons of paleomagnetic sites and detected magnetic polarities are shown beside the columnar sections. Solid (open) circle denotes normal (reverse) polarity, and open squares denote transitional polarities (see in the text). KF: horizons providing molluscan fossils referred to the Kadonosawa Fauna.

Formation [27]. K-Ar ages of 13.5–14.9 Ma were reported for rhyolite samples from the middle part of the Amino Formation [22]. The Tango Formation consists mainly of andesitic to dacitic pyroclastics, lavas, and intrusive rocks [2, 22], and K-Ar ages of 13.9 and 14.6 Ma were reported from intrusive rocks of the formation [22].

In the Yakuno-Wadayama area, volcanic sequences consisting of andesite lavas and dacitic pyroclastic rocks in the lower part of the Toyooka Formation and sedimentary ones consisting of conglomerate, sandstone, and alternation of sandstone and mudstone intercalating tuffs and andesite lavas in the upper part of the formation were exposed along two forest roads (**Figure 2**). From sandstone layers above a conglomerate bed overlying pyroclastic rocks at the boundary between the volcanic and sedimentary sequences, molluscan fossils referred to the Kadonosawa Fauna were found [24].

The Tottori Group in our sampling area is divided into two formations: the Yazu and Iwami Formations in ascending order [4, 5, 28]. The Yazu Formation consists mainly of thick piles of volcanic rocks, namely the Kawabara Volcanic Member, overlying a basal conglomerate of the Tottori Group (the Koge Conglomerate Member) [4, 5]. The Iwami Formation is subdivided into the following four members: the Moroga Conglomerate, Fuganji Mudstone, Oda Andesite, and Aragane Pyroclastic Members in ascending order [4, 5]. We performed paleomagnetic samplings on the Kawabara Volcanic Member of the Yazu Formation and on the Moroga Conglomerate and Fuganji Mudstone Members of the lower part of the Iwami Formation in the eastern area of Tottori Prefecture (**Figure 2**). The Kawabara Member is composed mainly of andesite lavas and andesitic to dacitic pyroclastics and the Moroga Member consists of alternating beds of conglomerate and sandstone [4, 5]. Abundant molluscan fossils of the Kadonosawa Fauna have been found from the Moroga Member [29]. The Fuganji Member is made mainly of black mudstone deposited in deep marine environment. Matsumoto and Seto [30] found large foraminiferal fossils identical with *Operculina complanata japonica* from the Fuganji Member, and they correlated the horizon with the fossils to the Blow [20]'s planktonic foraminiferal zones from N8 to the base of N9.

In the Yakuno-Wadayama area, we collected samples from the Toyooka Formation at 16 sites (**Figure 2**): dacitic pyroclastic rocks and andesite lavas in the lower part of the formation at six sites, and in the upper part, volcanic sand, tuffs, and mudstones at nine sites and andesite lava at one site. In the Tottori area, paleomagnetic samples were obtained from the Tottori Group at 17 sites (**Figure 2**): mudstones and tuffs in the Fuganji Member at 10 sites, andesite lava in the Moroga Member at one site, and dacitic pyroclastic deposits and andesite lavas in the Kawabara Member at six sites. Paleomagnetic samplings were carried out by using a gasoline-powered core drill or by hand sampling. Three to ten cores or block samples oriented with a magnetic compass were collected at each site. Two or more cylindrical specimens of paleomagnetic standard size (24 mm in diameter and 22 mm in height) were prepared for each sample. Tilting values of strata were measured at two to eight points in the sampling sites, and mean values in the sites were adopted for tilt correction. Tilting data from dacitic pyroclastic layer intercalating in andesite lava units at WD-1 were used for tilt corrections for four sites of andesite lavas (WD-2, 3, 4, 5) because we did not find any layers in the lava units for measuring the tilting of the lava units.

3. Paleomagnetic analyses

3.1. Experiments

The stability of natural remanent magnetization (NRM) was accessed by progressive thermal and alternating field (AF) demagnetization experiments. The NRM of each specimen was measured with a superconducting rock magnetometer (model 760R of 2G Enterprises) or a spinner magnetometer (model SMD-85 of Natsuhara Giken). Thermal demagnetization was performed in air using an electric furnace with in a four-layer μ-metal magnetic shield. The internal residual field in the furnace is less than 10 nT. AF demagnetization was performed using a 2G AF demagnetizer with a three-axis tumbler system.

Two specimens from each site were subjected to the progressive thermal and AF demagnetization experiments, respectively. When the pilot specimens of a site showed linear trends of the vector endpoints decaying the origin of the orthogonal vector demagnetization diagram [31], remaining specimens of the site were demagnetized progressively mainly by thermal method. AF demagnetization was used on the remaining specimens if the pilot AF and thermal demagnetization results provided similar stable components and AF method appeared to be more effective to isolate the components. The direction of the component displaying as a linear trend on the diagrams was determined by using the principal component analysis [32]. The determined direction with a maximum angular deviation less than 5° was adopted. The component showing the linear trend toward the origin was regarded as a characteristic remanent magnetization (ChRM). The best-fit line for the linear trend toward the origin was anchored on the diagram for determining the direction of ChRM.

Site-mean directions of ChRMs and associated statistical parameters were calculated after Fisher [33]. Site mean directions with α_{95} less than 20° were regarded as reliable means of ChRMs and were adopted in the further consideration.

3.2. Demagnetization results

Figure 3 shows typical demagnetization behaviors of progressive demagnetization experiments. In samples from 17 among 19 sites of mudstone, tuff, and volcanic sand, low-stability components with northerly declination and positive inclination were observed generally below 200–300°C and/or 10–20 mT, which are probably viscous overprints of the recent geomagnetic field direction. NRMs of samples at the remaining two sites were unstable at lower demagnetization steps in both thermal and AF methods. After removal of the low-stability components, ChRMs were isolated below about 400–480°C and/or 30–40 mT at 11 sites, followed by erratic behaviors of magnetization in higher demagnetization levels (**Figure 3a–c**), while no stable component was found in samples of three sites. ChRMs were observed in the demagnetization levels up to 560–600°C at two sites of tuff (YK-9 and TR-9: **Figure 3d** and **e**). Mudstone samples of WD-6 provided two stable components below about 360°C: one with northeasterly declination and positive inclination below about 240°C and the other with southwesterly declination and negative inclination between 240 and 360°C. After removal of the two components, ChRMs in samples of WD-6 were isolated between about 360 and 480°C.

Samples from six sites of andesite lavas provided ChRMs above 400–500°C and/or about 20 mT after removals of the low-stability components regarded as the viscous overprint of the recent geomagnetic field direction below about 300°C and/or 15 mT or removals of considerable overlaps of the low-stability components with ChRMs up to about 400°C and/or 30–40 mT (**Figure 3f**). The ChRMs in samples of WD-16 were isolated by AF methods more effectively.

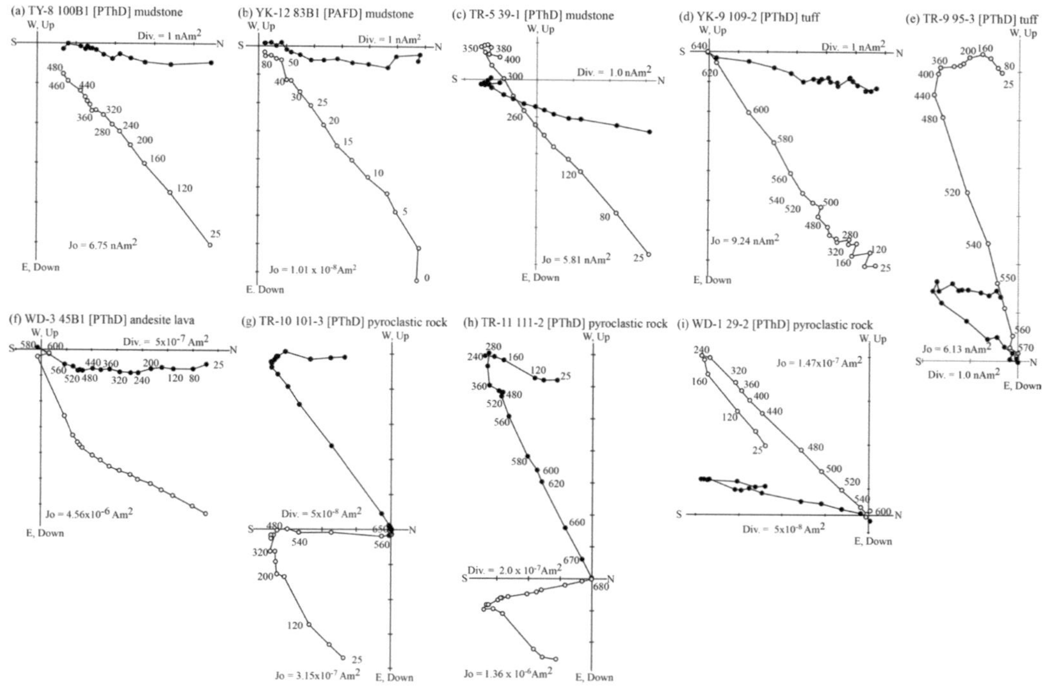

Figure 3. Typical results of progressive demagnetization experiments plotted on orthogonal vector demagnetization diagrams. Solid (open) symbols indicate projections of vector end-points on horizontal (vertical) planes in *in situ* coordinates. PThD: progressive thermal demagnetization. PAFD: progressive alternating field demagnetization.

Samples from six sites of pyroclastic rocks also showed the low-stability components below about 300°C and 15 mT (**Figure 3g–i**). The overlaps of the low-stability components with the higher stability components were also observed up to 400–500°C and 40 mT or more in four sites (TR-10, 11, 14 and 15: **Figures 3g** and **h**). At higher demagnetization steps, samples of four sites provided ChRMs below about 560°C for WD-1 (**Figure 3i**) and below 630–680°C for TR-11, 14, and 15 (**Figure 3h**). AF methods were not effective to isolate the ChRMs of TR-11, 14, and 15 because of high coercivity of the ChRMs: samples of the three sites were remained about 40–70% of initial intensity of NRM after AF demagnetization at 100 mT. Samples of TR-10 showed stable components shown as straight lines decaying not toward the origin on the demagnetization diagrams between 520 and 560°C before unstable magnetic behaviors in the higher demagnetization steps (**Figure 3g**). The directions of the stable components form TR-10 were close to those of ChRMs from TR-11 and14 as mentioned later.

3.3. Characteristic directions

Site mean directions of ChRMs with α_{95} less than 20° were obtained from seven sites of sedimentary rocks in the upper Toyooka Formation of the Hokutan Group (**Table 1** and **Figure 4**). The *in situ* directions with north declination and normal polarity are close to the directions of the present geomagnetic field and the geocentric axial dipole field (**Figure 4**) although the

Site	Locality		Lith	n/N	Demagnetization level		In situ		Tilt corrected		α_{95}	*k*	Bedding
	Lon (E)	Lat (N)					D (°)	I (°)	D (°)	I (°)			strike, dip
Hokutan Group													
Toyooka Formation (upper)													
YK-6	134°56′12″	35°24′07″	m	10/10	T	320–480°C	–4.8	49.4	–6.8	57.3	4.2	131.9	N70.7°W, 8°S
WD-6	134°55′59″	35°24′36″	t	9/10	T	360–500°C	–4.7	60.8	–15.4	70.8	6.4	65.6	N59.7°W, 11°S
YK-9	134°56′13″	35°24′12″	t	9/10	T	360–580°C	–2.5	49.5	–17.5	48.4	4.1	159.4	N9.3°E, 13°W
YK-7B	134°56′16″	35°24′11″	t	10/10	T	320–480°C	11.9	58.9	9.1	60.0	5.6	75.0	N9.7°W, 2°W
YK-7C	134°56′16″	35°24′11″	t	4/10	T	300–460°C	–1.3	57.9	–3.9	59.1	9.5	94.1	N25.7°W, 2°W
YK-8	134°56′23″	35°24′14″	m	9/10	T	320–480°C	0.3	52.6	–8.8	54.5	3.9	170.9	N5.7°W, 7°W
YK-12	134°56′25″	35°24′15″	m	10/10	A	15–40 mT	1.5	44.4	–3.2	48.5	3.6	179.1	N28.7°W, 6°W
		MEAN		*7*	*in-situ*		*–0.1*	*53.6*			*5.1*	*140.4*	
					tilt corrected				*–6.6*	*57.2*	*6.7*	*81.0*	
Toyooka Formation (lower)													
WD-2	134°55′53″	35°24′37″	al	10/10	T	480–570°C	43.2	52.1	48.1	49.2	5.5	77.2	N24.3°E, 5°E (*2)
WD-3	134°55′42″	35°24′41″	al	8/8	T	520–560°C	25.9	64.7	35.4	62.9	7.5	55.0	N24.3°E, 5°E (*2)
WD-4	134°55′32″	35°24′42″	al	10/10	T	320–600°C	17.2	47.4	22.5	46.5	4.7	105.5	N24.3°E, 5°E (*2)
WD-1	134°55′21″	35°24′48″	dp	4/4	T	280–550°C	–173.9	-44.5	–169.0	-44.6	2.5	1341.3	N24.3°E, 5°E
WD-5	134°55′24″	35°24′44″	al	3/3	T	360–560°C	–179.5	–50.6	–173.4	–51.1	5.8	452.5	N24.3°E, 5°E (*2)
		MEAN		*5*	*in-situ*		*17.5*	*52.8*			*12.4*	*39.2*	
					tilt corrected				*23.8*	*51.9*	*12.4*	*39.2*	

Site	Locality		Lith	n/N	Demagnetization level		In situ		Tilt corrected		α_{95}	*k*	Bedding
	Lon (E)	Lat (N)					D (°)	I (°)	D (°)	I (°)			strike, dip
Tottori Group													
Fuganji Mudstone Member													
TR-6M	134°23′40″	35°26′39″	m	6/9	T	200–400°C	11.0	71.6	15.3	55.7	11.7	33.8	N54.7°W, 16°N
TR-7	134°22′54″	35°26′45″	m	8/8	T	200–320°C	−155.3	−46.8	−161.2	−43.0	2.6	452.8	N69.3°E, 7°N
TR-9	134°22′22″	35°26′45″	t	7/10	T	440–570°C	−144.3	−61.9	−144.8	−50.9	7.5	66.0	N41.7°W, 11°E
TR-5	134°28′44″	35°21′15″	m	4/8	T	350–400°C	−179.9	−51.8	−178.8	−46.9	19.9	22.3	N63.7°W, 5°N
		MEAN		*4*	*in-situ*		*17.8*	*58.7*			*15.8*	*34.8*	
					tilt corrected				*17.4*	*49.8*	*12.1*	*58.6*	
Moroga Conglomerate Member													
TR-1	134°24′08″	35°23′22″	al	8/8	T	400–570°C	−85.3	−32.8	−93.7	−25.0	6.5	72.8	N47.7°W, 17°N
Kawabara Volcanic Member													
TR-10 [*1]	134°22′07″	35°26′36″	dp	10/10	T	520–580°C	−136.0	−0.5	−137.0	12.9	1.7	763.5	N1.3°E, 16°E
TR-11	134°22′02″	35°26′33″	dp	8/8	T	480–670°C	−124.9	6.2	−123.3	14.5	2.2	589.9	N65.7°W, 12°N
TR-14	134°21′56″	35°26′32″	dp	8/8	T	440–660°C	−119.6	0.1	−119.2	13.7	2.2	659.7	N28.7°W, 14°E
		MEAN		*3*	*in-situ*		*−126.8*	*1.9*			*14.0*	*78.5*	
					tilt corrected				*−126.4*	*13.8*	*14.0*	*78.9*	
TR-16	134°22′14″	35°24′33″	al	6/8	A	25–60 mT	−101.5	−39.9	−111.4	−49.8	11.6	34.2	N43.3°E, 14°E

Note: lith: lithology, m: mudstone, t: tuff, al: andesite lava, dp: dacitic pyroclastic rock. n/N: the number of samples used in a site mean calculation (n) and samples corrected at the sit (N), demagnetization: demagnetization levels of thermal (T) or alternating-field (A) methods showing the characteristic magnetic components, D and I: declination and inclination of the mean, respectively, α_{95}: the radius of the 95% circle of confidence around the site mean direction, *k*: Fisher precision parameter. Directions of declination and strike of bedding are corrected by the local geomagnetic declination (7.3°W) at each site. *1: the direction of TR-10 is a mean direction of stable components decaying not toward the origin (see in the text). *2: bedding data is that of WD-1.

Table 1. Site-mean directions of characteristic remanent components obtained from Hokutan and Tottori Groups.

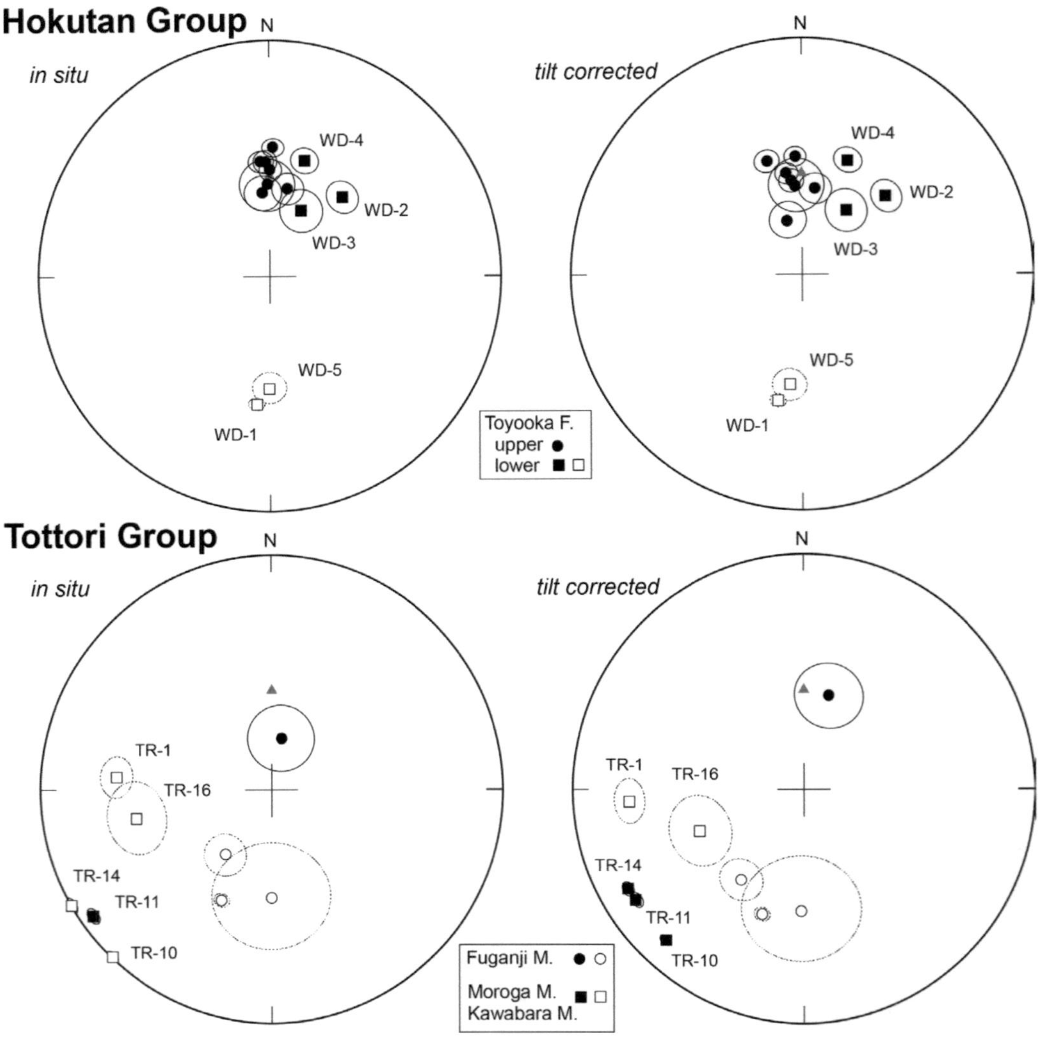

Figure 4. Site-mean directions of characteristic remanent magnetizations with 95% confidence circles plotted on equal-area nets before (left) and after (right) tilt corrections. Solid (open) symbols are on the lower (upper) hemisphere. A solid triangle on each net represents the direction of the geocentric axial dipole field expected at each sampling area.

low-stability components, which are regarded as viscous overprints of the recent geomagnetic field direction, were erased by the demagnetization experiments. Tilt-corrected directions do not show better grouping than the *in situ* directions (**Table 1**): α_{95} and the precision parameter are 5.1° and 140.4 for a mean of the *in situ* directions, while 6.7° and 81.0 for that of the tilt-corrected. It is probably inferred that the viscous overprints of the recent geomagnetic filed direction were not demagnetized perfectly and/or that tilt corrections were not effective because directional differences in tilting among the sites were little. The ChRMs of the sediment sites cannot be regarded as primary magnetic components acquired before tilting of the strata.

Site mean directions were obtained from five sites of volcanic sequence in the lower Toyooka Formation (**Table 1** and **Figure 4**). We performed tilt correction on the five *in situ* directions by tilting data from dacitic pyroclastic layer interbedded in andesite lava units at WD-1. Tilt-corrected directions show an antipodal relationship with a NNE-SSW trend approximately, and the lower two sites in the volcanic sequence are negative polarity while the upper three sites are positive polarity. A mean of the five tilt-corrected directions is D = 23.8°, I = 51.9°, and α_{95} = 12.4°, which is regarded as a paleomagnetic direction of the lower Toyooka Formation in the further consideration. Confidence limits on D and I associated with the value of α_{95}, namely Δ□ and ΔI, are 20.4° (Δ□ = sin¹[sin α_{95}/cos I]) and 12.4° (ΔI = α_{95}), respectively. The difference between the mean direction and the geocentric axial dipole field (GADF) direction expected at the Yakuno-Wadayama area (D_G = 0° and I_G = 54.9°) are D-D_G = 23.8° ± 20.4° in declination and I-I_G = −4.9° ± 12.4° in inclination. The mean direction is found to show a significant CW deflection in declination.

From the upper part of the Tottori Group, site mean directions are obtained at four sites of the Fuganji Mudstone Member, showing an antipodal relationship with a N-S trend (**Table 1**, **Figure 4**). The directions are grouped more tightly after tilt correction than in *in situ* coordinate: α_{95} and the precision parameter change from 15.1 to 12.1° and from 34.8 to 58.6, respectively (**Table 1**). The ChRMs from the four sites are considered as the components acquired before tilting. The geomagnetic polarity is negative for TR-5, 7, and 9 in the lower horizons of the Fuganji Member and positive for TR-6M in the upper horizon. A mean of the four tilt-corrected directions, D = 17.4°, I = 49.8°, and α_{95} = 12.1°, is regarded as a paleomagnetic direction of the Fuganji Member. Δ□ and ΔI are 19.0° and 12.1°, respectively. The difference between the paleomagnetic direction and the expected direction of the GADF at the Tottori area (D_G = 0° and I_G = 54.9°) is 17.4° ± 19.0° in declination and 5.1° ± 12.1° in inclination, indicating no statistical significance in the difference.

From the lower part of the Tottori Group, site mean directions of ChRMs were obtained at one site (TR-1) of andesite lava in the Moroga Conglomerate Member and at two sites of dacitic pyroclastic rocks (TR-11 and 14) and one site of andesite lava (TR-16) in the Kawabara Volcanic Member (**Table 1**, **Figure 4**). *In situ* and tilt-corrected directions of TR-1 and 16 show west-southwesterly declination with negative inclination, while those of TR-11 and 14 are characterized by very shallow inclination with southwesterly declination. The directions of TR-11 and 14 are similar to a mean direction of the stable components at TR-10 of pyroclastics (**Figure 4**). Mean directions from the three sites of pyroclastics before and after tilt correction have same values of α_{95} and precision parameter (**Table 1**). Based on the demagnetization results, the ChRMs of TR-11 and 14 are carried by magnetite and hematite and those of TR-10 are carried by magnetite. The stable components of the three sites seem to be characteristic components of pyroclastic rocks in the horizons of TR-10, 11, and 14, which are close stratigraphically (**Figure 2**) and might have been acquired at the formation time of the pyroclastics although the fold test is not positive for the directions of the three sites. Tilt-corrected directions of the three sites appear to be anomalous, possibly implying that the characteristic components are regarded as a record of anomalous direction of the geomagnetic field during a geomagnetic excursion or polarity change. A mean calculation and fold test for the only two directions of TR-1 and 16 are not meaningful. The stable ChRMs of the sites are considered to be free from the viscous overprint of the recent geomagnetic field based on the demagnetization results as mention above. Tilt-corrected directions of the two sites with negative inclinations are likely to indicate reverse polarities of the horizons (**Figure 2**).

4. Discussion

The paleomagnetic data from the Miocene sequences in the Tottori and Yakuno-Wadayama areas, the eastern part of the San'in district, imply a CW rotational motion after the volcanic activity prior to the marine transgression and no significant rotation after the formation of the marine sedimentary sequences for the eastern San'in district.

In the Yakuno-Wadayama area, a paleomagnetic direction was determined for the volcanic sequence in the lower Toyooka Formation and shows a significant CW deflection in declination (23.8° ± 20.4°: **Table 2**). A CW rotation of 24° is implied for this area after the volcanic activities forming the lower Toyooka Formation. Sakamoto [34] reported paleomagnetic data from the Toyooka Formation. Site-mean directions were obtained at three sites of sedimentary rocks (mudstone, sandstone, and tuff) and one site of volcanic rocks (rhyolite) showing horizontal bedding (**Figure 5**). An overall mean of the four sites (D = 4.6°, I = 46.4°, and α_{95} = 25.1°: **Table 2**) shows no significant deflection from the direction of the geocentric axial dipole field (GADF) expected at this area (D = 4.6° ± 38.0°), indicating no rotational motion after the formation of the Toyooka Formation [34]. Although sampling horizons of Sakamoto [34] in the Toyooka Formation are unknown, it is suggested that no significant rotational movement occurred after the formation of the sedimentary sequence in the upper Toyooka Formation. In the Tottori area, a paleomagnetic direction from the Fuganji Member shows a slightly CW deflection in declination relative to the expected direction of the GADF, but the deflection is not significant statistically (17.4° ± 19.0°: **Table 2**). It is indicated that the Tottori area has suffered no significant rotational motion after the formation of the Fuganji Member associated with the marine transgression. Two tilt-corrected directions from the Moroga and Kawabara Members show large CW deflections in declination (**Figure 4**). The CW deflected directions may also imply a possibility to represent a CW rotational motion before the formation of the marine sequence in the eastern San'in district.

Paleomagnetic and magnetobiostratigraphic investigations with radiometric age controls have been performed systematically on the Miocene sequences in the Hokuriku district (**Figure 1**), namely the Hokuriku Group and its corresponding strata, by Itoh and his corroborators [14, 35–37]. Itoh and Watanabe [36] revealed a magnetostratigraphy of the Yatsuo Group in the Yatsuo area (**Figures 1** and **5**), which is the lower unit of the Hokuriku Group. The Yatsuo Group consists of the Nirehara, Iwaine, Iozen, Kurosedani and Higashibessho Formations in ascending order [3] (**Figure 5**). The Kurosedani and Higashibessho Formations are marine sequences deposited under the marine transgression and subsequent rapid subsidence of sedimentary basins, and fossils of the Kadonosawa Fauna are found in the Kurosedani Formation [3]. A reverse polarity zone in the Iozen, Kurosedani and Higashibessho Formations is assigned to C5Br of the standard geomagnetic polarity scale of Berggren et al. [38] (**Figure 5** [36]). The age

Miocene sequences	n	D (°)	I (°)	α_{95} (°)	ΔD	Reference
Eastern San'in						
Hokutan Group						
Toyooka F.	4	4.6	46.4	25.1	38.0	[32]
lower part of Toyooka F.	5	23.8	51.9	12.4	20.4	This study
Tottori Group						
Fuganji Mudstone Member	4	17.4	49.8	12.1	19.0	This study
Hokuriku						
Hokuriku Group						
[Yastuo]						
Higashibessho F.	6	−4.5	56.4	10.0	18.3	[36]
Kurosedani F.	8	3.6	45.4	6.2	8.3	[35, 39, 40]
Iozen/Iwaine/Nirehara F.	6	20.9	49.4	6.7	10.3	[39, 40]
[Kanazawa-Iozen]						
Sunagozaka/Nanamagari F.	7	23.5	50.2	11.4	18.0	[14, 36]
Iozen F.	11	36.4	51.6	12.1	19.7	[34, 35]
Mt. Wasso Rhyolite	5	49.9	56.6	12.2	22.6	[13]
[Daishoji]						
Kasanomisaki F.	3	5.1	54.9	9.7	17.0	[34]
Kawaminami F.	4	33.5	52.8	4.4	7.3	[34]
Central San'in						
Iwami Group						
Omori F.	18	0.0	55.9	8.9	16.0	[8]
Kawai F.	8	38.4	39.8	15.1	19.8	[8]
Hata F.	6	69.4	48.2	14.9	22.7	[48]
Hata F.	5	3.5	42.9	9.9	13.6	[50]
Sada F.	6	52.5	43.2	13.1	18.1	[50]

Note: n: number of sites for calculating a mean, D/I: mean declination/inclination, α_{95}: radius of 95% confidence limit for a mean. ΔD is a confidence limit on D, estimated using the following equation: $\Delta D = \sin^{-1}[\sin \alpha_{95}/\cos I]$.

Table 2. Paleomagnetic directions from Miocene sequences in the Hokuriku and San'in districts.

range of C5Br is between 15.974 and 15.160 Ma based on Geological Time Scale (GTS) 2012 [39]. A horizon with normal polarity at the uppermost part of the Higashibessho Formation and a normal polarity zone in the lowermost part of the Iozen Formation and the Iwaine Formation of volcanic sequences is assigned to C5Bn and C5Cn, respectively [36]. The Nirehara Formation shows reverse polarity and is assigned to C5Cr [40].

In the Hokuriku district, paleomagnetic directions from the Kurosedani Formation and its corresponding strata [14, 35–37, 40, 41] show different amounts of CW deflection in declination (**Table 2**, **Figure 5**): the Sunagozaka/Nanamagari Formations at the Kanazawa-Iozen area and the Kasanomisaki Formation at the Daishoji area show CW deflections of 23.5° ± 18.0° and 33.5° ± 7.3°, respectively, while the Kurosedani Formation at the Yatsuo area shows no significant deflection (3.6° ± 8.8°). The difference in CW deflection among paleomagnetic declinations of the Kurosedani Formation and its corresponding strata has been attributed to the differential rotation in the eastern part of SW Japan caused by a collision of the Izu-Bonin arc just after the CW rotation of SW Japan as a coherent block [35, 42]. Itoh et al. [37] revealed that paleomagnetic directions from the strata assigned to NPD-4A zone of diatom biostratigraphy [43], namely the Higashibessho Formation at the Yatsuo area and the Kasanomisaki Formation at the Daishoji area (**Figure 1**), show no deflection in declination while CW deflections in paleomagnetic declinations are observed in the strata assigned to the diatom-biostratigraphic NPD-3A zone or the foraminiferal zone of N8 and older strata (**Table 2**, **Figure 5**). It is suggested that the CW rotation of SW Japan occurred at the latest early Miocene assigned to the diatom-biostratigraphic zone of NPD-3B [37]. Itoh and Kitada [14] revealed a successive decrease in CW deflection of paleomagnetic declinations among paleomagnetic directions from the Mt. Wasso Rhyolite of about 20 Ma (D =49.9° ± 22.6°), the Iozen Formation of volcanic sequences (D = 36.4° ± 19.7°) and the Sunagozaka/Nanamagari Formations in the Kanazawa-Iozen area (**Table 2**, **Figures 1** and **5**), and suggested that the CW rotation of SW Japan may have commenced in the early Miocene and that the rotation was accelerated after the initial stage of marine transgression.

In this study, samples of the Moroga Member and the lower part of the Fuganji Member at the Tottori area show reverse polarities (**Figures 2** and **5**). The Moroga Member includes fossils of the Kadonosawa Fauna. The Fuganji Member is assigned to the planktonic foraminiferal zone of N8 and the base of N9 [30]. Compared with the magnetostratigraphy of the Yatsuo Group [36], it is suggested that the reverse polarity horizons in the Moroga and Fuganji Members are probably correlated to C5Br and that a horizon with normal polarity at the upper part of the Fuganji Member (TR-6M) is possibly assigned to C5Bn. A reverse polarity of TR-16 below the horizons with transitional polarities (TR-10, 11 and 14) in the Kawabara Member (**Figure 2**) might have been correlated to one of reverse polarity subchrons in C5Cn or C5Cr. According to the lithology of the Toyooka Formation at the Yakuno-Wadayama area, sedimentary sequences in the upper part, which include fossils of the Kadonosawa Fauna at its basal part, and the volcanic sequences in the lower part are considered to be correlated to the Kurosedani Formation and the underlying volcanic sequences (the Iozen Formation and older volcanic units), respectively, in the Hokuriku Group. Horizons of normal and reverse polarities in the volcanic sequences at the lower Toyooka Formation (**Figure 2**) may be correlated to a normal polarity zone in the Iozen and Iwaine Formations, assigned to C5Cn [40] and to a reversal polarity one in the Nirehara Formation (C5Cr) at the Yatsuo area or possibly one of reverse polarity subchrons in C5Cn, respectively. According to the above-mentioned magnetostratigraphic correlation, it is implied that the CW rotation of 24° inferred for the eastern San'in district occurred at around 16 Ma after the early Miocene volcanic activity and before the middle Miocene marine transgression in the whole process of the CW rotation of SW Japan in Miocene time.

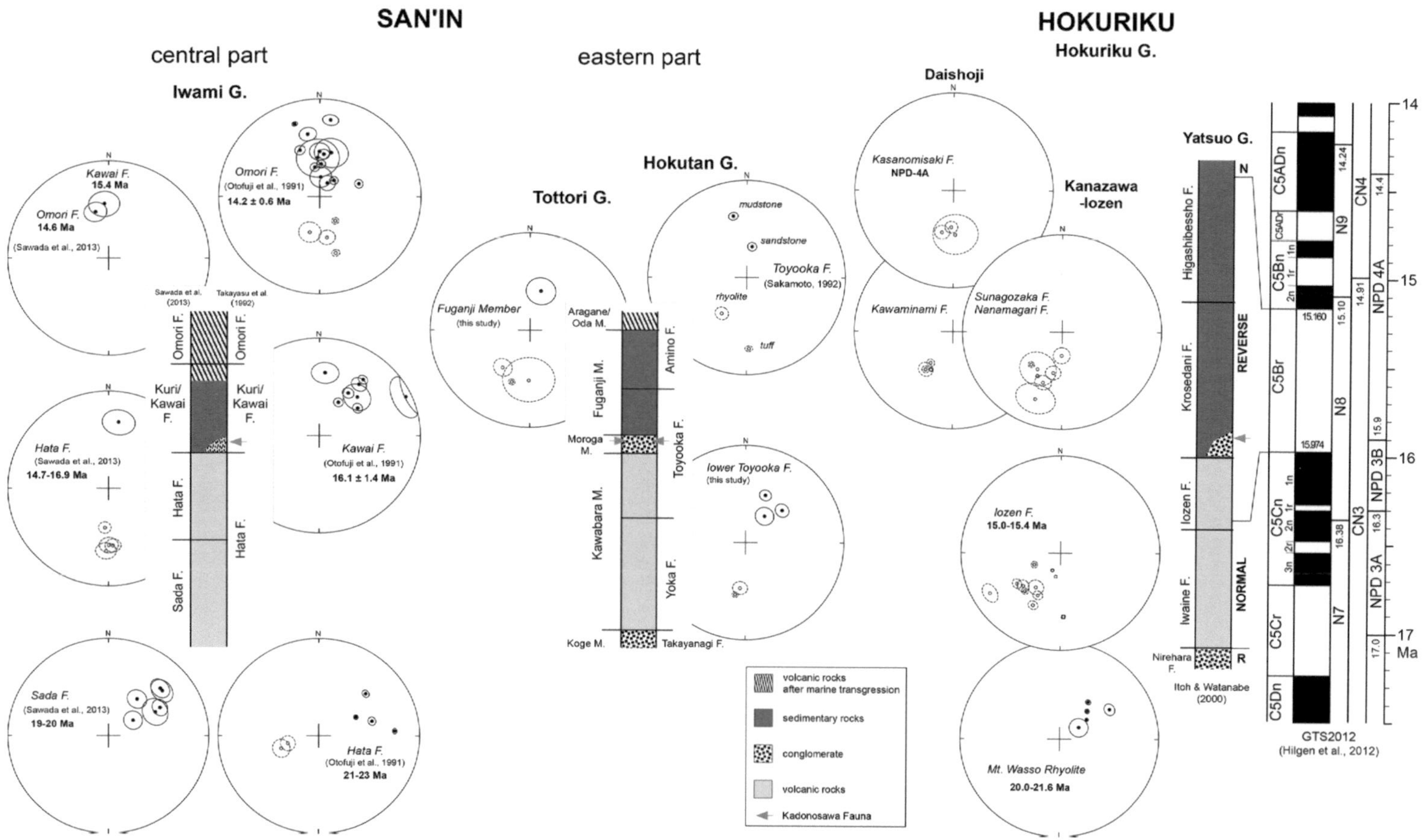

Figure 5. Comparison of paleomagnetic directions from Miocene sequences in the San'in and Hokuriku districts, listed in **Table 2**. Solid (open) symbols on equal-area nets are on the lower (upper) hemisphere. The geomagnetic polarity time scale and biostratigraphic zonations of planktonic foraminifera and calcareous nannofossils are after Geological time scale 2012 [39]. Ages for NPD zones [43] of diatom biostratigraphy are referred to Refs. [52, 53]. The generalized columnar section and magnetostratigraphy of the Yatsuo Group are modified from Ref. [36]. The generalized columnar section of the Hokutan and Tottori Groups is modified from Refs. [2, 29], and that of the Iwami Group is modified from Refs. [6, 51].

A paleomagnetic direction of the Fuganji Member shows no significant deflection in declination, implying that a rotational motion related to the CW rotation of SW Japan ceased before the formation of the Fuganji Member related to the marine transgression in the eastern San'in district. If so, it is also inferred that the horizons providing paleomagnetic data in the Fuganji Member, assigned to C5Br-C5Bn, may be correlated to the Higashibessho Formation and its corresponding strata of the Hokuriku Group assigned to the diatom zone of NPD-4A. A CW deflected site mean direction with reverse polarity from the Moroga Member (TR-1: **Figure 4**), which includes fossils of the Kadonosawa Fauna, may imply a possibility that a CW rotation occurred after the initiation of the marine transgression in the eastern San'in district. More detailed paleomagnetic investigations on the Fuganji Member, as well as the upper Toyooka Formation, may detect a rotational motion at the initial stage of the marine transgression in the eastern San'in district as suggested in the Hokuriku district [14].

In the central area of the San'in district (**Figure 1**), the Miocene sequences of the Iwami Group consist of the Hata, Kawai/Kuri, and Omori Formations in ascending order [6, 44] (**Figure 5**). The marine transgression is recorded in the Kawai/Kuri Formations, and fossils of the Kadonosawa Fauna were reported from the Kawai Formation [45, 46]. Hayashi et al. [47] found planktonic foraminiferal fossils and calcareous microfossils assigned to N8 and CN3 zones, respectively, from the Kuri Formation, and the overlap age of the two zones ranges between 16.38 and 15.10 Ma based on GTS2012 [39]. Otofuji and Matsuda [48] estimated a total amount of CW rotation of SW Japan of 44–47° with respect to East Asia associated with the opening of Japan Sea since about 20 Ma based on Cretaceous to Miocene paleomagnetic data from the western and central parts of the San'in district, including a CW deflected paleomagnetic direction from the Hata Formation (D = 69.4° ± 22.7°: **Table 2**) of 23–21 Ma [49]. Otofuji et al. [8] performed paleomagnetic and K-Ar dating analyses on volcanic rocks of the Kawai/Kuri and Omori Formations (**Table 2**, **Figure 5**): a paleomagnetic direction of the Kawai Formation with a mean age of 16.1 ± 1.4 Ma shows a CW-deflected declination (38.4 ± 19.8°), while that of the Omori Formation with a mean age of 14.2 ± 0.6 Ma has no significant deflection in declination (0.0 ± 16.0°). Otofuji et al. [8] suggest that more than 80% of the CW rotation of SW Japan occurred between 16 and 14 Ma.

All site mean directions from lavas of the Kawai Formation [8] have normal polarities (**Figure 5**), which may be correlated to the normal polarity zone in the Iozen and Iwaine Formations, assigned to C5Bn, based on the magnetostratigraphy of the Hokuriku Group [36]. A paleomagnetic direction of the volcanic sequences in the lower Toyooka Formation, which is also correlated to the Iozen and Iwaine Formations as mentioned above, has a CW deflected declination of 23.8° ± 20.4°. The deflected declination implies a CW rotation of 24° for the eastern San'in district. The rotation angle for the eastern San'in district is smaller than the total amount of the CW rotation for Southwest Japan estimated by Otofuji and Matsuda [48]. The amount of the CW deflection from the lower Toyooka Formation appears to be also smaller than that of the Kawai Formation, which might have inferred a differential rotation for the eastern San'in district during the CW rotation of SW Japan. For estimating the difference in declination between the Kawai Formation and the lower Toyooka Formation, expected directions at the representative point of SW Japan (35°N and 134°E) are calculated from the paleomagnetic directions of the two formations as follows: D = 38.5°, I = 40.7°, α_{95} = 15.1°, and ΔD = 20.1° for the Kawai Formation, and D = 23.9°, I = 51.9°, α_{95} = 12.3°, and ΔD = 20.2° for the lower Toyooka Formation. The difference in declination between the Kawai Formation and the lower Toyooka Formation (14.8° ± 28.5°) estimated as defined by Beck [50] is not statistically significant. It is probably implied that the CW rotation for the eastern San'in district at around 16 Ma represents a rotational motion of SW Japan at the climax stage or the final stage in the whole process of the CW rotation related to the Japan Sea opening as suggested by Otofuji et al. [8].

Based on the above-mentioned paleomagnetic data from the Miocene sequences in the San'in and Hokuriku districts, it is pointed out that magnetic polarities of the strata including fossils of the Kadonosawa Fauna are different among the Miocene sequences: the strata in the eastern San'in and Hokuriku districts, namely the Moroga Member and the Kurosedani Formation, show reverse polarities assigned to C5Br, while the Kawai Formation in the central San'in district has normal polarity probably assigned to C5Cn. It may be inferred that the marine transgression in the central San'in district occurred before the CW rotation of SW Japan, preceding the marine transgression in the eastern San'in and Hokuriku districts.

On the other hand, according to results of geochronological and paleomagnetic investigations of Sawada et al. [51] in the central San'in district (**Figure 5**), it is implied that a CW rotational motion in the central San'in district at about 16 Ma [51] ceased before the formation of the Kawai/Kuri Formations related to the marine transgression. Sawada et al. [51] indicated that a paleomagnetic direction of the Sada Formation with K-Ar ages of 20–19 Ma, which has been previously recognized as the Kawai Formation, shows a large CW deflection in declination (52.5° ± 18.1°: **Table 2** and **Figure 5**), while a paleomagnetic direction of the Hata Formation, of which K-Ar ages range between 16.9 and 14.7 Ma, has no significant deflected paleomagnetic declination (9.0° ± 12.0°). Site-mean directions from the Kawai Formation of 15.4° ± 0.6° Ma and the Omori Formation of 14.6° ± 0.8° Ma also show no deflected declinations with normal polarities (**Figure 5**). According to the results of Sawada et al. [51], the Kawai Formation with normal magnetic polarity may be assigned to C5Bn or younger normal magnetic chron. More detailed magnetostratigraphic investigations on the Miocene sequence in the central San'in district may be still indispensable in order to clarify the aspect of the Miocene marine transgression in SW Japan related to Japan Sea opening accompanied with the CW rotation of Southwest Japan.

5. Conclusions

We performed paleomagnetic investigations on volcanic and sedimentary rocks at 33 sites of Miocene sequences at the eastern part of the San'in district in SW Japan, namely the Tottori and Hokutan Group, and suggest the following conclusions:

1. Two paleomagnetic directions were determined for the Miocene sequences in the eastern San'in district as follows:

Volcanic sequence of the lower Toyooka Formation in the Hokutan Group: D = 23.8°, I = 51.9°, and α_{95} = 12.4°.

Sediments of the Fuganji Mudstone Member in the Tottori Group: D = 17.4°, I = 49.8°, and α_{95} = 12.1°.

The paleomagnetic direction of the lower Toyooka Formation has an easterly-deflected declination (D = 23.8° ± 20.4°), while that of the Fuganji Member shows no significant deflection in declination (D = 17.4° ± 19.0°).

2. Based on the comparison of lithology, magnetostratigraphy, and geochronological data in the Miocene sequences between the eastern San'in and Hokuriku districts, magnetic polarities in the lower Toyooka Formation and Fuganji Member are probably assigned to C5Cn and C5Br-C5Bn, respectively.

3. It is suggested that the eastern San'in district suffered a CW rotation of 24° at around 16 Ma after the early Miocene volcanic activity and before the middle Miocene marine transgression, which is attributed to the rotational motion at the climax stage in the whole process of the CW rotation of the SW Japan block related to the formation of the Japan Sea back-arc basin in Miocene time.

Acknowledgements

We thank K. Uno for his support and suggestions in the field and laboratory works. We are grateful to T. Kudo and the town office of Yakuno-cho, Kyoto Prefecture, for their help in our fieldwork. Discussions with K. Kaneko, M. Kato, N. Kawamura and other members in Division of Dynamics of Earth and Cosmos, Graduate school of Human and Environmental Studies, Kyoto University, were very helpful.

Author details

Naoto Ishikawa[1*], Takashi Suzuki[1] and Shiro Ishida[2]

*Address all correspondence to: ishikawa@gaia.h.kyoto-u.ac.jp

1 Graduate School of Human and Environment Studies, Kyoto University, Kyoto, Japan

2 Lake Biwa Museum, Siga, Japan

References

[1] Wadatsumi K, Matsumoto T. The stratigraphy of the Neogene formations in the northern Tazima—Study of the Neogene in the North-western part of the Kinki District—(Part 1). Journal of the Geological Society of Japan. 1958; **64**: 625-637 (in Japanese with English abstract).

[2] Wadatsumi K, Ikebe N, Matsumoto T. Correlation of the Neogene Formations of Northern Kinki, Southwest Japan—Studies of the Late Cenozoic Formations in Northern Kinki District, Part 3. Professor Susumu Matsushita Memorial Volume. 1966; 105-116 (in Japanese with English abstract).

[3] Hayakawa H, Takemura A. The Neogene system in the Yatsuo area, Toyama Prefecture, central Japan. Journal of the Geological Society of Japan. 1987; **93**: 717-732 (in Japanese with English abstract).

[4] Matsumoto T. Stratigraphy of the Miocene series in the area southeast of Tottori City, Tottori Prefecture, Southwest Japan. Journal of the Geological Society of Japan. 1986; **92**: 269-287 (in Japanese with English abstract).

[5] Matsumoto T. Stratigraphical study of the Miocene series in the eastern part of Tottori Prefecture, Southwest Japan. Journal of Science of the Hiroshima University. Series C. 1989; **9**:199-235.

[6] Takayasu K, Yamasaki H, Ueda T, Akagi S, Matsumoto T, Nomura R, Okada S, Sawada Y, Yamauchi S, Yoshitani A. Miocene stratigraphy and paleogeography of the San'in district, Southwest Japan. The Memoirs of the Geological Society of Japan. 1992; **37**: 97-116 (in Japanese with English abstract).

[7] Otofuji Y, Hayashida A, Torii M. When was the Japan Sea opened?: paleomagnetic evidence from Southwest Japan. In: Nasu N., Uyeda S., Kushiro I., Kobayashi K., Kagami H., editors. Formations of Active Ocean Margins: Terrapub; 1985. pp. 551-566.

[8] Otofuji Y, Itaya T, Matsuda T. Rapid rotation of southwest Japan-palaeomagnetism and K-Ar ages of Miocene volcanic rocks of southwest Japan. Geophysical Journal International. 1991; **105**: 397-405.

[9] Otofuji Y. Large tectonic movement of the Japan Arc in the late Cenozoic times inferred from paleomagnetism: review and synthesis. The Island Arc. 1996; **5**: 229-249.

[10] Hayashida A. Timing of rotational motion of Southweest Japan inferred from paleomagnetism of the Setouchi Miocene Series. Journal of Geomagnetism and Geoelectricity. 1986; **38**: 295-310.

[11] Nakajima T, Hayashi M, Nakagawa T. Two step motions of the clockwise rotation of Southwest Japan during the Miocene. Professor Shizuka Miura Memorial Volume. 1991; 105-119 (in Japanese with English abstract).

[12] Nakajima T, Sawada Y, Nakagawa T, Hayashi M, Itaya T. Paleomagnetic results and K-Ar dating on Miocene rocks in the northern part of Fukui Prefecture, Central Japan—with reference to the rotation of Southwest Japan. Journal of Mineralogy, Petrology and Economic Geology. 1990; **85**: 45-59 (in Japanese with English abstract).

[13] Itoh Y, Doshida S, Kitada K, Danhara T. Paleomagnetism and fission-track ages of the Mt. Wasso moonstone rhyolitic welded tuff in the Ishikawa Prefecture, central Japan. Bulletin of the Geological Survey of Japan. 2001; **52**: 573-579 (in Japanese with English abstract).

[14] Itoh Y, Kitada K. Early Miocene process in the eastern part of south-west Japan inferred from paleomagnetic studies. The Island Arc; 2003: **12**, 348-356.

[15] Otofuji Y, Matsuda T, Nohda S. Opening mode of the Japan Sea inferred from the paleomagnetism of the Japan Arc. Nature. 1985; **317**: 603-604.

[16] Oyama K. Studies of fossil molluskan biocoenosis, No. 1: biocoenological studies on the mangrove swamps, with descriptions of new species from Yatsuo Group. Reports, Geological Survey of Japan. 1950; **132**: 1-16.

[17] Chinzei K. Faunal succession and geographic distribution of Neogene molluscan faunas in Japan. In: Kotaka T, editors. Origin and Migration of the Japanese Cenozoic Mollusca. Paleontological Society of Japan. Special Papers. 1986; **29**: 17-32.

[18] Chinzei K. Opening of the Japan Sea and Marine Biogeography during the Miocene, Journal of Geomagnetism and Geoelectricity. 1986; **38**: 487-494.

[19] Chinzei K. Late Cenozoic zoogeography of the Sea of the Japan area, Episodes, 1991; **14**: 231-235.

[20] Blow WH. Late Middle Eocene to Recent planktonic foraminiferal biostratigraphy. In: Bronnimann P, Renz H.H, editors. Proceedings of the First International Conference on Planktonic Microfossils. Vol. 1: EJ Brill Leiden; 1969. pp. 199-422.

[21] Yamada N, Saito E, Murata Y. Computer-Generated Geologic Map of Japan (scale 1:200,000), 1:2,000,000 Map Series 22, Geological Survey of Japan, 1990.

[22] Yamamoto T, Hoshizumi H. Stratigraphy of the Neogene system in the Tango Peninsula, eastern part of the San-in district, Southwest Japan, and associated Middle Miocene volcanism. Journal of the Geological Society of Japan. 1988; **94**: 769-781 (in Japanese with English abstract).

[23] Editorial Committee of Engineering Geological Maps of Hyogo Prefecture. Geology of Hyogo—Explanatory text of geological maps (scale 1:100,000) of Hyogo Prefecture, geological part—Hyogo Construction Technology Center, Kobe. 1996. 361p (in Japanese).

[24] Ishida S. "Era of Island Arc"—Geological history of Yakuno Town since Neogene. In: Yakuno Town History editorial board, editors. History of Yakuno Town, Volume 1, Natural Sciences and Folk Traditions. 2005. pp. 68-134 (in Japanese).

[25] Furuyama K, Sawada Y, Itaya T, Miyake Y, Inoue Y, Kotaki A. K-Ar ages of some volcanic rocks from the Yoka Formation, Miocene Hokutan Group, northern Kinki district. Earth Science (Chikyu Kagaku). 1997; **51**: 452-457 (in Japanese).

[26] Matsubara T. Miocene shallow marine molluscs from the Hokutan Group in the Tajima area, Hyogo Prefecture, southwest Japan. Bulletin of the Mizunami Fossil Museum. 2011; **37**: 51-113.

[27] Nakagawa T, Chiji M, Miura S. Miocene geology and planktonic foraminifers of the Uchiura area, Fukui Prefecture, central Japan. Journal of the Geological Society of Japan. 1985; **91**: 389-402 (in Japanese with English abstract).

[28] Tottori Green Tuff Research Group. Studies on stratigraphy and structures of Tottori Group, Lower and Middle Miocene strata, distributed in the southeast area of Tottori City, Southwest Japan. Monograph of the Association for the Geological Collaboration in Japan. 1989; **36**: 85-104.

[29] Uemura F, Sakamoto T, Yamada N. Geology of the Wakasa District, Scale 1:50,000 Quadrangle Series, Geological Survey of Japan. 1979. 91p (in Japanese with English abstract).

[30] Matsumoto T, Seto K. New occurrence of Operculina from the Tottori Group, and its biostratigraphical and paleozoogeographical significance. Geological Report of Shimane University. 1994; **13**: 47-56 (in Japanese with English abstract).

[31] Zijderveid JDA. A. C. demagnetization of rocks: Analysis of results. In: Collinson, DW, Creer KM, Runcorn SK, editors. Methods in Palaeomagnetism. Amsterdam: Elsevier; 1967, pp. 254-268.

[32] Kirschvink JL. The least-squares line and plane and analysis of paleomagnetic data. Geophysical Journal of the Royal Astronomical Society. 1980; **62**: 699-718.

[33] Fisher RA. Dispersion on a sphere. Proceedings of the Royal Society of London. Series A. 1953; **217**: 295-305.

[34] Sakamoto M. Lava block rotation around vertical axis inferred from remanence directions of Miocene Yoka Formation, Southwest Japan. Journal of Geomagnetism and Geoelectricity. 1992; **44**: 55-63.

[35] Itoh Y, Ito Y. Confined ductile deformation in the Japan arc inferred from paleomagnetic studies. Tectonophysics. 1989; **167**: 57-73.

[36] Itoh Y, Watanabe M. Refined magnetostratigraphy of the Early Miocene sequence in the Yatsuo area, central Japan. Bulletin of the Geological Survey of Japan. 2000; **51**: 37-45 (in Japanese with English abstract).

[37] Itoh Y, Yamamoto A, Iwano H, Danhara T, Watanabe M. Paleomagnetism and fission-track ages of the Miocene sequence in the Kanazawa and Iozen areas, central Japan. Bulletin of the Geological Survey of Japan. 2000; **51**: 495-504 (in Japanese with English abstract).

[38] Berggren WA, Kent DV, Swisher CC-III, Aubry MP. Geochronology, Time Scales and Global Stratigraphic Correlation. In: Bergrenn WA, Kent DV, Aurbry MP, Hardenbol J, editors. Society for Sedimentary Geology, Tulsa, Oklahoma. SEPM Special Publication No. 54, 1995. pp. 129-212.

[39] Hilgen FJ, Lourens LJ, Van Dam JA. The neogene period. In: Gradstein FM, Ogg JG, Schmitz MD, Ogg GM, editors. The Geologic Time Scale 2012, vol. 2, 2012. pp. 923-978.

[40] Itoh Y, Hayakawa H. Magnetostratigraphy of Neogne rocks around the Yatsuo area in Toyama Prefecture, Japan. Journal of the Geological Society of Japan. 1988; **94**: 515-525 (in Japanese with English abstract).

[41] Itoh Y, Hayakawa H. On the boundary of polarity chron C5B/C5C in the Neogene section of the Yatsuo area in Toyama Prefecture, Japan. Journal of the Geological Society of Japan. 1989; **95**: 133-136 (in Japanese with English abstract).

[42] Itoh Y. Differential rotation of the eastern part of southwest Japan inferred from paleomagnetism of Cretaceous and Neogene rocks. Journal of Geophysical Research. 1988; **95**: 3401-3411.

[43] Yanagisawa Y, Akiba F. Refined Neogene diatom biostratigraphy for the northwest Pacific around Japan, with an introduction of code numbers for selected diatom biohorizons. Journal of the Geological Society of Japan. 1998; **104**: 395-414.

[44] Editorial Board of Geological Map of Shimane Prefecture. Geological Map of Shimane Prefecture (new version) scale 1:200,000. Naigai Map Co. Ltd., 1997.

[45] Okamoto K, Takahashi Y, Terauchi M. Molluscan assemblage from the Miocene Kawai Formation in Nima-chô, Shimane Prefecture. Professor Hisamichi Matsushita Memorial Volume. Part 2. 1971; 179-185 (in Japanese with English abstract).

[46] Takenouchi S, Tai Y, Kato M. Miocene microbiostratigraphy of Oda City, Shimane Prefecture, west Japan, with specieal reference to the strato-type sections of the Kawai and Kuri formations. Memoirs of the Faculty of Integrated Arts and Sciences, Hiroshima University. Series IV. 1982; **7**: 49-89 (in Japanese with English abstract).

[47] Hayashi H, Hashino S, Nomura R, Tanaka Y. Biostratigraphy of the Miocene Kuri Formation at the type locality in Oda City, Shimane Prefecture, southwestern Honshu, Japan. Journal of the Geological Society of Japan. 2013; **119**: 300-311 (in Japanese with English abstract).

[48] Otofuji Y, Matsuda T. Amount of clockwise rotation of Southwest Japan - fan shape opening of the southwest part of the Japan Sea. Earth and Planetary Science Letters. 1987; **85**: 289-301.

[49] Otofuji Y, Matsuda T. Timing of rotational motion of Southwest Japan inferred from paleomagnetism. Earth and Planetary Science Letters. 1984; **70**: 373-382.

[50] Beck ME Jr. Paleomagnetic record of plate-margin tectonic processes along the western edge of North America. Journal of Geophysical Research. 1980; **85**: 7115-7131.

[51] Sawada Y, Mishiro Y, Imaoka T, Yoshida K, Inada R, Hisai K, Kondo H, Hyodo M. K-Ar ages and paleomagnetism of the Miocene in the Izumo Basin, Shimane Prefecture. Journal of the Geological Society of Japan. 2013; **119**: 267-284 (in Japanese with English abstract).

[52] Jaeger JM, Gulick SPAS, LeVay LJ, the Expedition 341 Scientists. Methods. In: Proceedings of the Integrated Ocean Drilling Program. 2014; **341**:58. doi:10.2204/iodp.proc.341.102.2014

[53] Tada R, Murray RW, Alvarez Zarikian CA, the Expedition 346 Scientists. Methods. In: Proceedings of the Integrated Ocean Drilling Program. 2015; **346**: 71. doi:10.2204/iodp.proc.346.102.2015

Chapter 9

Oki-Dozen Dike Swarm: Effect of the Regional Stress Field on Volcano-Tectonic Orientations

Daisuke Miura, Kiyoshi Toshida, Ken-ichi Arai,
Takeshi Wachi and Yutaka Wada

Additional information is available at the end of the chapter

http://dx.doi.org/10.5772/67612

Abstract

This article presents new field, geochronological, and geochemical data for the Late Miocene Oki-dozen dike swarm (ODS), southwest Japan. This swarm is part of a volcanic suite comprising mafic and silicic dikes, sills, and pyroclastic cones from which we obtained structural measurements at a various genetic orders and scales. The mafic magmas generated three dike swarms with dikes oriented to NW-SE, N-S, and NE-SW. In comparison, the silicic intrusions do not have a preferred orientation but instead appear to radiate from the center of the volcanic suite. Comparison of the maximum thickness of 37 dikes with SiO_2 content (wt%) yielded a critical thickness (T_{cr}) value of $T_{cr} = 0.2 \times (SiO_2 - 40)$. These data indicate that the orientations of dikes were controlled by the magnitude of dike tip pressure and magma overpressure, both of which positively correlate with SiO_2 concentrations. The silicic units yield estimated pressures (up to 15–60 MPa) that are large enough to have counteracted the regional stress field, whereas the mafic dike swarm only yielded lower pressures. This result suggests that comparative analysis at a range of scales is essential for the accurate determination on the tectonic stress field by igneous rocks.

Keywords: dike, sill, mafic magma, silicic magma, stress field, volcanology

1. Introduction

Geological research into the formation of subvolcanic structures can provide insights into the long-term processes that contribute to magma plumbing and migration. One example of these processes is the formation of a dike swarm, where single dike formation involves a physical process in the form of crack opening [1, 2]. The ambient stress field during dike intrusion can be modeled as a mode 1 open crack that is associated with liquid magma injection [3]. This process occurs over timescales of hours to months, meaning that any physical contrasts in properties between the magma and the host rock are key controls on dike emplacement. The analysis of deformation structures within dike interiors can provide information on dike propagation processes prior to eruption [4–6]. However, dike swarms form as a result of the repeated injection of blade-like magma bodies into the crustal subsurface, a process that can take several ten of years, a thousand years, or more. Subvolcanic structures such as dike swarms form as a result of numerous competing geological factors, which include the following: the behavior of magma in stiff host rock(s); the physical properties of magma; local, regional, and tectonic stress fields; and the structure of the host rocks.

Changes in magma density and/or viscosity can provide insights into the behavior of magmas in deep-seated blade-like conduits. The rheological properties of magma usually govern the eruption rate, the kinetics of crystal and bubble formation, and the flow of lava on the Earth's surface [7]. Magmas can have a wide range of viscosities [8–11] that in turn lead to diverse eruption styles, structures, and evolution of eruptive episodes. The viscosity of a molten magma varies between 10^2 and 10^8 Pa s [12], and bulk viscosity dramatically increases with decreasing magma temperature and increasing crystal abundance. For example, a bulk magma with a crystal fraction of >0.6 may not be eruptible [13], and this limit of "eruptibility" is a critical threshold for the emplacement of viscous magmas into a crack-type conduit [11, 14].

The Oki-dozen volcanic complex (OVC) is a volcanic suite located 300 km from the present-day subduction trench in SW Japan (**Figure 1**) [15–21]. The complex contains numerous volcanic edifices that define the interior of the complex, including an outer-flank basaltic volcano with lava flows, pyroclastic cones, dikes, and laccolithic sills. These intrusive bodies are here termed the Oki-dozen dike swarm (ODS). The ODS includes ≥200 numbers of outcropping dikes and sills that allow the multiscale measurement of structural elements, in turn providing insights into the genetic relationships between the sills and dikes of the ODS.

Our new data from the ODS indicate that the propagation direction of dikes was controlled by pressure change in the dike interior and the state of the regional stress field. The presence of magma overpressure associated with magma density and magma chamber depth results in a distinctive pattern of dike orientation compared with the surrounding regional stress field (σ_3). This means that the development of a regional-scale dike swarm must involve the repetition of similar processes to produce dikes with a preferred orientation. This in turn indicates that a comparative analysis of the stress field at different scales is needed to infer the tectonic stress orientation in the volcanic rocks.

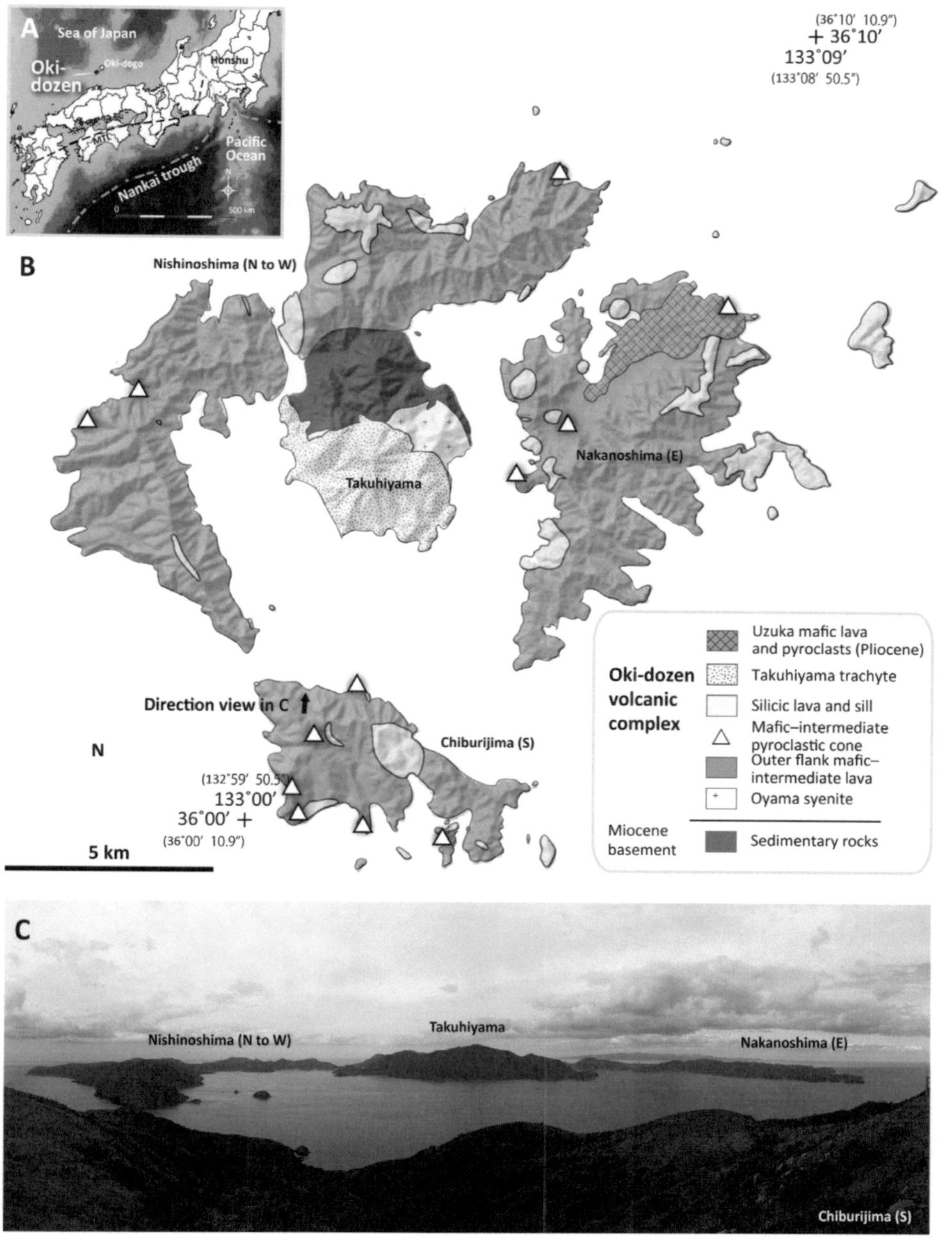

Figure 1. Maps showing the location (A) and geology (B) of the Oki-dozen volcanic complex (OVC) in southwest Japan (simplified after Refs. [16, 19]), and a panoramic view (C) of the topography of this region. The three main islands within the OVC are dominated by voluminous mafic lavas and laccolithic silicic intrusions that are associated with minor amounts of syenite and trachyte ash-flow tuff. Miocene sedimentary rocks, the Oyama syenite, and the Takuhiyama trachyte ash-flow tuff all crop out within the central OVC, suggesting this area is underlain by a fossil magma chamber. The open triangle in (B) indicates the location of a pyroclastic cone that represents a conduit for both mafic and silicic magma systems, as well as an area that concentrated mafic and silicic magma conduits.

2. General geology of the Oki-dozen volcanic complex

2.1. Geologic setting

The Oki-dozen islands of southwest Japan consist of the latest Miocene dissected OVC (**Figure 1**) [15–21], which is located in the southern Sea of Japan, ~50 km off the north coast of Shimane Peninsula, 15 km southwest of Oki-dogo island. The Oki-dozen islands comprise three main islands: Nishinoshima, Chiburijima, and Nakanoshima (**Figures 1** and **2**). These islands form part of the Oki-Shimane plateau within the Sea of Japan [22], and parts of these islands have been awarded UNESCO Global Geopark status as a result of the large number of excellent exposures of volcanic deposits [15].

The interior of the complex consists of the ODS, which comprises a suite of numerous volcanic edifices that include an outer-flank basaltic volcano, lava flows, pyroclastic cones, dikes,

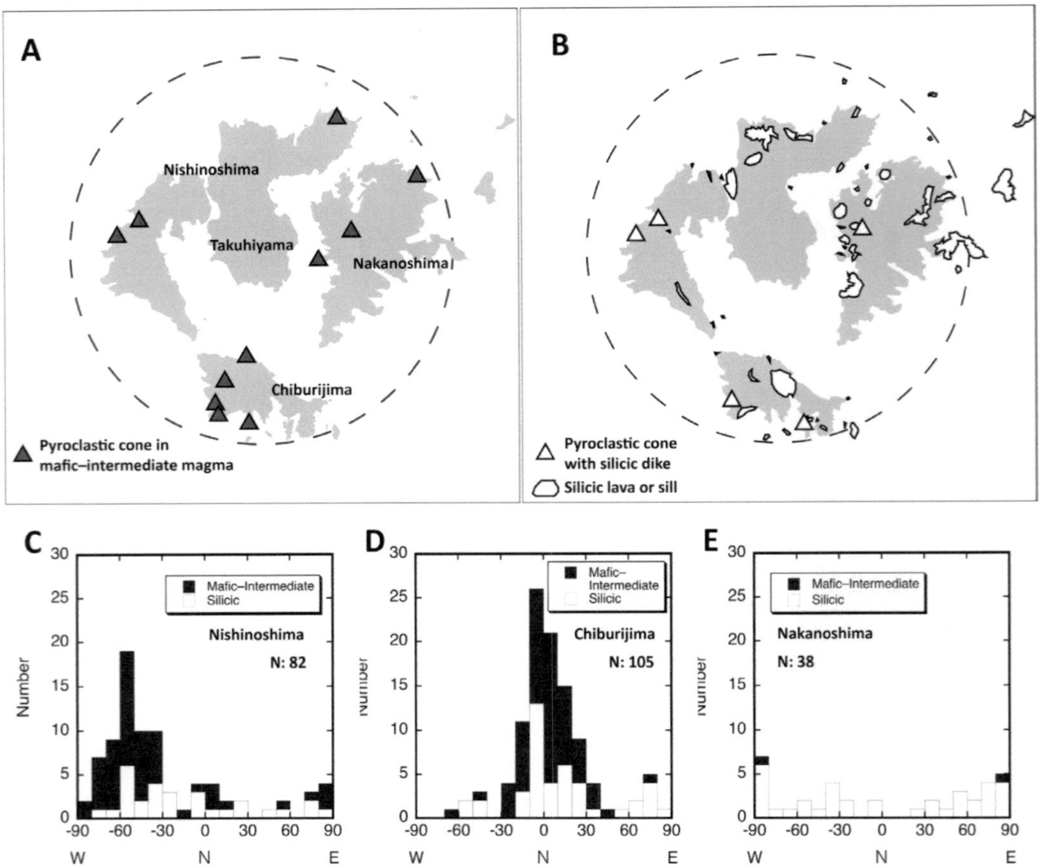

Figure 2. Overview map showing the location of magma source vents for the mafic (A) and silicic (B) magmas within the interior of the OVC, and histograms (D–F) showing the preferred orientation of the Oki-dozen dike swarm (ODS) (after Ref. [16]). These maps and diagrams indicate that the mafic magmatism in this area has NW-SE and N-S preferential orientations on Nishinoshima (D) and Chiburijima (E) islands, whereas the silicic magmatism in this area defines a radial pattern.

and laccolithic sills. The multiple mafic and silicic dike swarms of the ODS are located within the alkaline OVC and represent an ideal area for determining the processes involved in dike propagation and emplacement for the following reasons. (1) The ODS formed rapidly (within a million years), as evidenced by precise K-Ar ages of 5.9–5.5 Ma for the majority of the stratocone lavas and associated dikes and sills in this region [16]. In addition (2), the OVC is a volcanic suite that is located ~300 km from the present-day subduction trench in SW Japan, and (3) the coeval magmatism in this region formed ≥200 single, composite, and multiple dikes, pyroclastic cones, and sills [15–18]. Finally, (4) the ODS has a broad range of magma compositions with SiO_2 concentrations of 44–69 wt% [15–18].

This study presents multiscale structural data for a couple of hundred dike and sill outcrops, providing insights into the relationships between structures, stress, and dike and sill emplacement. The mafic trachybasalt and trachyandesite magmatism formed three swarms of parallel dikes oriented to NW-SE, N-S, and NE-SW. In contrast, the silicic trachytic intrusions show no preferential orientations but instead radiate from the central OVC. These differences suggest that the silicic magmas were emplaced into a volcano-related regional stress field in which the horizontal stress was similar in all directions.

2.2. Ages

Many radiometric dates have been reported for the OVC. Zircon U-Pb ages of 6.3–6 and 6.2–5.7 Ma have been reported for the subplutonic Oyama syenite and the Takuhiyama trachyte, respectively [15]; an anomalously old age of 7.4 Ma has been obtained for the Oyama syenite [18, 19]. Precise K-Ar dating of the majority of the stratocone lavas and associated ODS dikes and sills yields ages of 5.9–5.5 Ma (**Figure 3**) [16]. Another anomalously young age of ca. 2.8 Ma has been reported for a scoria cone and lava flows of the Uzuka basalt in the Nakanoshima area. These age data suggest that ODS was rapidly emplaced over <1 Myr, as evidenced by the precise K-Ar dates of 5.9–5.5 Ma outlined above [16].

The genetic order in a series of alkaline magmas was temporally and spatially coeval (**Figure 4**). For example, in **Figure 4A**, the silicic dike/sill has intruded into the mafic scoria cone, whereas the mafic trachyandesite dike has later intruded along the marginal part of the silicic dike at

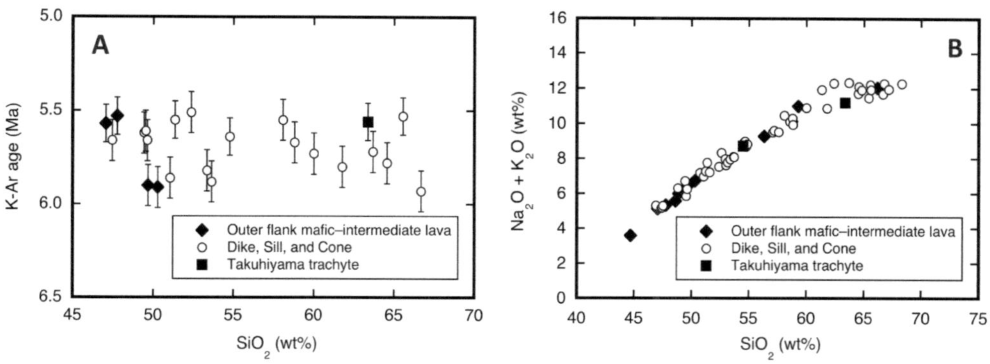

Figure 3. K-Ar age vs. SiO_2 concentration (A) and total alkali vs. silica $SiO_2 - Na_2O + K_2O$ (B) (modified after Ref. [16]) diagrams for the samples analyzed during this study.

center. This finding suggests that an arbitrary stress field, which was temporally coeval with Late Miocene tectonics, must have resulted in the variation for preferred orientations of the ODS.

2.3. Stress field

The ODS is thought to have been emplaced into a neutral or weak transpressional stress regime, primarily as the silicic intrusions of the ODS have a range of orientations (**Figure 2**) [16]. The mafic dikes that originate from the center of the OVC have at least NW-SE and N-S orientations, and additionally dikes oriented to NE-SW are presumed, although only poorly exposed sections of dike are present within the Nakanoshima area. The eastern OVC contains scoria pyroclastic cones that have a NE-SW orientation (**Figure 2A**), suggesting that the mafic conduits in this area are aligned in three different directions. These orientations may directly reflect the nature of the tectonic stress field over an area of >100 km wide [20, 23]. This contrasts with research into back-arc opening and counter compression regime in SW Japan, which identified N-S compression and related folding of Late Miocene units [22]. The presence of diverse dike orientations within the ODS is inconsistent with this interpretation of the regional tectonic

Figure 4. Examples of the relationships between pyroclastic cones and stratified lavas within the ODS. (A) A red-colored (a darker grey in greyscale) mafic scoria cone that forms a famous natural monument called Chibu-no-Sekiheki. The cone is cross-cut by trachyandesite and trachyte dikes and inclined sheets, with the thickness of a trachyte dike clearly varying as a result of host-rock stiffness, where the dike has a thickness of 4 m within stiff lavas and 17 m within soft pyroclastic ejecta [30]. (B) A spatter-rich pyroclastic cone called Byobu-iwa, where a yellow-colored (a pale in greyscale) trachytic dike-sill cross-cuts the cone. The images shown in (A) and (B) indicate that different magma conduits coalesced during the formation of the OVC. (C) A trachytic laccolith sill emplaced into stratified trachybasalt lava flows.

stress field. The magnitude and longevity of the stress field associated with a magmatic intrusion depend on the diversity in length, or area of distribution, of stress indicators such as dikes. Magma-generated stress fields are commonly short-lived [24]. In fact the ODS case, the massive syenite body aged in 6–7 Ma, acting as a core of volcanic suite, could have modified the regional stress regime from tectonics origin. This is a sense coincident with the modeling of geochemical study [15]. The model suggested that the basaltic magma injections were arrested by the syenite core in the center, whereas it could be readily reached to the outer region surfaces. Therefore, it is unclear whether the stress indicator of magma bodies in the study area reflects the size and long-term state of the tectonic stress field.

This article uses the following terminology for the size of the stress field at different scales: local stress field: 10^{-2} to 10^{3} m^{2} (i.e., equivalent to a square with sides of 0.1–31.6 m in length); regional stress field: 10^{4} to 10^{7} m^{2} (100–3162 m); and tectonic stress field: 10^{8} to 10^{10} m^{2} (10–100 km).

2.4. Dissection depth at ODS

The ODS has undergone shallow dissection, with the center of the OVC containing outcropping sedimentary basement that appears to be at present-day sea level. The volcanic edifices in the study area and the sedimentary basement of the Ichibu and Shimazushima formations have not undergone any significant tilting [18]. In addition, the presence of numerous pyroclastic cones provides evidence of the paleosurface during eruption. The topography of the study area is the highest at Nishinoshima (451 m above sea level (asl)), with other peaks at Chiburijima (325 m asl) and Nakanoshima (256 m asl). These topographic highs mark the approximate land-surface level at the time of magmatic activity. Borehole data from Nakanoshima show a basal breccia unit within the OVC at a depth of >345 m below sea level [21]. This means that the depth of these units (D) is estimated to be 250–600 m from the original surface, yielding a lithostatic pressure of 6.9×10^{6} to 1.6×10^{7} Pa for a host rock density of 2800 kg m^{-3}.

2.5. Analytical methods

We measured the orientation, thickness, offset, and overlap between dike segments and collected samples from the ODS during this study. Sample preparation and analysis were undertaken at the Central Research Institute of Electric Power Industry (CRIEPI), Abiko, Japan, with thin sections of representative samples prepared for petrographic analysis. Modal phenocryst proportions were determined using an automated point counter and ≥2500 counts per thin section. Whole-rock major element concentrations on 54 lava flow, dike, and sill samples of >2 cm in diameter (i.e., excluding fine-grained layers) were determined using X-ray fluorescence (XRF) spectrometry employing a Shimazu XRF-1500 spectrometer with a rhodium tube [25]. Prior to analysis, washed and dried rock samples were powdered using an agate mill before being fused to form glass disks for analysis. These XRF analyses were calibrated using Geological Survey of Japan JR-3, JB-2, JGb-2, JH-1, JSy-1, and JSd-2 standards. Representative geochemical data of samples from the ODS are given in **Figure 3**.

3. Characteristics of the Oki-dozen dikes

3.1. Geochemistry

Volcanic rocks of the Late Miocene OVC have a broad range of alkaline compositions and contain 44–69 wt% SiO_2 [15, 16, 18]. The ODS can be divided into three types of alkaline rocks on the basis of a total alkali-silica diagram (**Figure 3**) [26], namely trachybasalt, trachyandesite, and trachyte groups. The trachybasaltic and trachyandesitic magmas formed piles of lava flows, spatter-rich pyroclastic cones and maars, pyroclastic surges, and ballistic ejecta, all of which were associated with strombolian and phreatomagmatic eruptions. The lavas and pyroclastic deposits are crosscut by numerous mafic, intermediate, and silicic intrusions, the majority of which are subvertical dikes although some trachytic rocks occur as subhorizontal or inclined laccolithic sills. The ODS also contains a trachytic ash-flow tuff and a subplutonic syenite body, both of which crop out within a small caldera in the Takuhiyama area of the central OVC [21].

3.2. Phenocryst abundances

Phenocrysts make up <26 vol% of the samples from the study area (**Figure 5**), suggesting that the crystal volume fraction of these magmas had a negligible effect on viscosity [13, 27]. This in turn indicates that the viscosity of the ODS magmas was controlled by SiO_2 content rather than crystal abundance.

3.3. Viscosity

The viscosity of mafic-intermediate magmas (i.e., magmas containing 44.8–56.3 wt% SiO_2) within the ODS has been previously measured and estimated [28, 29], yielding viscosities of 1.90–3.00 log poise at 1300–1122°C, corresponding to 0.90–2.00 log Pa s. Assuming these magmas were Newtonian fluids, they are inferred to have had low viscosity and low internal pressure (P_{om}). This characteristic generally results in thinner dikes as a result of the low magma overpressure against the ambient stress σ_3 at the dike wall.

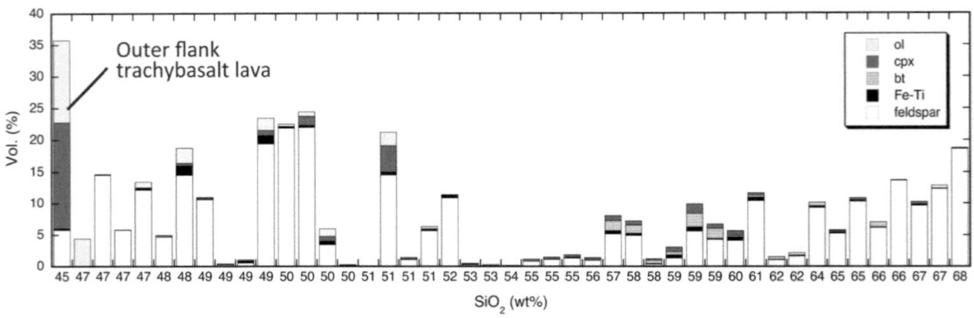

Figure 5. Variations in phenocryst type and abundance with changing SiO_2 concentration within the ODS rocks (modified after Ref. [16]). Abbreviations: ol: olivine; cpx: clinopyroxene; bt: biotite; Fe-Ti: Fe-Ti oxide minerals; feldspar: plagioclase and/or potassium feldspar.

http://dx.doi.org/10.5772/67612

4. Field observations

Extensive field measurements were made of the dikes and pyroclastic cones within the ODS, including location, orientation, and thickness measurements on more than 200 dikes and sills, their offsets, and overlaps where possible. The thickness associations with whole-rock SiO_2 concentrations have been discussed at the end of this section. The locations of pyroclastic cone were mapped (**Figure 2**) and the emplacement order of the associated dikes and sills was determined using structural relationships.

4.1. Pyroclastic cones

Typical relationships within the ODS are exemplified by pyroclastic cones and stratified lavas. One of these relationships forms the famous Chibu-no-Sekiheki natural monument, a red-colored mafic scoria cone with an interior that is crosscut by later trachyandesite and trachyte dikes and inclined sheets (**Figure 4A**). The trachyte dikes shown in **Figure 4A** vary in thickness as a result of variations in host rock stiffness, where the dike emplaced into stiff lavas is 4 m thick but bulges to a thickness of 17 m within soft pyroclastic ejecta. This difference is similar to that between feeder and nonfeeder dikes [30], although the difference observed in the study area reflects changing host rock stiffness rather than any variations in the flow of magma. The thicker trachyte dike was most likely caused by soft deformation of the host ejecta, a process that might have arrested dike ascent near the surface. The spatter-rich Byobu-iwa pyroclastic cone also contains a yellow-colored trachyte dike that was emplaced into stiff trachybasalt lavas before forming a laccolithic sill within the softer pyroclastic ejecta (**Figure 4B**). On the other hand, the outer-flank trachybasalt unit has a lot of laccolithic sills intercalated into the stratified lavas (**Figure 4C**). Those relevant trachyte forms massive intrusive bodies within the outer-flank trachybasalt lavas and are widely distributed across the OVC. The magmatic feeder system for trachyte sills in this area is evident as inclined sheets or subvertical dikes near the sill shown in **Figure 4**, indicating that many conduits containing magmas of various compositions coalesced during the formation of the OVC, regardless of their orientation.

4.2. Ideal dike segment shapes

The shape of a dike segment is a function of the physical processes involved in the intrusion of a magma as well as the propagation of a dike through the surrounding host rocks. A couple of distinctively shaped trachybasaltic dikes in the study area provide examples of different ideal dike shapes (**Figure 6**). The simplest case is exemplified by a teardrop-shaped trachybasalt dike (**Figure 6A** and **C**) that formed dike head, body, and tail segments as a result of magma flow. The bulging head of the dike formed as a result of the concentration of stress caused by excess magma pressure, contrasting with the narrow body and tail of the dike. The magma within the dike flowed from right to left in **Figure 6**, with pale orange-colored pyroclastic material filling a cavity at the tip of the head of the dike. Another example is given in **Figure 6B** and **D**, which shows a bifurcating trachybasalt dike that is adjacent to a teardrop-shaped dike. The bifurcating head of this dike splits into two orthogonal sections whereas the main body projects straight toward the narrow tail of the dike. All of these observations indicate the direction of dike propagation toward the next dike, with the magma in this image of **Figure 6B** and **D** flowing from right to left, as for the teardrop-shaped dike. The bifurcating dike itself has a dike fingertip that is continuous toward the tail of the next teardrop-shaped dike. The surrounding brecciated host rock consists of numerous clasts within matrix material formed by hydrofracturing as a result of the injection of hot magma.

4.3. Structural relationships between dike segments

Structural relationships at dike tips are an important control on the form of connections between dike segments, which are a fundamental part of dike propagation. Here, we describe well-exposed outcrops that reveal the nature of connections between trachybasalt, basaltic trachyandesite, and silicic trachyandesite dike segments.

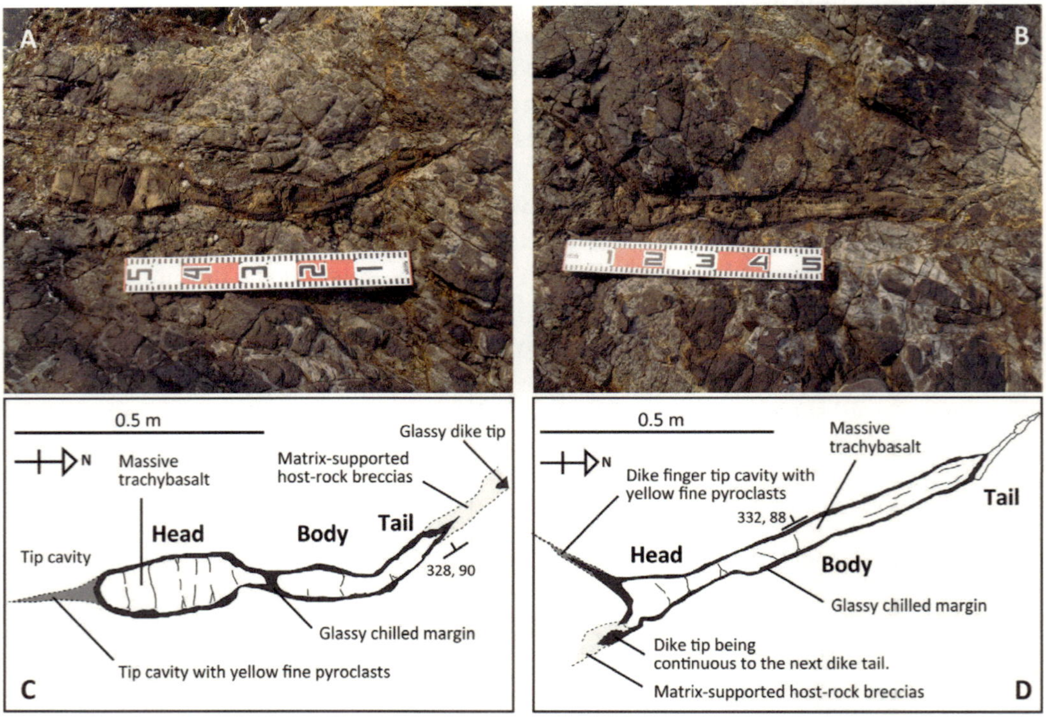

Figure 6. Examples of the ideal dike shape for trachybasalt magmas. (A) and (C) are a photograph and accompanying sketch showing a teardrop-shaped trachybasalt dike that provides evidence of a magma flow process that separated the magma into distinctive dike head, body, and tail sections. The bulging head shape formed as a result of the concentration of stress caused by excess magma pressure, with the inferred flow direction from right to left. Note the pale orange-colored (the grey in greyscale) pyroclastic material that fills the tip cavity space adjacent to the dike head. (B) and (D) are a photograph and accompanying sketch showing a bifurcating trachybasalt dike that flowed from right to left.

The coalescence of trachybasaltic dike segments in the Shiratori area of Chiburijima Island is shown in **Figure 7**, with these magmas recording the coalescence of two segments to form a hook- or pull-apart-shaped linking structure around the dike tips. The maximum thickness of the dike is 0.12 m, with offset and overlap distances between the two dike segments of 0.24 and 0.52 m, respectively. The trachybasalts in coalescing areas record the stepwise deformation of dike segments, with the host rock enclosed in these structures undergoing significant brecciation to form numerous large and small clasts and associated matrix material, most likely due to concentrated stress at the segment tip resulting in hydrofracturing. The host rock in this area is a stiff, massive, and uniform silicic laccolith, with soft-state deformation or brecciation possibly the result of episodic pressure increases caused by stress at the dike tip. This soft-state deformation caused the coalescence of the two dike segments.

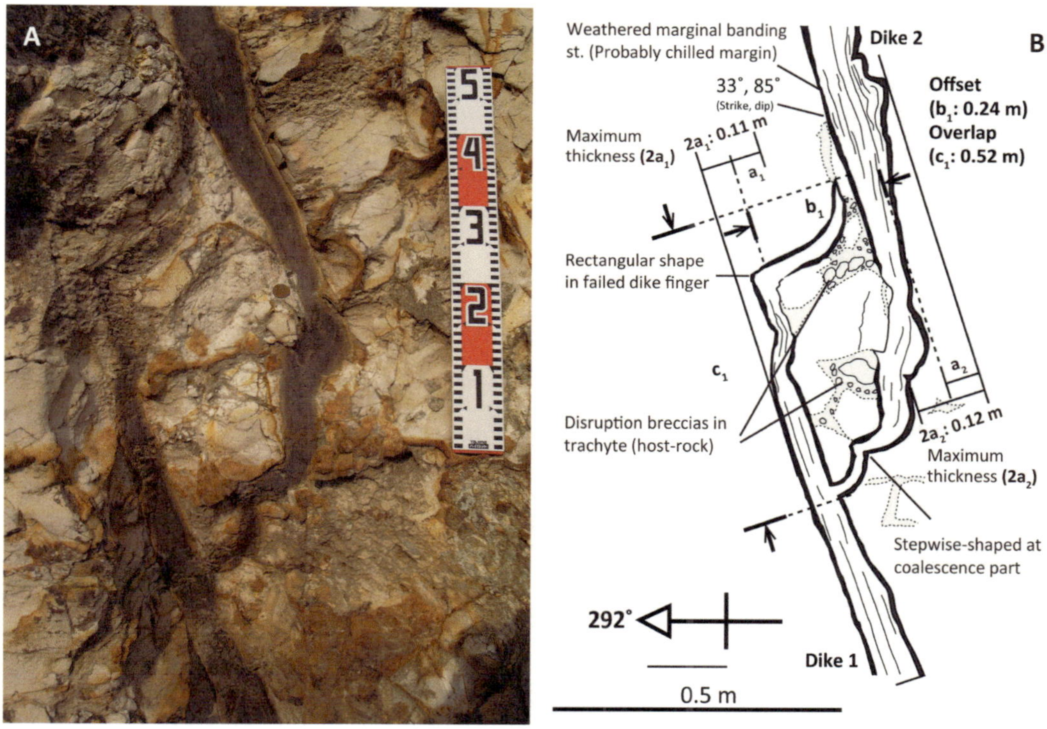

Figure 7. A hook-shaped structure linking dike segments within a trachybasalt. (A) Photograph and (B) explanatory sketch of the trachybasalt at Shiratori on Chiburi-jima island, where two segments of magma have coalesced to form a hook-shaped link (i.e., pull-apart shaped) structure around the dike tips. The host-rock enclosed in the structure is intensively brecciated and consists of numerous large and small clasts hosted by matrix material, all of which are thought to have formed by the concentration of stress at the segment tip and hydraulic fracturing (hydro-fracturing). The soft-state deformation around the dike tip induced the coalescence of the two segments. This dike is hosted by a stiff host-rock in the form of a less fractured trachyte intrusion. Abbreviations are as follows: a_1, a_2 = half maximum thickness, b_1 = offset length between dikes 1 and 2, c_1 = overlap length between dikes 1 and 2.

An offsetting of dike segments is also present within a basaltic trachyandesite dike in the study area (**Figure 8**), where an offset is present between two paper-knife-shaped dike tip segments. The hosting trachyte is brecciated and disrupted and has mingled with the dike magma, possibly related to the concentration of stress at the dike tip as well as hydrofracturing. The maximum thickness of this dike is 1.17 m, with offset and overlap distances between the two dike segments of 0.65 and 0.41 m, respectively. The glassy chilled margin of the dike contains a gently plunging lineation that suggests the magma in this area flowed horizontally.

Another offset relationship is visible within a silicic trachyandesite dike (**Figure 9**). This thick dike contains an offset in the form of a separate finger and formerly adjoined tail within two adjacent dike segments. The host rock is a stiff laccolith, and only weakly fractured trachybasalt lava and the dike have a maximum thickness of 1.33 m, with offset and overlap distances between the two tips of 1.96 and 1.18 m, respectively. These distances are larger

Figure 8. Basaltic trachyandesite dike segment offsets shown in a photograph (A) and accompanying explanatory sketch (B) of an outcrop at Shiratori on Chiburi-jima Island. Here, the magmas have been offset across two paper-knife-shaped tips located in different dike segments. The location contains brecciated, disrupted, and mingled trachyte host-rock and trachyandesite dike material. The surface of the glassy chilled margin in this area shows a gently plunging lineation, suggesting the magma in this area moved horizontally. Abbreviations are as follows: a_3, a_4 = half maximum thickness, b_2 = offset length between dikes 3 and 4, c_2 = overlap length between dikes 3 and 4.

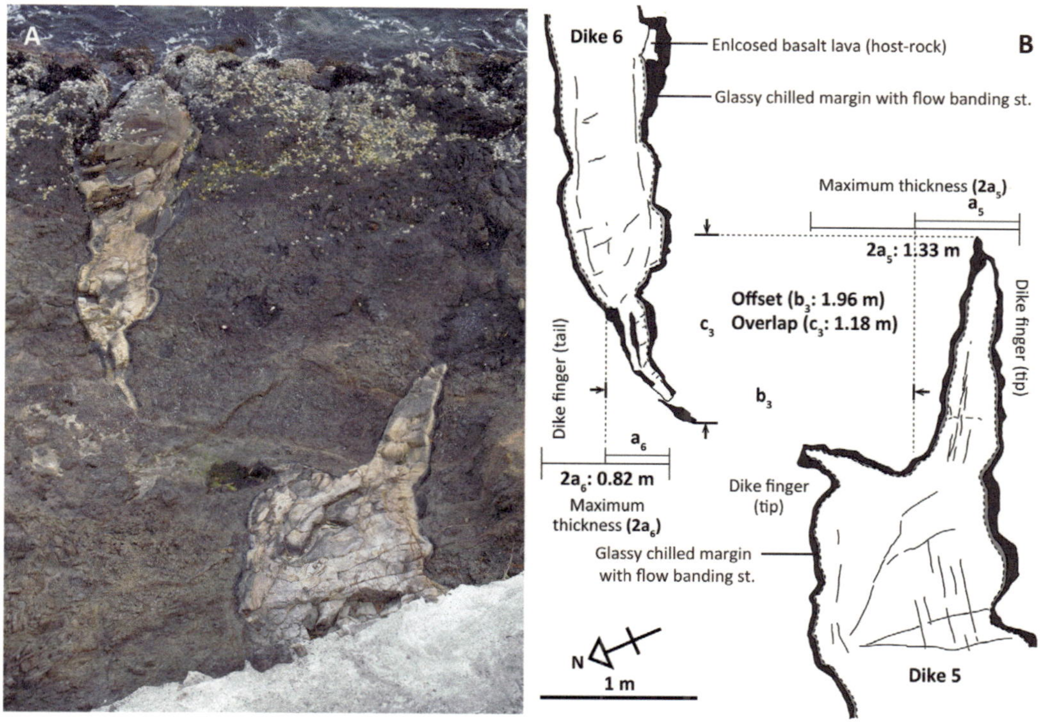

Figure 9. Offset relationship within silicic trachyandesite dike segments shown in a photograph (A) and an accompanying explanatory sketch (B) of an outcrop at Shimanehana on Nishinoshima Island. The thick trachyandesite dike is offset across the tips of the dikes visible in the figures. This dike was emplaced into a stiff and less-fractured trachybasalt lava. Abbreviations are as follows: a_5, a_6 = half of maximum thickness, b_3 = offset length between dikes 5 and 6, c_3 = overlap length between dikes 5 and 6.

than those determined for the trachybasalt and basaltic trachyandesite dikes discussed above (**Figures 7** and **8**), suggesting that the silicic trachyandesite dike experienced a greater concentration of stress than the other two dikes.

4.4. Relationship between critical dike thickness and SiO_2 concentration

The maximum thicknesses of the 37 ODS dikes were measured to determine a critical thickness value that correlates with SiO_2 concentration (**Figure 10A**). All of the dike sections measured during this study were hosted by stiff host lavas rather than soft pyroclastic ejecta. The relationship between dike thickness and SiO_2 concentrations yields a critical thickness (T_{cr}) threshold line that can be approximately modeled using $T_{cr} = 0.2 \times (SiO_2 - 40)$. This suggests that ODS dikes within stiff host rocks containing 75 and 80 wt% SiO_2 have maximum thicknesses of 7 and 8 m, respectively, contrasting with a trachytic dike emplaced into a pyroclastic cone that has a thickness of 17 m (**Figure 4**). The emplacement of this dike into the cone may have resulted in the soft-state deformation of the hosting ejecta, thereby arresting the ascent of the trachyte dike near the surface.

5. Discussion

5.1. Relationships among dike thickness, dike offset, and the overlap of dike segments

The ideal dike shape, dike tip relationships, dike thicknesses, dike offsets, and dike overlaps all provide information on the processes that operate during the propagation of a single dike in a given direction. A dike has a shape defined by its propagation direction and is subdivided into head, body, and tail segments (**Figure 6**). The dike head and tip bulge as a result of the concentration of stress (**Figure 6A** and **C**), a process that leads to dike propagation (**Figure 6B** and **D**). Offset and overlap structures form as a result of the magnitude of stress in a given part of the dike, where the concentrated stress is usually four times larger than the dike magma overpressure [2]. This means that the direction of dike propagation can change in areas with concentrated stress, although propagation outside of these areas is generally controlled by local and regional stress fields.

The relationship between offset (b) and overlap (c) yields a linear curve with a relatively gentle gradient $c = 0.7b$ (**Figure 10B**), whereas areas beyond $b = 0.6$ show a thickness ($2a$) to offset (b) relationship that yields a gentle curve, $a = 0.06b$. This reflects a difference in stress concentration between magma overpressure ($2a$) and the pressure at the dike tip (b), with the concentrated pressure at the dike tip drastically increasing with only a minor increase in magma overpressure.

5.2. Magma overpressures within mafic and silicic dikes

Dike magma overpressure can be estimated using fracture mechanics such as hydrofracturing [2], a process that is an important control on dike propagation and the formation of a dike

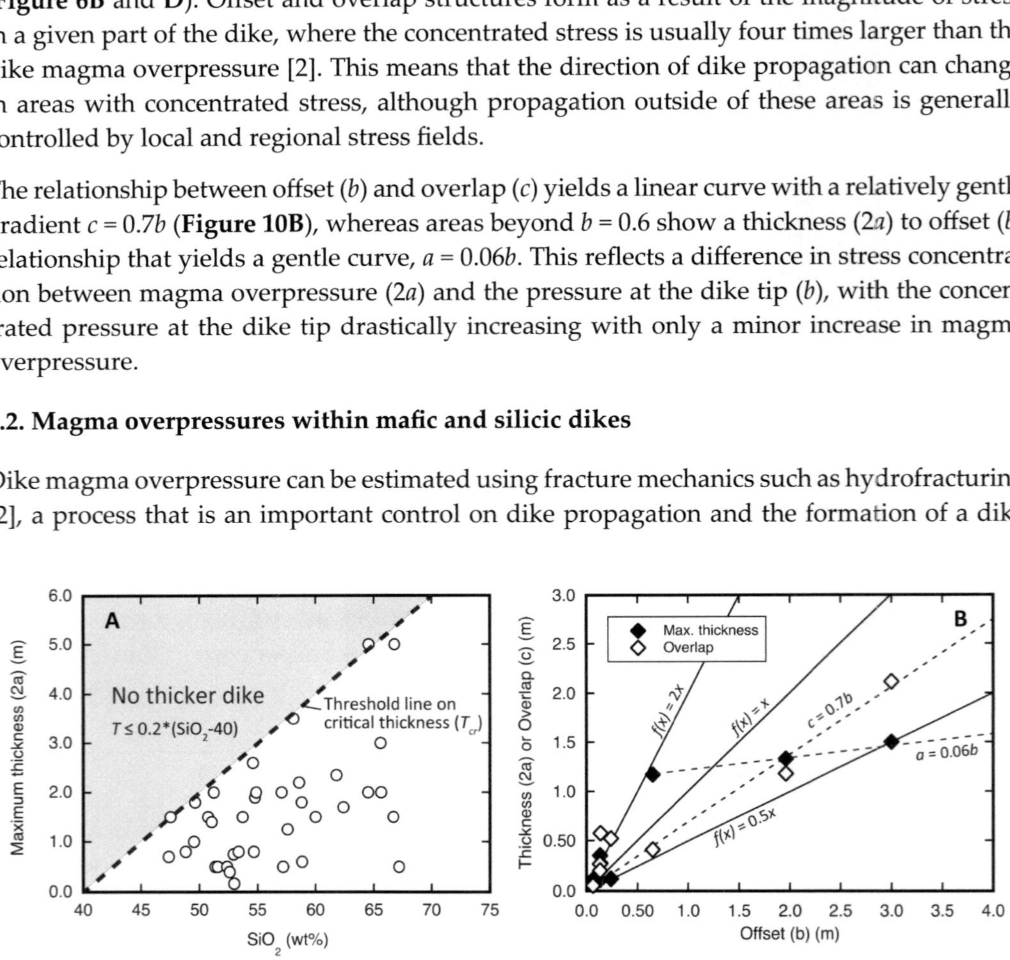

Figure 10. Relationship between SiO_2 concentration and maximum dike thickness (A), and offset distances to either overlap length or maximum dike thickness (B). (A) shows a threshold critical thickness line (T_{cr}) for the ODS magmas, where an empirical curve fit to T_{cr} is expressed as $T_{cr} = 0.2 \times (SiO_2 - 40)$. Assuming a thickness/length ratio of 1000 [2], we obtain dike lengths from 1.0×10^2 to 5.0×10^3 m, with these lengths relating to the regional stress field. (B) indicates that the offset (b) to overlap (c) relationship is roughly proportional to a linear curve fit where $c = 0.7b$, whereas the maximum thickness ($2a$) forms two distinct regions within a curved slope, with $0.6 \geq b > 0$ yielding a steep curve for $a \sim b$ and $b > 0.6$ yielding a gently sloping curve where $a \sim 0.06b$.

swarm. Generating a vertical hydrofracture from a source chamber at a given depth below the original surface involves the rupturing of a reservoir, with hydrofracturing initiating when the following condition is satisfied [2]:

$$p_l + p_e = \sigma_3 + T_0, \tag{1}$$

where p_l is lithostatic stress at the rupture site, $p_e = p_t - p_l$ is the difference between total fluid pressure p_t in the reservoir and the lithostatic stress p_l at the time of reservoir rupture, σ_3 is the minimum compressive principle stress, and T_0 is the local in situ tensile strength at the rupture site. The propagation of hydrofracturing leads to the magma overpressure p_o becoming

$$p_o = p_e + (\rho_r - \rho_m) gh + \sigma_d, \tag{2}$$

where p_e is the fluid excess pressure at the source, ρ_r is host rock density, ρ_m is magma density, g is acceleration due to gravity, h is the height of the given part of the hydrofracture, and σ_d is the differential stress at the level where the hydrofracture is being examined [2]. Hydrofracturing initiates when p_e becomes equal to the tensile strength T_0, which is usually T_0 = 0.5–6 MPa, yielding an average value of 3.25 MPa. Here, we simplify this problem by assuming that the stress difference σ_d is also equal to the tensile strength of the host rock, T_0. The host rock is the stiff trachybasalt lava, meaning that the average host rock density ρ_r and the average trachyte magma density ρ_m are 2800 and 2350 kg m^{-3}, respectively.

Clinopyroxene thermobarometric analysis of the OVC rocks yielded pressures of formation of 40–270 MPa for trachyte magma, a value that is consistent with the results of amphibole barometry for these units (<50 MPa) [15]. This means that the trachyte formed in a magma chamber at a pressure of 40–50 MPa, roughly equivalent to a depth of 1.5–2 km. The magma overpressure p_{os} of the trachyte magma can be given by

$$p_{os} = 3.25 \times 10^6\,\text{Pa} + \left(2800\,\text{kg m}^{-3} - 2350\,\text{kg m}^{-3}\right) \times 9.8\,\text{m s}^{-2} \times 2000\,\text{m} + 3.25 \times 10^6\,\text{Pa} = 1.53 \times 10^7\,\text{Pa} = 15.3\,\text{MPa}. \tag{3}$$

Similarly, the magma overpressure of the trachybasalt magma p_{om} is given by

$$p_{om} = 6.5 \times 10^6\,\text{Pa} = 6.5\,\text{MPa}. \tag{4}$$

These results suggest that the trachybasalt magma reflects the ambient state of the regional stress, whereas the trachyte magma overpressure may override the regional stress field. We assume a dike length (along strike) to thickness ratio of 1000 [2] or 1500 [31], which are common values obtained from dike measurements worldwide [31]. A single dike has a length of several kilometers, a scale that is influenced by the regional stress field. The presence of regional horizontal differential stress ($\sigma_{Hmax} - \sigma_{Hmin}$) at a shallow level may cause the trachybasalt dike to propagate with a preferred orientation aligned with the orientation of the regional σ_{Hmax}. This is consistent with the presence of three different orientations of mafic dike swarms in the study area, each of which is different from the others and corresponds with the size of the islands in the study area. Fracture mechanics also suggest that the trachyte magma has more than double the magma overpressure p_{os} than the trachybasalt magma p_{om}. The magma overpressure and dike tip stress associated with silicic magmas are large enough to counteract the pressure, in which the dike follows to propagate in a preferred orientation governed by the 6.9–16 MPa regional stress field (see Section 2.4).

5.3. Implications for volcano-tectonic stress orientation

The present results suggest that the orientation of intrusions within the ODS reflects the relationship between dike tip pressure and magma overpressure and the surrounding regional stress field. Dike tip pressure and magma overpressure both positively correlate with the SiO_2 concentration in magma. Fracture mechanics suggests that these characteristics are determined by a balance between magma and host rock densities and magma chamber depth, all of which decrease with increasing SiO_2. The development of a dike swarm also requires similar processes to be repeated in such a way to form a regional-scale preferred orientation defined by the intrusion of multiple dikes.

The preferred orientation of dikes and sills within the ODS correlates with magma composition, as reflected by SiO_2 concentrations. This clear difference in preferred orientation represents a difference in magma overpressure between the mafic and silicic magmas. This in turn means that a comparative analysis of stress fields at different scales could provide insights into the orientation of volcano-tectonic stress in an area, indicating that dike orientations may not fully reflect the tectonic stress field of an area during emplacement.

6. Conclusions

A detailed field, geochronological, and geochemical study of the Late Miocene Oki-dozen dike swarm (ODS) of southwest Japan provided the following insights into the effects of the regional stress field on volcano-tectonic orientations.

Swarms of parallel trachybasalt and trachyandesite dikes in the study area can be grouped by orientation to NW-SE, N-S, and NE-SW sets. In addition, more evolved trachytic magmas form dikes and sills that radiate from the central Oki-dozen volcanic complex, suggesting that these more silicic magmas were emplaced when the magnitude of horizontal stress was similar in all directions.

The maximum thickness of 37 dikes was compared with their SiO_2 concentrations, yielding a critical thickness (T_{cr}) relationship that can be approximately expressed as $T_{cr} = 0.2 \times (SiO_2 - 40)$. Our data suggest that the orientation of intrusions within the ODS is a function of dike tip pressure and magma overpressure, both of which vary with SiO_2 concentration. These characteristics are in turn determined by magma and host rock densities and the depth of the parental magma chamber.

The preferred orientation of dikes and sills within the ODS is related to the SiO_2 content of the magma, with mafic magmas being intruded along three locally parallel preferential orientations, whereas the silicic dikes and sills form a radial arrangement. This clear difference in preferred orientation is the result of differences in magma overpressure between the mafic and silicic magmas, where the estimated magma overpressure of ~15 MPa and the dike tip pressure of ~60 MPa within the silicic magmas were large enough to counteract the effects of the 6.9–16 MPa regional stress. In contrast, the mafic dike swarm is aligned as a result of the effect of the regional stress field, albeit with small amounts of differential stress. The development

http://dx.doi.org/10.5772/67612

of a regional-scale dike swarm requires these processes to be repeated to ensure the dikes of the swarm have a common orientation. As such, the comparative analysis of different scales of stress fields would provide significant insights into the orientation of regional volcano-tectonic stress.

Acknowledgements

We thank Prof. Y. Itoh for providing the opportunity to undertake this research and for encouragement in preparing the manuscript. DM, KT, KA, and TW also thank T. Chiba for assistance during fieldwork.

Author details

Daisuke Miura[1*], Kiyoshi Toshida[1], Ken-ichi Arai[2], Takeshi Wachi[2†] and Yutaka Wada[3]

*Address all correspondence to: dmiura@criepi.denken.or.jp

1 Geosphere Sciences, Central Research Institute of Electric Power Industry, Abiko, Japan

2 Asia Air Survey, Kawasaki, Japan

3 Nara University of Education, Nara, Japan

† Present address: In Situ Solutions, Tokyo, Japan

References

[1] Rubin AM. Propagation of magma-filled cracks. Annual Review of Earth and Planetary sciences. 1995;23:287-336. doi:10.1146/annurev.ea.23.050195.001443

[2] Gudmundsson A. Rock fractures in geological processes: Cambridge University Press. Cambridge, UK. 2011. 578 p. doi:10.1017/CBO9780511975684

[3] Takada A. The influence of regional stress and magmatic input on styles of monogenetic and polygenetic volcanism. Journal of Geophysical Research. 1994;99:13563-13573. doi:10.1029/94JB00494

[4] Seaman SJ, Scherer EE, Standish J. Multistage magma mingling and the origin of flow banding in the Aliso lava dome, Tumacacori Mountains, southern Arizona. Journal of Geophysical Research. 1995;100:8381-8398. doi:10.1029/94JB03260

[5] Tuffen H, Dingwell D: Fault textures in volcanic conduits: evidence for seismic trigger mechanisms during silicic eruptions. Bulletin of Volcanology. 2005;67:370-387. doi:10.1007/s00445-004-0383-5

[6] Pallister JS, Cashman KV, Hagstrum JT, Beeler NM, Moran SC, Denlinger RP. Faulting within the Mount St. Helens conduit and implications for volcanic earthquakes. Geological Society of America Bulletin. 2013;125:359-376. doi:10.1130/B30716.1

[7] Cashman KV, Sparks RSJ. How volcanoes work: A 25 year perspective. Geological Society of America Bulletin. 2013;125:664-690. doi:10.1130/B30720.1

[8] Cruden AR. Flow and fabric development during the diacritic rise of magma. Journal of Geology. 1990;98:681-698. doi:10.1086/629433

[9] Webb SL, Dingwell DB. Non-Newtonian rheology of igneous melts at high stresses and strain rates: Experimental results for rhyolite, andesite, basalt, and nephelinite. Journal of Geophysical Research. 1990;95:15695-15701. doi:10.1029/JB095iB10p15695

[10] Cashman K, Blundy J. Degassing and crystallization of ascending andesite and dacite. Philosophical Transactions of the Royal Society of London A, Mathematical, Physical, and Engineering sciences. 2000;358:1487-1513. doi:10.1098/rsta.2000.0600

[11] Blundy J, Cashman K. Ascent-driven crystallisation of dacite magmas at Mount St Helens, 1980-1986. Contributions to Mineralogy and Petrology. 2001;140:631-650. doi:10.1007/s004100000219

[12] Takeuchi S. Preeruptive magma viscosity: An important measure of magma eruptibility. Journal of Geophysical Research. 2011;106:B10201. doi:10.1029/2011JB008243

[13] Marsh BD. On the crystallinity, probability of occurrence, and rheology of lava and magma. Contributions to Mineralogy and Petrology. 1981;78:85-98. doi:10.1007/BF00371146

[14] Scaillet B, Holtz F, Pichavant M. Phase equilibrium constraints on the viscosity of silicic magmas. 1. Volcanic-plutonic comparison. Journal of Geophysical Research. 1998;103: 27257-27266. doi:10.1029/98JB02469

[15] Brenna M, Nakada S, Miura D, Toshida K, Ito H, Hokanishi N, Nakai S. A trachyte–syenite core within a basaltic nest: filtering of primitive injections by a multi-stage magma plumbing system (Oki-Dozen, south-west Japan). Contributions to Mineralogy and Petrology. 2015;170:22. doi:10.1007/s00410-015-1181-0

[16] Toshida K, Miura D, Hataya R. Proposition of assessment method for lateral magma migration – Effects of magma geochemistry studied from the distribution of conduits at Oki Dozen volcano– (in Japanese with English abstract). CRIEPI Report. 2006; N05026;24 p.

[17] Morris PA, Itaya T, Iizumi S, Kagami H, Watling RJ, Murakami H. Age relations and petrology of alkalic igneous rocks from Oki Dozen, Southwest Japan. Geochemical Journal. 1997;31:135-154. doi:10.2343/geochemj.31.135

[18] Tiba T, Kaneko N, Kano K. Geology of the Urago District—with 1:50,000 geological sheet map (in Japanese with English abstract). Geological Survey of Japan; 2000. 74 p.

[19] Kaneko N, Tiba T. Occurrence and K–Ar age of an alkali olivine basalt from Nakanoshima in Oki-Dozen, Shimane Prefecture, Southwest Japan (in Japanese with English abstract). The Journal of the Geological Society of Japan. 1998;104:419-422. doi:10.5575/geosoc.104.419

[20] Wada Y, Itaya T, Ui T. K–Ar ages of Oki-Dozen and Tango dike Swarms, Western Honshu, Japan (in Japanese with English abstract). Bulletin of Volcanological Society of Japan. 1990;35:217-229.

[21] Kano K, Kaneko N, Ishizuka O, Tiba T, Yanagisawa Y. The beginning and lifetime of Dozen volcano, Oki Islands, SW Japan (in Japanese with English abstract). Bulletin of the Volcanological Society of Japan. 2014;59:77-88.

[22] Itoh Y, Nakajima T, Takemura A. Neogene deformation of the back-arc shelf of Southwest Japan and its impact on the palaeoenvironments of the Japan Sea. Tectonophysics. 1997;281:71-82. doi:10.1016/S0040-1951(97)00159-5

[23] Nakamura K. Volcanoes as possible indicators of tectonic stress orientation—principle and proposal. Journal of Volcanology and Geothermal Research. 1977;2:1-16. doi:10.1016/0377-0273(77)90012-9

[24] Mogi K. Relations between the eruptions of various volcanoes and the deformations of the ground surfaces around them. Bulletin of the Earthquake Research Institute, University of Tokyo. 1958;36:99-134.

[25] Nakata E. The proposal of chemical analysis method of sedimentary rocks by X ray fluorescence – The combination analysis of pressed powder method and glass bead method (in Japanese with English abstract). CRIEPI Report. 2006;N05063;24 p.

[26] Le Bas MJ, Le Maitre RW, Streckeisen A, Zanettin BA. Chemical classification of volcanic rocks based on the total alkali–silica diagram. Journal of Petrology. 1986;27:745-750. doi:10.1093/petrology/27.3.745

[27] Pinkerton H, Steavenson RJ. Methods of determining the rheological properties of magmas at sub-liquidus temperatures. Journal of Volcanology and Geothermal Research. 1992;53:47-66. doi:10.1016/0377-0273(92)90073-M

[28] Taniguchi H. On the volume dependence of viscosity of some magmatic silicate melts. Mineralogy and Petrology. 1993;49:13-25. doi:10.1007/BF01162923

[29] Wada Y. On the relationship between dyke width and magma viscosity. Journal of Geophysical Research. 1994;99:17743-17755. doi:10.1029/94JB00929

[30] Geshi N, Kusumoto S, Gudmundsson A. Geometric difference between non-feeders and feeder dykes. Geology. 2010;38:195-198. doi:10.1130/G30350.1

[31] Browning J, Drymoni K, Gudmundsson A. Forecasting magma-chamber rupture at Santorini volcano, Greece. Scientific Reports. 2015;5:15785. doi:10.1038/srep15785.

Chapter 10

Effectiveness for Determination of Depositional Age by Detrital Zircon U–Pb Age in the Cretaceous Shimanto Accretionary Complex of Japan

Tetsuya Tokiwa, Makoto Takeuchi, Yusuke Shimura, Kazuho Shobu, Akari Ota, Koshi Yamamoto and Hiroshi Mori

Additional information is available at the end of the chapter

http://dx.doi.org/10.5772/67982

Abstract

Detrital zircon U–Pb ages indicate the crystallization age. Therefore, it is necessary to evaluate the effectiveness of determining the age of deposition using zircon age data. We carried out U–Pb dating of detrital zircons from sandstone at eight sites in the Cretaceous Shimanto accretionary complex on Kii Peninsula, Japan, with the aim of evaluating the accuracy of U–Pb zircon ages as indicators of the depositional age of sedimentary rocks by comparing zircon ages with radiolarian ages. Our results reveal zircons of late Cretaceous age, and the youngest peak ages are in good agreement with depositional ages inferred from radiolarian fossils. In addition, the youngest peak ages become younger as tectono-structurally downwards, and this tendency is clearer for the zircon ages than for the radiolarian ages. These results indicate that newly crystalized zircons were continuously supplied to the sediment by constant igneous activity during the late Cretaceous and that zircon ages provide remarkably useful information for determining the age of deposition in the Cretaceous Shimanto accretionary complex.

Keywords: detrital zircon, U–Pb age, accretionary complex, Shimanto, Cretaceous

1. Introduction

Determination of the depositional age of sedimentary rocks is essential for understanding tectonic processes, and microfossils have generally been used for this purpose. In particular, radiolarian ages have been played an important role in understanding accretionary tectonics, because the order of accretion of strata is inferred from radiolarian ages (for example, see Refs. [1, 2]). However, it is very difficult to determine the depositional age of coarse-grained sedimentary rocks such as sandstone and conglomerate, because radiolarian fossils are rare in such rocks, whereas they are commonly found in fine-grained rocks such as mudstone and chert. In addition, it is impossible to identify the species of radiolarian fossils in metamorphosed sedimentary rocks.

Recent progress in analytical techniques has enabled rapid and accurate U–Pb isotopic age determination of zircons using inductively coupled plasma–mass spectrometry with laser ablation sampling (LA–ICP–MS) (for example, see Refs. [3, 4]). This method has been widely used to determine zircon ages in coarse-grained and weakly metamorphosed sedimentary rock (for example, see Refs. [5–7]). However, zircon ages do not directly indicate the depositional age; instead, they indicate the crystallization age. Thus, it is possible that a large age gap exists between the depositional ages and zircon ages. Therefore, to determine the depositional age from detrital zircon U–Pb ages, it is necessary to evaluate the effectiveness of this method; e.g., by comparing detrital zircon ages with fossil ages.

In this study, we performed U–Pb dating on detrital zircons from sandstone in the Cretaceous Shimanto accretionary complex on Kii Peninsula, Japan, and we clarify the relationship between the detrital zircon ages and radiolarian ages in order to develop the zircon dating method to obtain accurate depositional ages.

2. Geological setting

The Shimanto accretionary complex is a well-studied ancient accretionary complex exposed along the Pacific side of Southwest Japan (for example, see Ref. [1]). The Shimanto accretionary complex in the Kii Peninsula is divided into the Kouyasan Sub-belt (Cretaceous), the Hidakagawa Sub-belt (Cretaceous) and the Hikigawa Sub-belt (Paleogene) (**Figure 1**; [8]). This study considers the Yukawa complex (Albian–Cenomanian), Miyama complex (Turonian–Campanian) and Ryujin complex (Campanian–Maastrichtian) in the Hidakagawa Sub-belt. These complexes are further subdivided into several units, and these complexes and units are in thrust contact with each other [8]. The unit names are shown in **Figure 2**. Many radiolarian studies have been carried out in this study area [9–17], and the results indicate that the depositional ages have a tendency to become younger as tectono-structurally downwards (north to south).

3. Samples

Sandstone samples for U–Pb dating were collected from eight sites. The sites were named Site 1 to Site 8 from tectono-structurally upwards to downwards (north to south), and the relationship between sampling sites and geological units is shown in **Figure 2**.

Site 1 (sample no. 150503-06): This sample was collected from a medium-grained massive sandstone of the Yanase unit (Yu1), Yukawa complex (GPS: N34°6′13.28″, E135°30′32.82″) (**Figures 3a** and **4a**). The sandstone is composed of quartz (23.3%), plagioclase (27.3%), K-feldspar (8.6%), rock fragments (15.9%), matrix (17.1%) and the others (7.8%) (**Figure 5**).

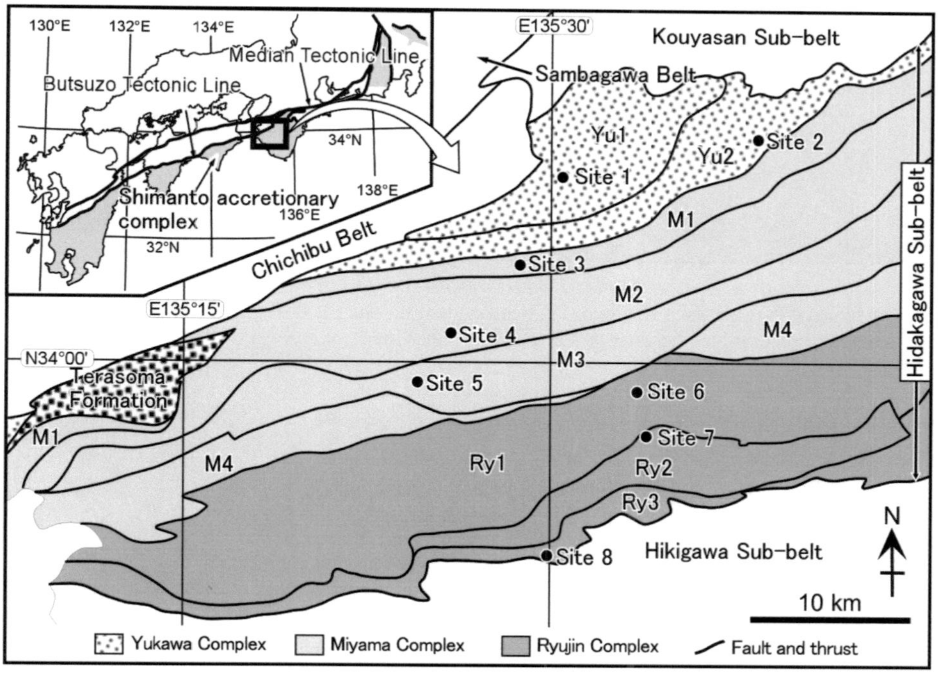

Figure 1. Geological distribution of the Shimanto accretionary complex, Kii Peninsula, southwest Japan. (Modified form Kishu Shimanto Research Group [8].) Sampling sites for zircon U–Pb dating are shown in this figure. Yu1: Yanase Unit, Yu2: Kitamata Unit, M1: Chikai Unit, M2: Gomadanzan Unit, M3: Hattomaki Unit, M4: Ubuyukawa Unit, Ry1: Sohgawa Unit, Ry2: Komatagawa Unit, and Ry3: Yunohara Unit.

Tectono-stratigraphic division			Sampling site	Sample no.
Hidakagawa Sub-belt	Yukawa Complex	Yanase Unit (Yu1)	Site 1	150503-06
		Kitamata Unit (Yu2)	Site 2	151014-02
	Miyama Complex	Chikai Unit (M1)	Site 3	160705-06
		Gomadanzan Unit (M2)	Site 4	160705-02
		Hattomaki Unit (M3)	Site 5	160705-01
		Ubuyukawa Unit (M4)	–	–
	Ryujin Complex	Sohgawa Unit (Ry1)	Site 6	160704-05
		Komatagawa Unit (Ry2)	Site 7	160704-04
		Yunohara Unit (Ry3)	Site 8	160704-03

Tectono-structurally upward
northward
southward
Tectono-structurally downward

Figure 2. Relationship between tectono-structural division and sampling site. (Division is modified form the Kishu Shimanto Research Group [8]).

Site 2 (sample no. 151014-02): This sample was collected from a medium-grained massive sandstone of the Kitamata unit (Yu2), Yukawa complex (GPS: N34°7′32.36″, E135°38′32.48″).

Figure 3. Photographs showing the occurrence of the sandstone at the Site 1 in the Yu1 of the Yukawa Complex (a), Site 3 in the M1 of the Miyama Complex (b), and Site 7 in the Ry2 of the Ryujin Complex (c).

Figure 4. Photomicrograph of the sandstone at the Site 1 in the Yu1 of the Yukawa Complex (a), Site 3 in the M1 of the Miyama Complex (b), and Site 7 in the Ry2 of the Ryujin Complex (c). Sandstone under crossed polarized light.

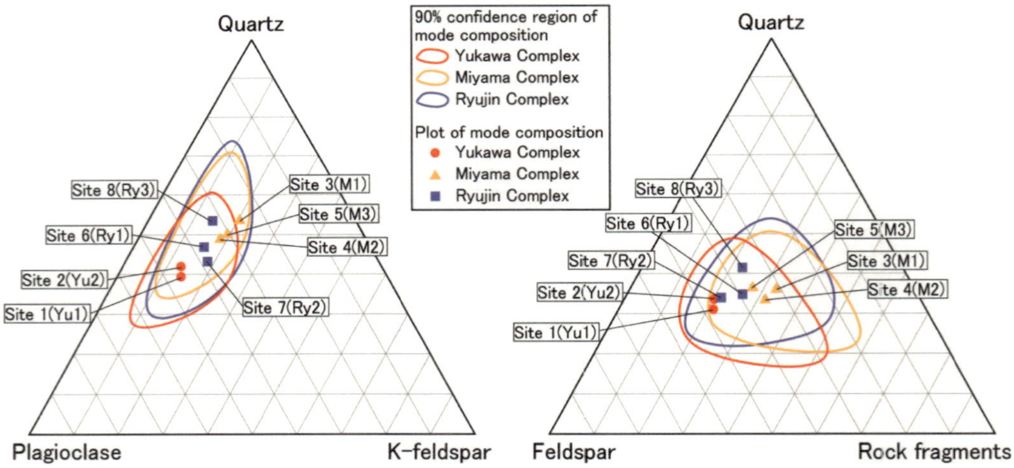

Figure 5. Modal composition of sandstones in the each site. Closed lines show the confidence regions of 90% in each of the complexes. (Modified form Kumon et al. [18].)

The sandstone is composed of quartz (29.9%), plagioclase (32.1%), K-feldspar (9.4%), rock fragments (17.9%), matrix (10.3%) and the others (0.4%).

Site 3 (sample no. 160705-06): This sample was collected from a medium-grained massive sandstone of the Chikai unit (M1), Miyama complex (GPS: N34°3′22.37″, E135°28′47.3″)

(**Figures 3b** and **4b**). The sandstone is composed of quartz (29.1%), plagioclase (13.9%), K-feldspar (11.0%), rock fragments (26.2%), matrix (16.5%) and the others (3.3%).

Site 4 (sample no.160705-02): This sample was collected from a medium-grained massive sandstone of the Gomadanzan unit (M2), Miyama complex (GPS: N34°1′8.81″, E135°25′41.19″). The sandstone is composed of quartz (25.3%), plagioclase (16.7%), K-feldspar (9.4%), rock fragments (23.6%), matrix (23.2%) and the others (1.8%).

Site 5 (sample no.160705-01): This sample was collected from a medium-grained massive sandstone in the Hattomaki unit of (M3), Miyama complex (GPS: N33°59′4.15″, E135°24′40.99″). The sandstone is composed of quartz (28.9%), plagioclase (17.3%), K-feldspar (11.1%), rock fragments (21.3%), matrix (18.2%) and the others (3.2%).

Site 6 (sample no.160704-05): This sample was collected from a medium-grained sandstone of the alternating beds of sandstone and mudstone in the Sohgawa unit (Ry1), Ryujin complex (GPS: N33°59′3.65″, E135°33′23.67″). The sandstone is composed of quartz (25.0%), plagioclase (19.9%), K-feldspar (8.5%), rock fragments (18.6%), matrix (19.9%) and the others (8.1%).

Site 7 (sample no. 160704-04): This sample was collected from a medium-grained sandstone of the alternating beds of sandstone and mudstone in the Komatagawa unit (Ry2), Ryujin complex (GPS: N33°57′35.99″, E135°33′50.36″) (**Figures 3c** and **4c**). The sandstone is composed of quartz (25.9%), plagioclase (23.0%), K-feldspar (11.1%), rock fragments (16.5%), matrix (19.3%) and the others (4.2%).

Site 8 (sample no. 160704-03): This sample was collected from a medium-grained sandstone of the alternating beds of sandstone and mudstone in the Yunohara unit (Ry3), Ryujin complex (GPS: N33°53′42.86″, E135°29′40.63″). The sandstone is composed of quartz (29.7%), plagioclase (17.8%), K-feldspar (8.1%), rock fragments (16.1%), matrix (19.5%) and the others (8.8%).

On ternary composition diagrams, the compositions of all the sandstones samples, within the 90% confidence region of the modal composition of each complex (**Figure 5**; [18]).

4. Analytical techniques

Zircons were separated by conventional techniques including crushing, sieving, water-based panning and magnetic separation. Zircons were randomly handpicked under a stereoscopic microscope, mounted in epoxy resin on a glass slide, and polished to approximately half of their original thickness. In order to investigate the internal structure and zonation patterns of zircons, cathodoluminescence (CL) images were obtained using a scanning electron microscope (SEM) JSM-6510 (JOEL Ltd) at Shinshu University of Japan. Target spots for U–Pb dating analyses were identified from the CL images (**Figure 6**).

The U–Pb zircon dating analyses carried out using a laser ablation–inductively coupled plasma mass spectrometer (LA–ICP–MS) at Nagoya University of Japan. The ICP–MS part is Agilent 7700x (Agilent Technologies), and the LA part is NWR213 (Electro Scientific Industries). Nancy zircon 91500 [19] was utilized for normalization of NIST SRM 610, and the latter was used as the external standard for age determinations. Analyses were carried out with an

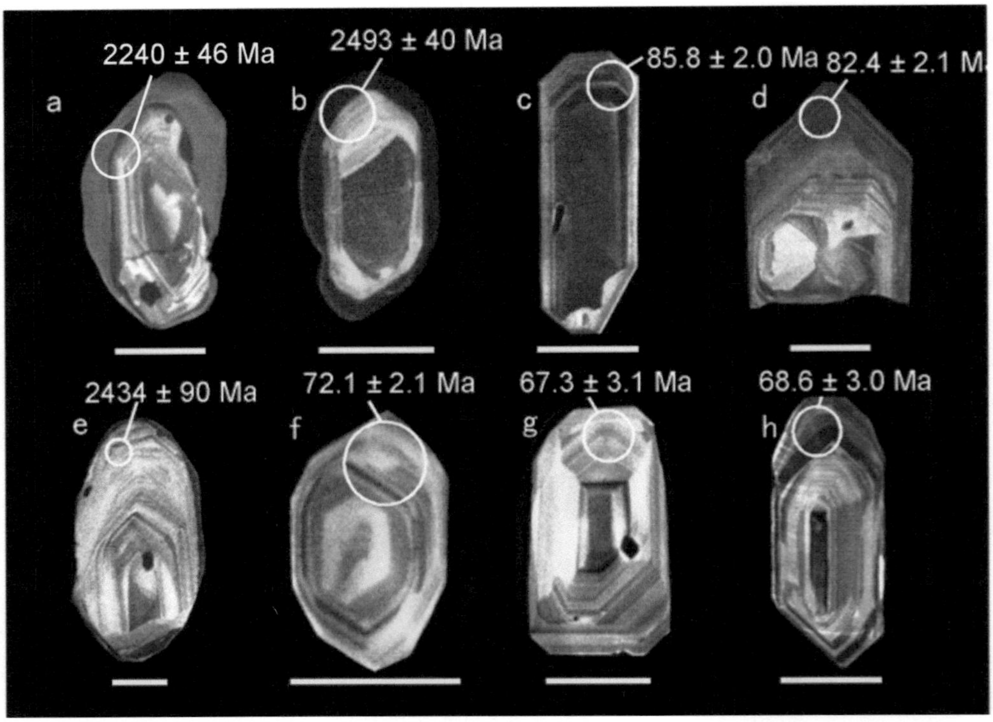

Figure 6. Cathodoluminescence images of selected zircon grains. (a) Site 1 in the Yu1, grain no. Yu-041. (b) Site 2 in the Yu2, grain no. Yu2-095. (c) Site 3 in the M1, grain no. M1-046. (d) Site 4 in the M2, grain no. M2-003. (e) Site 5 in the M3, grain no. M3-018. (f) Site 6 in the Ry1, grain no. Ry1-022. (g) Site 7 in the Ry2, grain no. Ry2-037. (h) Site 8 in the Ry3, grain no. Ry3-021. Scale bars are 50 μm. Ages indicate ^{238}U– ^{206}Pb ages in Ma.

ablation pit size of 25 μm, energy density of 11.7 J/cm^2 and pulse repetition rate of 10 Hz. Detailed descriptions of the LA–ICP–MS analysis are provided by Orihashi et al. [20] and Kouchi et al. [21].

In order to obtain accurate age data, we deleted the missed shot spot that is indicative of an anomalous value. In addition, based on the judgment of discordance showed by the many previous studies (for example, see Refs. [22–24]), the U–Pb zircon ages with discordances of >10% were rejected in data interpretation in this paper. Age calculations were performed using Isoplot/Ex 4.15 [25]. All ages indicate the ^{238}U–^{206}Pb ages, and uncertainties are given at the 2σ level.

5. Zircon U–Pb ages

Analyses were carried out on more than 160 spots in each sample. Results are shown in histograms and probability density plots (**Figures 7** and **8**). The probability density plots show the cumulative Gaussian probability curve for a collection of single-valued data and errors [25]. All data are shown in the Appendix.

Site 1 (sample no. 150503-06): 200 spots on 200 zircon grains were analysed, and 142 spots (discordances of <10%) were selected for statistical interpretations. The zircon U–Pb ages consist mainly of two age groups; ca. 90–320 Ma (66%) and ca. 1500–2600 Ma (31%). The youngest age is 98.8 ± 2.5 Ma, and the highest peak age on the probability density plot is at ca. 190 Ma. The zircons that yield ages of ca. 1500–2600 Ma are rounded and they show distinct cores and rims in the CL images (**Figure 6a**). The other zircons are euhedral and exhibit clear oscillatory zoning.

Site 2 (sample no. 151014-02): 200 spots on 200 zircon grains were analysed, and 140 spots (discordances of <10%) were selected for statistical interpretations. The zircon U–Pb ages consist mainly of two age groups: ca. 100–310 Ma (62%) and ca. 1500–2500 Ma (36%). The youngest age is 100.1 ± 2.8 Ma and the highest peak age on the probability density plot is at ca. 178 Ma. The zircons that yield ages of ca. 1500–2500 Ma are rounded and they show distinct cores and rims in the CL images (**Figure 6b**). The other zircons are euhedral and show clear oscillatory zoning.

Site 3 (sample no. 160705-06): 180 spots on 180 zircon grains were analysed, and 86 spots (discordances of <10%) were selected for statistical interpretations. The zircon U–Pb ages consist mainly of one age group at ca. 80–140 Ma (91%). The youngest age is 85.6 ± 3.4 Ma and the highest peak age on the probability density plot is at ca. 100 Ma. Most of the zircons are euhedral and display clear oscillatory zoning (**Figure 6c**).

Site 4 (sample no. 160705-02): 180 spots on 180 zircon grains were analysed, and 78 spots (discordances of <10%) were selected for statistical interpretations. The zircon U–Pb ages consist mainly of one age group at ca. 70–130 Ma (88%). The youngest age is 75 ± 2.9 Ma and the highest peak age on the probability density plot is at ca. 98 Ma. Most of the zircons are euhedral and exhibit clear oscillatory zoning (**Figure 6d**).

Site 5 (sample no.160705-01): 180 spots on 174 zircon grains were analysed, and 119 spots (discordances of <10%) were selected for statistical interpretations. The zircon U–Pb ages consist mainly of three age groups at ca. 70–110 Ma (7%), ca. 160–240 Ma (17%), and ca. 1300–2600 Ma (74%). The youngest age is 77.3 ± 2.1 Ma and the highest peak age on the probability density plot is at ca. 184 Ma. The zircons that yield ages of ca. 1300–2600 Ma are rounded and show cores and rims (**Figure 6e**). The other zircons are euhedral and display clear oscillatory zoning.

Site 6 (sample no.160704-05): 180 spots on 180 zircon grains were analysed, and 42 spots (discordances of <10%) were selected for statistical interpretations. The zircon U–Pb ages consist mainly of one age group at ca. 60–120 Ma (79%). The youngest age is 62.9 ± 5.7 Ma and the highest peak age on the probability density plot is at ca. 86 Ma. Most of the zircons are euhedral and indicate show clear oscillatory zoning (**Figure 6f**).

Site 7 (sample no. 160704-04): 180 spots on 180 zircon grains were analysed, and 58 spots (discordances of <10%) were selected for statistical interpretations. The zircon U–Pb ages consist mainly of one age group at ca. 60–120 Ma (78%). The youngest age is 62 ± 8.4 Ma and the highest peak age on the probability density plot is at ca. 67 Ma. Most of the zircons are euhedral and exhibit clear oscillatory zoning (**Figure 6g**).

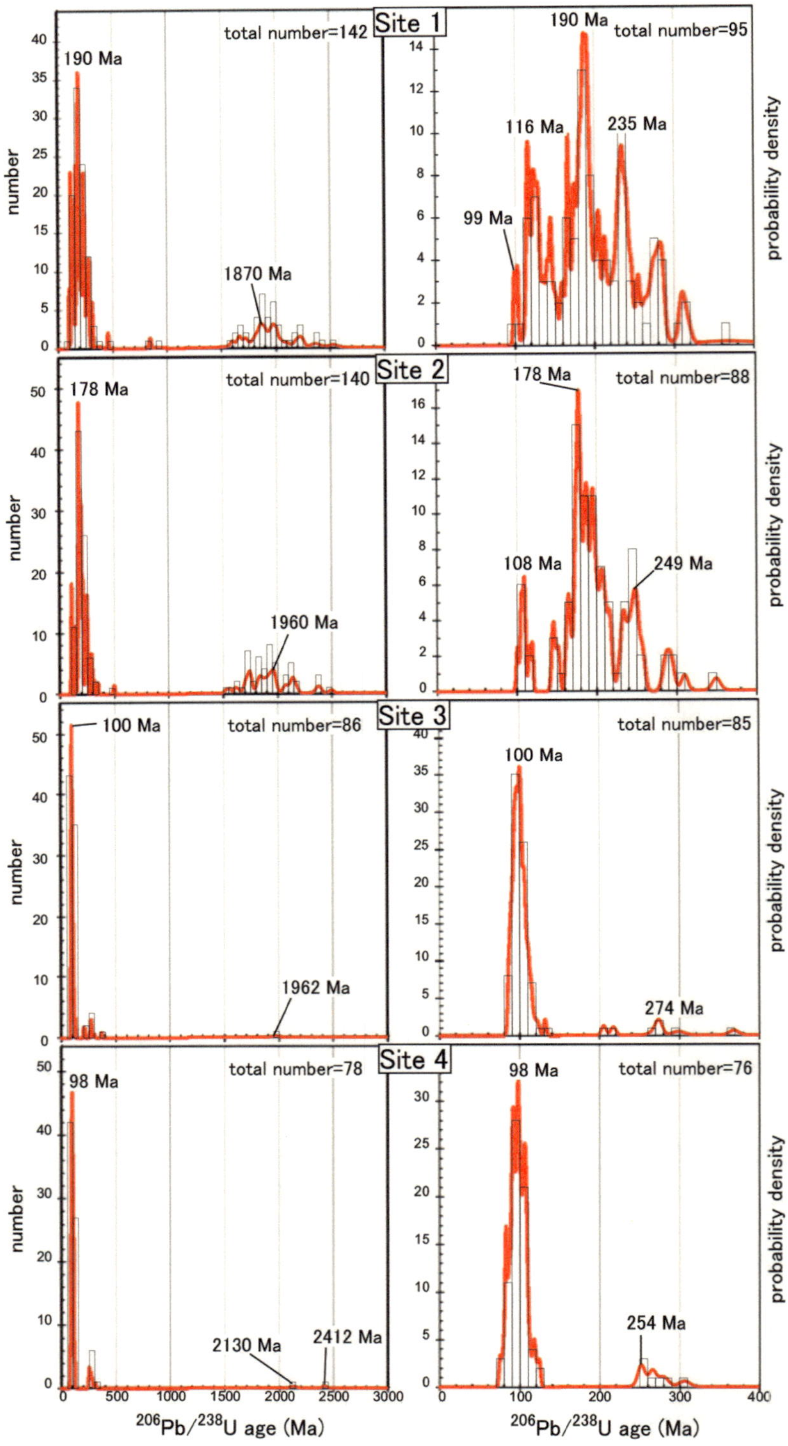

Figure 7. Histogram and probability density plot of Site 1 to Site 4. Left side and right side diagrams show all data and 0–400 Ma data, respectively.

http://dx.doi.org/10.5772/67982

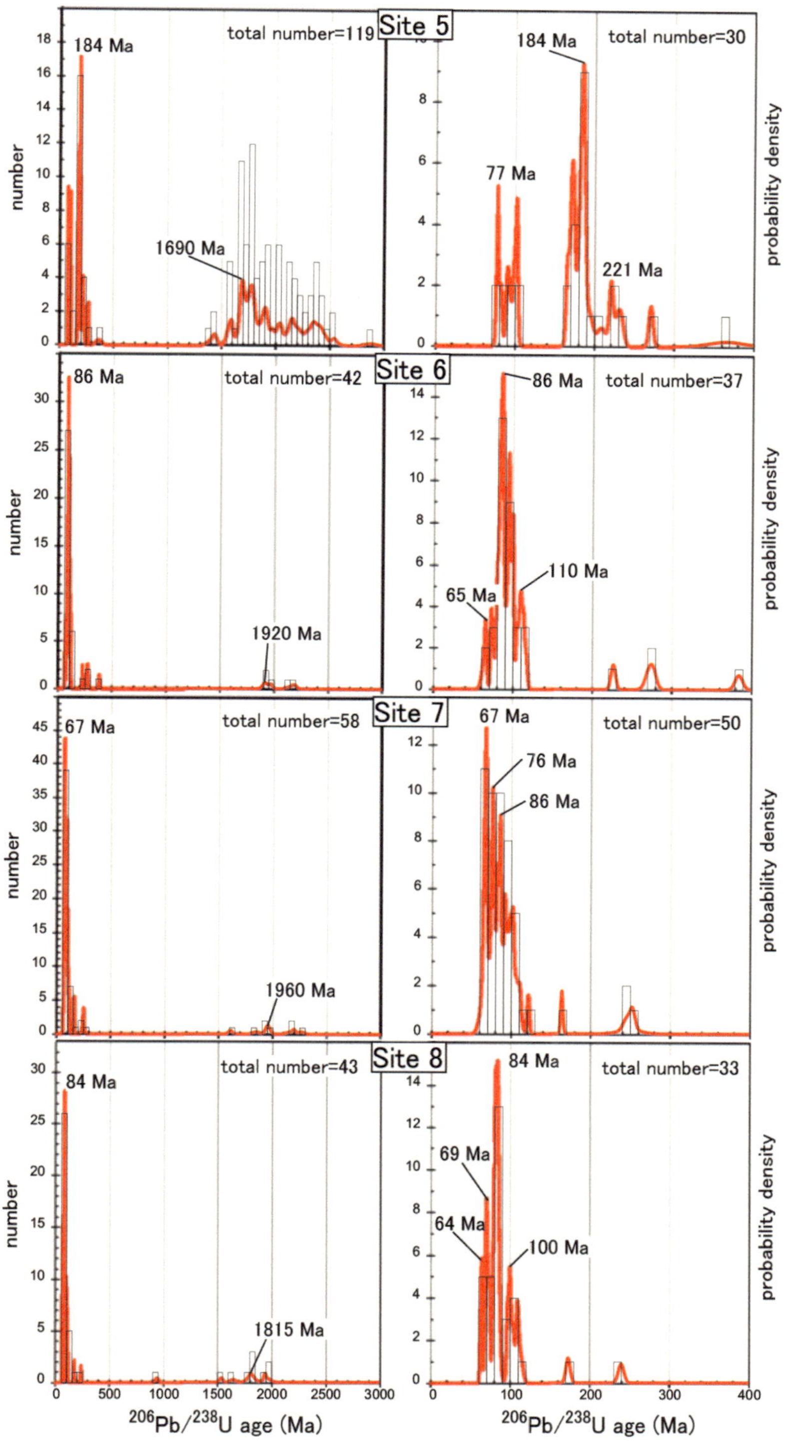

Figure 8. Histogram and probability density plot of Site 5 to Site 8. Left side and right side diagrams show all data and 0–400 Ma data, respectively.

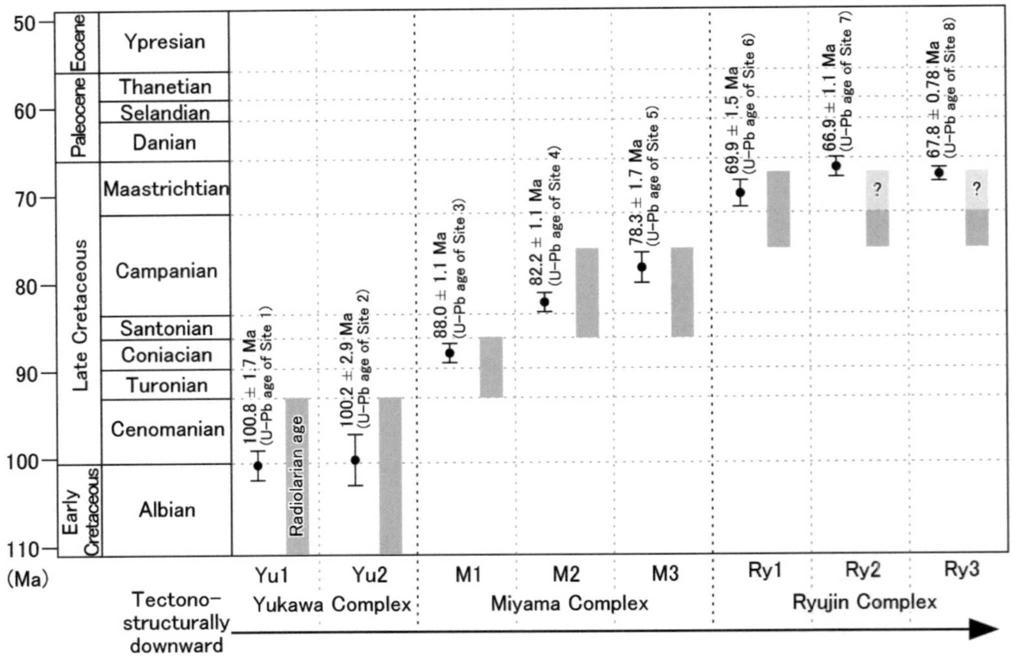

Figure 9. Summary of age distribution of U–Pb zircon ages and radiolarian ages. See text for the radiolarian ages of the each unit.

Site 8 (sample no. 160704-03): 180 spots on 180 zircon grains were analysed, and 43 spots (discordances of <10%) were selected for statistical interpretations. The zircon U–Pb ages consist mainly of one age group at ca. 60–120 Ma (72%). The youngest age is 62.7 ± 2.1 Ma and the highest peak age on the probability density plot is at ca. 84 Ma. Most of the zircons are euhedral and display clear oscillatory zoning (**Figure 6h**).

6. Relationship between detrital zircon U–Pb age and radiolarian fossil age

In this paper, we have not used the youngest single age but the youngest peak age in comparing detrital zircon U–Pb ages with radiolarian fossil ages because it is possible that the zircon U–Pb ages indicate younger than the crystallization age as a result of lead loss [26]. We applied the 'mixture modelling' proposed by Sambridge and Compston [27] to determine the youngest peak age. The 'mixture modelling' is in common usage in estimating the youngest peak age (for example, see Refs. [28–30]). The results indicate that the youngest peak ages are 100.8 ± 1.7 Ma (fraction = 0.09) at Site 1 (Yu1), 100.2 ± 2.9 Ma (fraction = 0.13) at Site 2 (Yu2), 88.0 ± 1.1 Ma (fraction = 0.13) at Site 3 (M1), 82.2 ± 1.1 Ma (fraction = 0.15) at Site 4 (M2), 78.3 ± 1.7 Ma (fraction = 0.37) at Site 5 (M3), 69.9 ± 1.5 Ma (fraction = 0.12) at Site 6 (Ry1), 66.9 ± 1.1 Ma (fraction = 0.27) at Site 7 Ry2) and 67.8 ± 0.78 Ma (fraction = 0.26) at Site 8 (Ry3).

Radiolarian fossils indicate Albian to Cenomanian ages for Yu1 and Yu2 [9, 11], Turonian to Coniacian ages for M1 [9, 11], Santonian to early Campanian ages for M2 and M3 [9], late

Campanian to middle Maastrichtian ages for Ry1 (**Figure 9**; [10, 13, 14, 16, 17]). In Ry2 and Ry3, the radiolarian fossils indicate a late Campanian age [15]; however, a previous study point out the possibility that Ry2 and Ry3 contain radiolarian fossils of middle Maastrichtian age [12].

Detrital zircon U–Pb ages are in good agreement with depositional ages inferred from radiolarian fossils (**Figure 9**). In addition, the youngest peak ages become younger as tectono-structurally downwards; this trend is clearer in the U–Pb ages than in the radiolarian ages. In general, the detrital zircon U–Pb ages do not directly indicate the depositional age, because the ages indicate their crystallization ages. However, igneous activity occurred continuously during the late Cretaceous along the Eastern Asian continental margin where the proto-Japan arc was situated (for example, see Refs. [31–33]). In addition, it is reported that the sandstone compositions of the Cretaceous Shimanto accretionary complex provide evidence for the igneous activity (for example, see Refs. [18, 34, 35]). Most of detrital zircons of Cretaceous age from this study exhibit clear oscillatory zonings. Such zoning reflects crystal growth in a magma chamber, and the zoning is commonly altered to other structure such as core–rim textures and homogeneous textures as the result of metamorphism-induced recrystallization and hydrothermal alteration [36]. Therefore, these detrital Cretaceous zircons in this study are of igneous origin. Thus, it is considered that newly crystallized zircons were continuously supplied by constant igneous activity during the late Cretaceous period, which led to the concordance between the depositional age and zircon age.

From above, we can conclude that detrital zircon U–Pb ages can provide remarkably useful information for determining of the depositional age in the Cretaceous Shimanto accretionary complex.

7. Conclusions

Detrital zircons from sandstone in the Cretaceous Shimanto accretionary complex were subjected to U–Pb dating. The youngest peak ages in the samples from tectono-structurally upwards to downwards were 100.8 ± 1.7 Ma, 100.2 ± 2.9 Ma, 88.0 ± 1.1 Ma, 82.2 ± 1.1 Ma, 78.3 ± 1.7 Ma, 69.9 ± 1.5 Ma, 66.9 ± 1.1 Ma and 67.8 ± 0.78 Ma. These youngest peak ages are in good agreement with the depositional ages inferred from radiolarian fossils. In addition, the youngest peak ages become younger as tectono-structurally downwards. This trend is clearer in the U–Pb age data than in the radiolarian age data and is typical of an accretionary complex. Therefore, U–Pb dating of detrital zircons in sandstone is an effective way to determine the depositional age of strata in the Cretaceous Shimanto accretionary complex.

Acknowledgements

This study was partially supported by a grant from the Fukada Geological Institute (Fukada Grant-in-Aid, 2015). Constructive review comments by Y. Itoh greatly improved the manuscript.

Appendix 1. LA-ICP-MS U–Pb isotopic data.

Grain No.	$^{206}Pb/^{238}U$	$^{207}Pb/^{235}U$	$^{238}U–^{206}Pb$ age (Ma)	$^{235}U–^{207}Pb$ age (Ma)	Th/U	Grain No.	$^{206}Pb/^{238}U$	$^{207}Pb/^{235}U$	$^{238}U–^{206}Pb$ age (Ma)	$^{235}U–^{207}Pb$ age (Ma)	Th/U
Site 1 (Sample 150503-06) in Yanase Unit (Yu1) of Yukawa Complex						Yu1-086	0.02259 ± 0.00044	0.1524 ± 0.0095	144.0 ± 2.8	144.0 ± 9.0	1.46
Yu1-001	0.29308 ± 0.00675	4.5101 ± 0.1505	1656.9 ± 38.2	1732.8 ± 57.8	0.22	Yu1-087	0.35608 ± 0.00588	5.6450 ± 0.1725	1963.6 ± 32.4	1922.9 ± 58.7	0.43
Yu1-002	0.34581 ± 0.00893	5.2821 ± 0.2485	1914.6 ± 49.5	1865.9 ± 87.8	1.32	Yu1-088	0.03080 ± 0.00070	0.2042 ± 0.0170	195.6 ± 4.4	188.6 ± 15.7	0.52
Yu1-003	0.30564 ± 0.00692	4.7023 ± 0.1446	1719.2 ± 38.9	1767.6 ± 54.3	0.24	Yu1-089	0.03006 ± 0.00054	0.2074 ± 0.0101	190.9 ± 3.4	191.4 ± 9.3	0.22
Yu1-004	0.03773 ± 0.00095	0.2675 ± 0.0163	238.8 ± 6.0	240.7 ± 14.7	0.55	Yu1-090	0.04459 ± 0.00127	0.3274 ± 0.0171	281.2 ± 8.0	287.6 ± 15.0	0.55
Yu1-005	0.01834 ± 0.00053	0.1319 ± 0.0119	117.2 ± 3.4	125.8 ± 11.3	0.37	Yu1-091	0.40599 ± 0.01116	8.2032 ± 0.2864	2196.6 ± 60.4	2253.6 ± 78.7	1.06
Yu1-006	0.37125 ± 0.00840	6.5750 ± 0.1993	2035.3 ± 46.1	2055.9 ± 62.3	0.10	Yu1-092	0.33660 ± 0.01008	5.2328 ± 0.2733	1870.3 ± 56.0	1857.9 ± 97.0	0.78
Yu1-007	0.04174 ± 0.00112	0.3220 ± 0.0224	263.6 ± 7.1	283.5 ± 19.7	0.38	Yu1-093	0.01801 ± 0.00057	0.1327 ± 0.0111	115.1 ± 3.7	126.5 ± 10.6	1.23
Yu1-008	0.03643 ± 0.00094	0.2475 ± 0.0143	230.7 ± 6.0	224.6 ± 13.0	0.78	Yu1-094	0.04283 ± 0.00128	0.3119 ± 0.0206	270.3 ± 8.1	275.7 ± 18.2	0.70
Yu1-009	0.01851 ± 0.00067	0.1359 ± 0.0178	118.2 ± 4.3	129.4 ± 16.9	0.43	Yu1-095	0.02943 ± 0.00532	0.2034 ± 0.0387	187.0 ± 33.8	188.0 ± 35.8	0.40
Yu1-010	0.35598 ± 0.00883	5.5068 ± 0.2313	1963.1 ± 48.7	1901.6 ± 79.9	0.17	Yu1-096	0.15109 ± 0.02729	1.6411 ± 0.3016	907.1 ± 163.9	986.1 ± 181.2	0.25
Yu1-011	0.03426 ± 0.00117	0.2247 ± 0.0274	217.2 ± 7.4	205.8 ± 25.1	0.65	Yu1-097	0.02644 ± 0.00478	0.1850 ± 0.0350	168.2 ± 30.4	172.3 ± 32.6	0.21
Yu1-012	0.31714 ± 0.00871	4.9983 ± 0.2712	1775.8 ± 48.8	1819.0 ± 98.7	0.63	Yu1-098	0.02783 ± 0.00504	0.2003 ± 0.0383	176.9 ± 32.0	185.4 ± 35.4	0.91
Yu1-013	0.03570 ± 0.00120	0.2509 ± 0.0286	226.1 ± 7.6	227.3 ± 25.9	0.29	Yu1-099	0.03132 ± 0.00567	0.2146 ± 0.0408	198.8 ± 36.0	197.4 ± 37.6	0.39
Yu1-014	0.01836 ± 0.00258	0.1258 ± 0.0272	117.3 ± 16.5	120.3 ± 26.0	0.51	Yu1-100	0.02617 ± 0.00474	0.1800 ± 0.0347	166.5 ± 30.1	168.0 ± 32.4	0.57
Yu1-015	0.33863 ± 0.04701	5.4033 ± 1.0076	1880.1 ± 261.0	1885.3 ± 351.6	0.27	Yu1-101	0.05890 ± 0.01065	0.4458 ± 0.0830	368.9 ± 66.7	374.3 ± 69.7	0.55
Yu1-016	0.04315 ± 0.00602	0.3141 ± 0.0619	272.3 ± 38.0	277.3 ± 54.6	0.36	Yu1-102	0.02856 ± 0.00092	0.1962 ± 0.0110	181.5 ± 5.8	181.9 ± 10.2	0.33
Yu1-017	0.28236 ± 0.03921	4.4232 ± 0.8256	1603.2 ± 222.6	1716.6 ± 320.4	0.13	Yu1-103	0.32550 ± 0.01024	5.6463 ± 0.2400	1816.6 ± 57.2	1923.1 ± 81.7	0.14
Yu1-018	0.46045 ± 0.06404	10.6060 ± 1.9925	2441.5 ± 339.6	2489.1 ± 467.6	0.40	Yu1-104	0.36460 ± 0.01144	5.6560 ± 0.2381	2003.9 ± 62.9	1924.6 ± 81.0	0.21
Yu1-019	0.29075 ± 0.04040	4.6481 ± 0.8710	1645.3 ± 228.6	1757.9 ± 329.4	0.30	Yu1-105	0.01937 ± 0.00075	0.1425 ± 0.0162	123.7 ± 4.8	135.2 ± 15.3	0.89
Yu1-020	0.03105 ± 0.00435	0.2244 ± 0.0468	197.1 ± 27.6	205.6 ± 42.9	0.48	Yu1-106	0.02620 ± 0.00083	0.1786 ± 0.0091	166.7 ± 5.3	166.8 ± 8.5	0.79
Yu1-021	0.02130 ± 0.00055	0.1329 ± 0.0094	135.8 ± 3.5	126.7 ± 8.9	0.82	Yu1-107	0.41387 ± 0.00921	7.7769 ± 0.2551	2232.6 ± 49.7	2205.5 ± 72.4	0.71
Yu1-022	0.03934 ± 0.00117	0.2624 ± 0.0263	248.7 ± 7.4	236.6 ± 23.8	0.42	Yu1-108	0.02493 ± 0.00060	0.1770 ± 0.0093	158.7 ± 3.8	165.5 ± 8.7	0.42
Yu1-023	0.28086 ± 0.00648	4.2625 ± 0.1459	1595.7 ± 36.8	1686.1 ± 57.7	0.61	Yu1-109	0.02608 ± 0.00076	0.1816 ± 0.0156	166.0 ± 4.8	169.4 ± 14.5	0.71

Grain No.	$^{206}Pb/^{238}U$	$^{207}Pb/^{235}U$	$^{238}U–^{206}Pb$ age (Ma)	$^{235}U–^{207}Pb$ age (Ma)	Th/U	Grain No.	$^{206}Pb/^{238}U$	$^{207}Pb/^{235}U$	$^{238}U–^{206}Pb$ age (Ma)	$^{235}U–^{207}Pb$ age (Ma)	Th/U
Yu1-024	0.32488 ± 0.00732	5.1435 ± 0.1535	1813.5 ± 40.8	1843.3 ± 55.0	0.14	Yu1-110	0.04332 ± 0.00124	0.3020 ± 0.0248	273.4 ± 7.8	268.0 ± 22.0	1.64
Yu1-025	0.01912 ± 0.00056	0.1386 ± 0.0131	122.1 ± 3.6	131.8 ± 12.5	0.55	Yu1-111	0.01928 ± 0.00054	0.1380 ± 0.0103	123.1 ± 3.4	131.2 ± 9.8	0.84
Yu1-026	0.34328 ± 0.00775	6.8643 ± 0.2042	1902.4 ± 42.9	2094.0 ± 62.3	0.30	Yu1-112	0.48029 ± 0.01290	10.6520 ± 0.3390	2528.5 ± 67.9	2493.2 ± 79.4	0.71
Yu1-027	0.02191 ± 0.00048	0.1451 ± 0.0123	139.7 ± 3.0	137.6 ± 11.7	0.65	Yu1-113	0.40824 ± 0.01088	8.6211 ± 0.2664	2206.9 ± 58.8	2298.7 ± 71.0	0.15
Yu1-028	0.14017 ± 0.00262	1.3545 ± 0.0770	845.6 ± 15.8	869.5 ± 49.4	0.26	Yu1-114	0.02075 ± 0.00067	0.1535 ± 0.0138	132.4 ± 4.3	145.0 ± 13.0	0.53
Yu1-029	0.04021 ± 0.00062	0.2900 ± 0.0117	254.1 ± 3.9	258.5 ± 10.4	0.81	Yu1-115	0.40300 ± 0.01105	7.2154 ± 0.2565	2182.8 ± 59.8	2138.3 ± 76.0	0.72
Yu1-030	0.03881 ± 0.00063	0.2780 ± 0.0131	245.4 ± 4.0	249.0 ± 11.8	0.70	Yu1-116	0.04410 ± 0.00157	0.3287 ± 0.0262	278.2 ± 9.9	288.5 ± 23.0	0.47
Yu1-031	0.03343 ± 0.00064	0.2430 ± 0.0161	212.0 ± 4.1	220.9 ± 14.6	0.43	Yu1-117	0.02962 ± 0.00079	0.2079 ± 0.0121	188.2 ± 5.0	191.8 ± 11.2	0.24
Yu1-032	0.03529 ± 0.00103	0.2489 ± 0.0303	223.6 ± 6.5	225.7 ± 27.5	0.44	Yu1-118	0.02726 ± 0.00079	0.1952 ± 0.0127	173.4 ± 5.0	181.0 ± 11.7	0.34
Yu1-033	0.04460 ± 0.00109	0.3061 ± 0.0266	281.3 ± 6.9	271.1 ± 23.5	0.46	Yu1-119	0.03759 ± 0.00142	0.2654 ± 0.0227	237.9 ± 9.0	239.0 ± 20.4	1.30
Yu1-034	0.04953 ± 0.00169	0.3975 ± 0.0542	311.6 ± 10.6	339.9 ± 46.4	0.40	Yu1-120	0.29430 ± 0.00647	4.4864 ± 0.1474	1663.0 ± 36.5	1728.4 ± 56.8	0.11
Yu1-035	0.01817 ± 0.00056	0.1126 ± 0.0150	116.1 ± 3.6	108.4 ± 14.4	0.47	Yu1-121	0.01976 ± 0.00098	0.1302 ± 0.0138	126.2 ± 6.3	124.3 ± 13.1	0.45
Yu1-036	0.04902 ± 0.00107	0.3934 ± 0.0257	308.5 ± 6.7	336.8 ± 22.0	0.24	Yu1-122	0.45026 ± 0.02077	9.0129 ± 0.4866	2396.4 ± 110.5	2339.2 ± 126.3	0.46
Yu1-037	0.03767 ± 0.00092	0.2814 ± 0.0235	238.4 ± 5.8	251.8 ± 21.1	0.63	Yu1-123	0.36066 ± 0.01660	5.8082 ± 0.3114	1985.3 ± 91.4	1947.6 ± 104.4	0.18
Yu1-038	0.33022 ± 0.01275	5.8938 ± 0.3428	1839.5 ± 71.0	1960.3 ± 114.0	0.12	Yu1-124	0.04334 ± 0.00133	0.3122 ± 0.0254	273.5 ± 8.4	275.9 ± 22.4	0.67
Yu1-039	0.29209 ± 0.01966	4.6595 ± 0.4578	1651.9 ± 111.2	1759.9 ± 172.9	0.42	Yu1-125	0.38653 ± 0.01040	6.0288 ± 0.2674	2106.7 ± 56.7	1979.9 ± 87.8	0.20
Yu1-040	0.03799 ± 0.00081	0.2703 ± 0.0138	240.3 ± 5.1	242.9 ± 12.4	0.82	Yu1-126	0.37922 ± 0.01092	6.0652 ± 0.3330	2072.6 ± 59.7	1985.2 ± 109.0	0.67
Yu1-041	0.41540 ± 0.00843	7.6183 ± 0.2621	2239.6 ± 45.5	2186.9 ± 75.2	0.53	Yu1-127	0.30472 ± 0.00817	4.9258 ± 0.2157	1714.7 ± 46.0	1806.6 ± 79.1	0.19
Yu1-042	0.02020 ± 0.00053	0.1384 ± 0.0120	128.9 ± 3.4	131.6 ± 11.4	0.55	Yu1-128	0.02287 ± 0.00087	0.1675 ± 0.0223	145.8 ± 5.6	157.2 ± 21.0	0.77
Yu1-043	0.33439 ± 0.00682	5.8951 ± 0.2073	1859.6 ± 37.9	1960.4 ± 68.9	0.85	Yu1-129	0.44038 ± 0.01189	9.4131 ± 0.4175	2352.3 ± 63.5	2379.0 ± 105.5	0.81
Yu1-044	0.35117 ± 0.00690	5.4899 ± 0.1716	1940.2 ± 38.1	1898.9 ± 59.4	0.39	Yu1-130	0.03684 ± 0.00111	0.2702 ± 0.0210	233.2 ± 7.0	242.8 ± 18.9	0.54
Yu1-045	0.03749 ± 0.00124	0.2875 ± 0.0353	237.3 ± 7.9	256.6 ± 31.5	0.53	Yu1-131	0.03704 ± 0.00056	0.2709 ± 0.0139	234.4 ± 3.6	243.4 ± 12.5	1.21
Yu1-046	0.02987 ± 0.00128	0.2137 ± 0.0140	189.8 ± 8.1	196.7 ± 12.9	0.16	Yu1-132	0.03232 ± 0.00051	0.2238 ± 0.0124	205.1 ± 3.2	205.1 ± 11.4	0.47
Yu1-047	0.03435 ± 0.00147	0.2399 ± 0.0162	217.7 ± 9.3	218.3 ± 14.7	0.51	Yu1-133	0.02941 ± 0.00045	0.2089 ± 0.0110	186.9 ± 2.9	192.7 ± 10.1	0.25
Yu1-048	0.02971 ± 0.00130	0.2194 ± 0.0174	188.7 ± 8.3	201.4 ± 15.9	0.26	Yu1-134	0.01604 ± 0.00036	0.1128 ± 0.0105	102.6 ± 2.3	108.5 ± 10.1	1.39

Grain No.	$^{206}Pb/^{238}U$	$^{207}Pb/^{235}U$	^{238}U–^{206}Pb age (Ma)	^{235}U–^{207}Pb age (Ma)	Th/U	Grain No.	$^{206}Pb/^{238}U$	$^{207}Pb/^{235}U$	^{238}U–^{206}Pb age (Ma)	^{235}U–^{207}Pb age (Ma)	Th/U
Yu1-049	0.03771 ± 0.00164	0.2595 ± 0.0197	238.6 ± 10.4	234.2 ± 17.8	0.62	Yu1-135	0.36392 ± 0.00439	5.7826 ± 0.1354	2000.7 ± 24.1	1943.7 ± 45.5	0.21
Yu1-050	0.02956 ± 0.00149	0.2023 ± 0.0291	187.8 ± 9.5	187.0 ± 26.9	0.49	Yu1-136	0.02008 ± 0.00058	0.1388 ± 0.0177	128.2 ± 3.7	132.0 ± 16.9	1.32
Yu1-051	0.02731 ± 0.00052	0.1961 ± 0.0115	173.7 ± 3.3	181.8 ± 10.6	0.72	Yu1-137	0.02618 ± 0.00034	0.1794 ± 0.0069	166.6 ± 2.2	167.5 ± 6.4	0.36
Yu1-052	0.02997 ± 0.00075	0.2045 ± 0.0196	190.4 ± 4.8	188.9 ± 18.1	0.53	Yu1-138	0.33782 ± 0.01119	5.1525 ± 0.3154	1876.2 ± 62.2	1844.7 ± 112.9	0.21
Yu1-053	0.33314 ± 0.00526	5.3443 ± 0.1403	1853.6 ± 29.3	1875.9 ± 49.2	0.25	Yu1-139	0.04484 ± 0.00161	0.3597 ± 0.0313	282.7 ± 10.1	312.0 ± 27.1	0.60
Yu1-054	0.01974 ± 0.00058	0.1343 ± 0.0160	126.0 ± 3.7	127.9 ± 15.3	0.85	Yu1-140	0.03036 ± 0.00111	0.2023 ± 0.0200	192.8 ± 7.1	187.1 ± 18.5	0.28
Yu1-055	0.03344 ± 0.00085	0.2207 ± 0.0219	212.0 ± 5.4	202.5 ± 20.1	0.57	Yu1-141	0.02922 ± 0.00099	0.1980 ± 0.0146	185.7 ± 6.3	183.4 ± 13.5	0.31
Yu1-056	0.33998 ± 0.00557	5.3254 ± 0.1553	1886.6 ± 30.9	1872.9 ± 54.6	0.24	Yu1-142	0.03201 ± 0.00128	0.2190 ± 0.0273	203.1 ± 8.1	201.1 ± 25.1	0.31
Yu1-057	0.02363 ± 0.00083	0.1679 ± 0.0189	150.6 ± 5.3	157.6 ± 17.8	1.57	Site 2 (Sample 151014-02) in Kitamata Unit (Yu2) of Yukawa Complex					
Yu1-058	0.04092 ± 0.00125	0.2635 ± 0.0216	258.5 ± 7.9	237.5 ± 19.5	0.72	Yu2-001	0.33832 ± 0.00590	5.2966 ± 0.1520	1878.6 ± 32.8	1868.3 ± 53.6	0.20
Yu1-059	0.03644 ± 0.00105	0.2855 ± 0.0171	230.7 ± 6.6	255.0 ± 15.3	0.40	Yu2-002	0.35570 ± 0.00620	5.5867 ± 0.1598	1961.7 ± 34.2	1914.0 ± 54.7	0.57
Yu1-060	0.36254 ± 0.00967	5.7486 ± 0.1897	1994.2 ± 53.2	1938.6 ± 64.0	0.34	Yu2-003	0.03432 ± 0.00082	0.2325 ± 0.0194	217.5 ± 5.2	212.2 ± 17.7	0.34
Yu1-061	0.03201 ± 0.00111	0.2158 ± 0.0247	203.1 ± 7.1	198.4 ± 22.7	0.84	Yu2-004	0.03113 ± 0.00082	0.2276 ± 0.0218	197.6 ± 5.2	208.2 ± 19.9	0.39
Yu1-062	0.02243 ± 0.00082	0.1583 ± 0.0197	143.0 ± 5.2	149.3 ± 18.6	1.35	Yu2-005	0.03666 ± 0.00078	0.2574 ± 0.0164	232.1 ± 5.0	232.5 ± 14.8	0.30
Yu1-063	0.02852 ± 0.00061	0.2017 ± 0.0109	181.3 ± 3.9	186.5 ± 10.1	0.37	Yu2-006	0.03179 ± 0.00091	0.2020 ± 0.0231	201.7 ± 5.8	186.8 ± 21.4	0.91
Yu1-064	0.01544 ± 0.00038	0.1101 ± 0.0087	98.8 ± 2.5	106.0 ± 8.4	1.55	Yu2-007	0.01798 ± 0.00042	0.1218 ± 0.0096	114.8 ± 2.7	116.7 ± 9.2	0.39
Yu1-065	0.02914 ± 0.00059	0.2053 ± 0.0090	185.2 ± 3.8	189.6 ± 8.3	0.40	Yu2-008	0.30263 ± 0.00564	4.6860 ± 0.1588	1704.3 ± 31.8	1764.7 ± 59.8	0.54
Yu1-066	0.03651 ± 0.00089	0.2677 ± 0.0200	231.2 ± 5.6	240.9 ± 18.0	0.63	Yu2-009	0.02842 ± 0.00058	0.1942 ± 0.0112	180.6 ± 3.7	180.2 ± 10.4	0.18
Yu1-067	0.01792 ± 0.00056	0.1253 ± 0.0147	114.5 ± 3.6	119.9 ± 14.1	0.90	Yu2-010	0.03316 ± 0.00093	0.2142 ± 0.0235	210.3 ± 5.9	197.0 ± 21.6	0.60
Yu1-068	0.03020 ± 0.00066	0.2118 ± 0.0122	191.8 ± 4.2	195.1 ± 11.2	0.41	Yu2-011	0.32489 ± 0.00576	5.0978 ± 0.1488	1813.6 ± 32.1	1835.7 ± 53.6	0.16
Yu1-069	0.02882 ± 0.00086	0.2142 ± 0.0165	183.1 ± 5.4	197.1 ± 15.2	0.44	Yu2-012	0.39869 ± 0.00743	8.1638 ± 0.2640	2163.0 ± 40.3	2249.3 ± 72.7	0.59
Yu1-070	0.02803 ± 0.00076	0.1966 ± 0.0107	178.2 ± 4.9	182.2 ± 9.9	0.12	Yu2-013	0.02781 ± 0.00057	0.1894 ± 0.0109	176.8 ± 3.6	176.1 ± 10.1	0.11
Yu1-071	0.34981 ± 0.00907	6.6285 ± 0.2377	1933.7 ± 50.2	2063.1 ± 74.0	0.17	Yu2-014	0.03043 ± 0.00084	0.2118 ± 0.0217	193.2 ± 5.3	195.0 ± 20.0	0.39
Yu1-072	0.03185 ± 0.00091	0.2468 ± 0.0164	202.1 ± 5.8	223.9 ± 14.9	0.64	Yu2-015	0.03922 ± 0.00160	0.2920 ± 0.0329	248.0 ± 10.1	260.1 ± 29.3	0.62
Yu1-073	0.35782 ± 0.00945	5.6855 ± 0.2238	1971.8 ± 52.1	1929.1 ± 75.9	0.36	Yu2-016	0.34878 ± 0.01190	6.6032 ± 0.2986	1928.8 ± 65.8	2059.7 ± 93.1	0.41

http://dx.doi.org/10.5772/67982

Grain No.	$^{206}Pb/^{238}U$	$^{207}Pb/^{235}U$	$^{238}U–^{206}Pb$ age (Ma)	$^{235}U–^{207}Pb$ age (Ma)	Th/U	Grain No.	$^{206}Pb/^{238}U$	$^{207}Pb/^{235}U$	$^{238}U–^{206}Pb$ age (Ma)	$^{235}U–^{207}Pb$ age (Ma)	Th/U
Yu1-074	0.03027 ± 0.00084	0.2128 ± 0.0129	192.2 ± 5.4	195.9 ± 11.9	0.19	Yu2-017	0.35085 ± 0.01194	6.5776 ± 0.2944	1938.7 ± 66.0	2056.3 ± 92.0	0.34
Yu1-075	0.33721 ± 0.00885	5.9032 ± 0.2239	1873.2 ± 49.2	1961.6 ± 74.4	0.47	Yu2-018	0.02964 ± 0.00105	0.1987 ± 0.0131	188.3 ± 6.7	184.0 ± 12.1	0.20
Yu1-076	0.03619 ± 0.00110	0.2845 ± 0.0241	229.2 ± 7.0	254.3 ± 21.5	0.56	Yu2-019	0.02811 ± 0.00106	0.1926 ± 0.0171	178.7 ± 6.7	178.8 ± 15.9	0.46
Yu1-077	0.02847 ± 0.00078	0.1858 ± 0.0120	180.9 ± 5.0	173.0 ± 11.1	0.24	Yu2-020	0.03045 ± 0.00088	0.1953 ± 0.0234	193.4 ± 5.6	181.1 ± 21.7	0.80
Yu1-078	0.05002 ± 0.00165	0.3717 ± 0.0387	314.6 ± 10.4	320.9 ± 33.4	0.42	Yu2-021	0.33004 ± 0.00591	5.1093 ± 0.1858	1838.6 ± 32.9	1837.6 ± 66.8	0.19
Yu1-079	0.32841 ± 0.00834	5.0641 ± 0.2009	1830.7 ± 46.5	1830.0 ± 72.6	0.14	Yu2-022	0.01857 ± 0.00039	0.1307 ± 0.0085	118.6 ± 2.5	124.7 ± 8.1	0.76
Yu1-080	0.02611 ± 0.00071	0.1719 ± 0.0106	166.2 ± 4.5	161.1 ± 9.9	0.27	Yu2-023	0.03348 ± 0.00096	0.2360 ± 0.0269	212.3 ± 6.1	215.1 ± 24.5	0.85
Yu1-081	0.02959 ± 0.00083	0.2063 ± 0.0145	188.0 ± 5.3	190.5 ± 13.3	0.59	Yu2-024	0.30775 ± 0.00699	4.7734 ± 0.1655	1729.6 ± 39.3	1780.2 ± 61.7	0.15
Yu1-082	0.02760 ± 0.00057	0.1854 ± 0.0128	175.5 ± 3.6	172.7 ± 11.9	0.58	Yu2-025	0.34762 ± 0.00903	5.5670 ± 0.2948	1923.2 ± 49.9	1910.9 ± 101.2	1.02
Yu1-083	0.35905 ± 0.00580	6.9546 ± 0.1954	1977.7 ± 32.0	2105.6 ± 59.2	0.15	Yu2-026	0.39270 ± 0.00930	7.3412 ± 0.2909	2135.3 ± 50.6	2153.8 ± 85.3	0.28
Yu1-084	0.04527 ± 0.00086	0.3571 ± 0.0191	285.4 ± 5.4	310.1 ± 16.6	0.68	Yu2-027	0.02803 ± 0.00072	0.1874 ± 0.0130	178.2 ± 4.6	174.4 ± 12.1	0.20
Yu1-085	0.07313 ± 0.00159	0.5711 ± 0.0411	455.0 ± 9.9	458.7 ± 33.0	0.68	Yu2-028	0.08018 ± 0.00221	0.5913 ± 0.0481	497.2 ± 13.7	471.7 ± 38.3	0.76

Appendix 2.

Grain No.	$^{206}Pb/^{238}U$	$^{207}Pb/^{235}U$	^{238}U–^{206}Pb age (Ma)	^{235}U–^{207}Pb age (Ma)	Th/U	Grain No.	$^{206}Pb/^{238}U$	$^{207}Pb/^{235}U$	^{238}U–^{206}Pb age (Ma)	^{235}U–^{207}Pb age (Ma)	Th/U
Yu2-029	0.02707 ± 0.00072	0.2032 ± 0.0151	172.2 ± 4.6	187.9 ± 14.0	0.40	Yu2-115	0.33730 ± 0.00709	5.2624 ± 0.1937	1873.7 ± 39.4	1862.7 ± 68.6	0.36
Yu2-030	0.02598 ± 0.00071	0.1866 ± 0.0149	165.3 ± 4.5	173.7 ± 13.9	0.64	Yu2-116	0.03074 ± 0.00057	0.2235 ± 0.0100	195.2 ± 3.6	204.8 ± 9.2	0.39
Yu2-031	0.03135 ± 0.00087	0.2396 ± 0.0193	199.0 ± 5.5	218.1 ± 17.6	0.58	Yu2-117	0.02859 ± 0.00056	0.1962 ± 0.0107	181.7 ± 3.6	181.9 ± 9.9	0.54
Yu2-032	0.34609 ± 0.00630	5.4136 ± 0.1610	1915.9 ± 34.9	1886.9 ± 56.1	0.09	Yu2-118	0.02263 ± 0.00058	0.1416 ± 0.0139	144.2 ± 3.7	134.5 ± 13.2	1.25
Yu2-033	0.02281 ± 0.00066	0.1722 ± 0.0185	145.4 ± 4.2	161.3 ± 17.3	0.80	Yu2-119	0.02817 ± 0.00055	0.1920 ± 0.0102	179.1 ± 3.5	178.3 ± 9.5	0.22
Yu2-034	0.03945 ± 0.00096	0.2761 ± 0.0225	249.4 ± 6.0	247.5 ± 20.2	0.38	Yu2-120	0.44076 ± 0.00782	9.2948 ± 0.2490	2354.1 ± 41.8	2367.4 ± 63.4	0.41
Yu2-035	0.02933 ± 0.00064	0.1961 ± 0.0127	186.4 ± 4.0	181.8 ± 11.8	0.38	Yu2-121	0.05568 ± 0.00164	0.4115 ± 0.0460	349.3 ± 10.3	349.9 ± 39.1	0.47
Yu2-036	0.44876 ± 0.00839	9.8060 ± 0.3082	2389.7 ± 44.7	2416.6 ± 76.0	0.74	Yu2-122	0.44581 ± 0.00792	9.7228 ± 0.2274	2376.6 ± 42.2	2408.8 ± 56.3	2.01
Yu2-037	0.03227 ± 0.00070	0.2225 ± 0.0143	204.7 ± 4.4	204.0 ± 13.1	0.50	Yu2-123	0.03261 ± 0.00072	0.2525 ± 0.0739	206.9 ± 4.5	228.6 ± 66.9	0.81
Yu2-038	0.01707 ± 0.00037	0.1116 ± 0.0074	109.1 ± 2.4	107.4 ± 7.1	0.79	Yu2-124	0.39375 ± 0.00745	6.9113 ± 0.2252	2140.2 ± 40.5	2100.0 ± 68.4	0.17
Yu2-039	0.30182 ± 0.00578	4.7164 ± 0.1742	1700.3 ± 32.5	1770.1 ± 65.4	0.16	Yu2-125	0.34194 ± 0.00856	5.2821 ± 0.1963	1896.0 ± 47.5	1865.9 ± 69.3	0.93
Yu2-040	0.31219 ± 0.00632	4.8866 ± 0.2077	1751.5 ± 35.4	1799.9 ± 76.5	0.61	Yu2-126	0.02890 ± 0.00075	0.2057 ± 0.0108	183.6 ± 4.8	190.0 ± 10.0	0.34
Yu2-041	0.34474 ± 0.00859	5.6217 ± 0.3504	1909.4 ± 47.6	1919.4 ± 119.6	1.37	Yu2-127	0.03037 ± 0.00080	0.2107 ± 0.0122	192.9 ± 5.1	194.1 ± 11.2	0.60
Yu2-042	0.27325 ± 0.00543	4.2696 ± 0.1744	1557.3 ± 31.0	1687.5 ± 68.9	0.66	Yu2-128	0.01636 ± 0.00046	0.1065 ± 0.0082	104.6 ± 2.9	102.7 ± 7.9	0.72
Yu2-043	0.39994 ± 0.00800	8.0902 ± 0.3250	2168.8 ± 43.4	2241.1 ± 90.0	1.31	Yu2-129	0.03593 ± 0.00102	0.2394 ± 0.0186	227.5 ± 6.5	217.9 ± 16.9	0.69
Yu2-044	0.02641 ± 0.00072	0.1787 ± 0.0187	168.0 ± 4.6	166.9 ± 17.5	0.54	Yu2-130	0.03702 ± 0.00073	0.2573 ± 0.0155	234.3 ± 4.6	232.5 ± 14.0	0.45
Yu2-045	0.03908 ± 0.00093	0.2979 ± 0.0233	247.1 ± 5.9	264.8 ± 20.7	0.24	Yu2-131	0.02595 ± 0.00058	0.1789 ± 0.0145	165.2 ± 3.7	167.1 ± 13.6	0.35
Yu2-046	0.02794 ± 0.00056	0.1864 ± 0.0105	177.6 ± 3.6	173.6 ± 9.8	0.20	Yu2-132	0.02711 ± 0.00058	0.1890 ± 0.0138	172.4 ± 3.7	175.8 ± 12.9	0.61
Yu2-047	0.02973 ± 0.00057	0.2126 ± 0.0099	188.8 ± 3.6	195.7 ± 9.1	0.31	Yu2-133	0.01659 ± 0.00043	0.1211 ± 0.0122	106.1 ± 2.7	116.0 ± 11.7	0.39
Yu2-048	0.30971 ± 0.00546	4.7676 ± 0.1320	1739.3 ± 30.6	1779.1 ± 49.3	0.42	Yu2-134	0.03081 ± 0.00094	0.2306 ± 0.0308	195.6 ± 6.0	210.7 ± 28.1	0.49
Yu2-049	0.30912 ± 0.00546	4.8220 ± 0.1345	1736.4 ± 30.7	1788.7 ± 49.9	0.17	Yu2-135	0.01711 ± 0.00045	0.1138 ± 0.0093	109.4 ± 2.9	109.4 ± 9.0	0.41
Yu2-050	0.03933 ± 0.00145	0.2908 ± 0.0169	248.7 ± 9.2	259.2 ± 15.1	0.74	Yu2-136	0.01566 ± 0.00044	0.1106 ± 0.0100	100.1 ± 2.8	106.5 ± 9.7	0.37
Yu2-051	0.30009 ± 0.01089	4.6834 ± 0.2219	1691.8 ± 61.4	1764.2 ± 83.6	0.12	Yu2-137	0.02376 ± 0.00061	0.1585 ± 0.0124	151.4 ± 3.9	149.4 ± 11.7	0.36

http://dx.doi.org/10.5772/67982

Grain No.	$^{206}Pb/^{238}U$	$^{207}Pb/^{235}U$	$^{238}U–^{206}Pb$ age (Ma)	$^{235}U–^{207}Pb$ age (Ma)	Th/U	Grain No.	$^{206}Pb/^{238}U$	$^{207}Pb/^{235}U$	$^{238}U–^{206}Pb$ age (Ma)	$^{235}U–^{207}Pb$ age (Ma)	Th/U
Yu2-052	0.02328 ± 0.00090	0.1589 ± 0.0129	148.3 ± 5.8	149.7 ± 12.2	1.52	Yu2-138	0.02929 ± 0.00070	0.2223 ± 0.0132	186.1 ± 4.4	203.8 ± 12.1	0.19
Yu2-053	0.34617 ± 0.01257	5.4714 ± 0.2605	1916.3 ± 69.6	1896.1 ± 90.3	0.65	Yu2-139	0.03119 ± 0.00074	0.2229 ± 0.0132	198.0 ± 4.7	204.4 ± 12.1	0.40
Yu2-054	0.32788 ± 0.01187	5.2129 ± 0.2439	1828.1 ± 66.2	1854.7 ± 86.8	0.11	Yu2-140	0.03850 ± 0.00115	0.2541 ± 0.0272	243.5 ± 7.3	229.9 ± 24.6	0.10
Yu2-055	0.03070 ± 0.00123	0.2291 ± 0.0208	194.9 ± 7.8	209.5 ± 19.0	0.57	Site 3 (Sample 160705-06) in Chikai Unit (M1) of Miyama Complex					
Yu2-056	0.03823 ± 0.00113	0.2984 ± 0.0335	241.9 ± 7.1	265.2 ± 29.8	1.19	M1-001	0.01528 ± 0.00068	0.1127 ± 0.0163	97.8 ± 4.4	108.4 ± 15.7	0.50
Yu2-057	0.35667 ± 0.00688	5.7215 ± 0.2771	1966.4 ± 37.9	1934.5 ± 93.7	0.39	M1-002	0.01605 ± 0.00068	0.1162 ± 0.0153	102.7 ± 4.4	111.6 ± 14.7	0.45
Yu2-058	0.02911 ± 0.00060	0.1904 ± 0.0121	185.0 ± 3.8	176.9 ± 11.2	0.33	M1-003	0.01457 ± 0.00062	0.1042 ± 0.0138	93.2 ± 4.0	100.7 ± 13.3	0.96
Yu2-059	0.39177 ± 0.00742	6.7510 ± 0.3166	2131.0 ± 40.3	2079.3 ± 97.5	0.25	M1-004	0.01337 ± 0.00054	0.0970 ± 0.0108	85.6 ± 3.4	94.0 ± 10.5	0.50
Yu2-060	0.38692 ± 0.00727	6.7445 ± 0.3125	2108.5 ± 39.6	2078.4 ± 96.3	0.20	M1-005	0.01512 ± 0.00064	0.1065 ± 0.0136	96.7 ± 4.1	102.7 ± 13.1	0.40
Yu2-061	0.04515 ± 0.00120	0.3293 ± 0.0324	284.7 ± 7.6	289.0 ± 28.4	0.65	M1-006	0.01854 ± 0.00069	0.1256 ± 0.0096	118.4 ± 4.4	120.1 ± 9.2	0.25
Yu2-062	0.03751 ± 0.00095	0.2706 ± 0.0248	237.4 ± 6.0	243.2 ± 22.3	0.68	M1-007	0.01505 ± 0.00067	0.1097 ± 0.0146	96.3 ± 4.3	105.7 ± 14.1	0.44
Yu2-063	0.35662 ± 0.01148	5.6792 ± 0.2601	1966.1 ± 63.3	1928.1 ± 88.3	0.36	M1-008	0.01472 ± 0.00066	0.1057 ± 0.0147	94.2 ± 4.2	102.0 ± 14.2	0.61
Yu2-064	0.01673 ± 0.00067	0.1218 ± 0.0147	107.0 ± 4.3	116.7 ± 14.1	0.25	M1-009	0.01400 ± 0.00063	0.0870 ± 0.0130	89.6 ± 4.0	84.7 ± 12.6	0.58
Yu2-065	0.02961 ± 0.00053	0.2053 ± 0.0115	188.1 ± 3.4	189.6 ± 10.6	0.21	M1-010	0.01486 ± 0.00060	0.0996 ± 0.0141	95.1 ± 3.9	96.4 ± 13.7	0.42
Yu2-066	0.03410 ± 0.00064	0.2298 ± 0.0144	216.1 ± 4.1	210.0 ± 13.1	0.41	M1-011	0.01554 ± 0.00060	0.0992 ± 0.0130	99.4 ± 3.8	96.1 ± 12.5	0.49
Yu2-067	0.35519 ± 0.00583	5.5668 ± 0.2050	1959.3 ± 32.1	1910.9 ± 70.4	0.36	M1-012	0.01573 ± 0.00061	0.1008 ± 0.0132	100.6 ± 3.9	97.5 ± 12.7	0.83
Yu2-068	0.31045 ± 0.00501	5.4161 ± 0.1927	1742.9 ± 28.1	1887.3 ± 67.2	0.14	M1-013	0.01406 ± 0.00060	0.0875 ± 0.0141	90.0 ± 3.9	85.1 ± 13.7	0.55
Yu2-069	0.30484 ± 0.00584	4.9608 ± 0.1395	1715.3 ± 32.9	1812.6 ± 51.0	0.39	M1-014	0.01668 ± 0.00050	0.1135 ± 0.0107	106.6 ± 3.2	109.1 ± 10.3	0.34
Yu2-070	0.03304 ± 0.00077	0.2387 ± 0.0167	209.5 ± 4.9	217.3 ± 15.2	0.75	M1-015	0.04354 ± 0.00130	0.3058 ± 0.0291	274.7 ± 8.2	270.9 ± 25.7	0.91
Yu2-071	0.28792 ± 0.00562	4.3254 ± 0.1344	1631.1 ± 31.9	1698.2 ± 52.8	0.93	M1-016	0.05879 ± 0.00182	0.4456 ± 0.0441	368.3 ± 11.4	374.2 ± 37.0	0.72
Yu2-072	0.02802 ± 0.00065	0.1938 ± 0.0137	178.1 ± 4.1	179.8 ± 12.8	0.32	M1-017	0.01503 ± 0.00042	0.1015 ± 0.0084	96.1 ± 2.7	98.1 ± 8.1	0.29
Yu2-073	0.02954 ± 0.00060	0.2093 ± 0.0099	187.6 ± 3.8	192.9 ± 9.1	0.40	M1-018	0.01674 ± 0.00083	0.1120 ± 0.0199	107.0 ± 5.3	107.8 ± 19.2	0.47
Yu2-074	0.04916 ± 0.00127	0.3650 ± 0.0296	309.4 ± 8.0	315.9 ± 25.7	0.68	M1-019	0.01680 ± 0.00075	0.1124 ± 0.0163	107.4 ± 4.8	108.2 ± 15.7	0.56
Yu2-075	0.38176 ± 0.00812	6.0917 ± 0.2339	2084.5 ± 44.3	1989.0 ± 76.4	0.34	M1-020	0.01422 ± 0.00058	0.0914 ± 0.0107	91.0 ± 3.7	88.8 ± 10.4	0.70
Yu2-076	0.33106 ± 0.00688	5.0870 ± 0.1829	1843.5 ± 38.3	1833.9 ± 65.9	0.30	M1-021	0.01605 ± 0.00079	0.0990 ± 0.0176	102.6 ± 5.0	95.9 ± 17.1	0.53

Grain No.	$^{206}Pb/^{238}U$	$^{207}Pb/^{235}U$	$^{238}U–^{206}Pb$ age (Ma)	$^{235}U–^{207}Pb$ age (Ma)	Th/U	Grain No.	$^{206}Pb/^{238}U$	$^{207}Pb/^{235}U$	$^{238}U–^{206}Pb$ age (Ma)	$^{235}U–^{207}Pb$ age (Ma)	Th/U
Yu2-077	0.02569 ± 0.00060	0.1703 ± 0.0114	163.5 ± 3.8	159.7 ± 10.7	0.53	M1-022	0.01376 ± 0.00062	0.0937 ± 0.0133	88.1 ± 3.9	91.0 ± 12.9	0.55
Yu2-078	0.03186 ± 0.00108	0.2241 ± 0.0303	202.2 ± 6.9	205.3 ± 27.7	0.70	M1-023	0.01612 ± 0.00068	0.1043 ± 0.0131	103.1 ± 4.3	100.8 ± 12.7	0.32
Yu2-079	0.35789 ± 0.00740	5.6833 ± 0.1997	1972.2 ± 40.8	1928.8 ± 67.8	0.36	M1-024	0.01588 ± 0.00074	0.0984 ± 0.0185	101.6 ± 4.7	95.3 ± 18.0	0.47
Yu2-080	0.02532 ± 0.00054	0.1745 ± 0.0083	161.2 ± 3.4	163.3 ± 7.8	0.70	M1-025	0.35579 ± 0.01026	5.7542 ± 0.2561	1962.2 ± 56.6	1939.5 ± 86.3	0.31
Yu2-081	0.03225 ± 0.00100	0.2416 ± 0.0245	204.6 ± 6.3	219.7 ± 22.3	0.75	M1-026	0.04741 ± 0.00238	0.3672 ± 0.0699	298.6 ± 15.0	317.6 ± 60.4	0.32
Yu2-082	0.02796 ± 0.00069	0.1935 ± 0.0115	177.7 ± 4.4	179.6 ± 10.7	0.24	M1-027	0.01367 ± 0.00051	0.0952 ± 0.0116	87.5 ± 3.3	92.3 ± 11.2	0.78
Yu2-083	0.02793 ± 0.00069	0.1866 ± 0.0113	177.6 ± 4.4	173.7 ± 10.5	0.55	M1-028	0.01493 ± 0.00074	0.0903 ± 0.0189	95.5 ± 4.7	87.8 ± 18.4	0.88
Yu2-084	0.02743 ± 0.00071	0.1972 ± 0.0134	174.4 ± 4.5	182.8 ± 12.5	0.47	M1-029	0.01507 ± 0.00064	0.1000 ± 0.0157	96.4 ± 4.1	96.7 ± 15.2	0.70
Yu2-085	0.03023 ± 0.00103	0.1985 ± 0.0595	192.0 ± 6.5	183.9 ± 55.1	0.85	M1-030	0.01461 ± 0.00060	0.1018 ± 0.0148	93.5 ± 3.8	98.5 ± 14.3	0.73
Yu2-086	0.04627 ± 0.00144	0.3125 ± 0.0712	291.6 ± 9.1	276.1 ± 62.9	0.71	M1-031	0.01648 ± 0.00081	0.1134 ± 0.0171	105.4 ± 5.2	109.0 ± 16.5	0.51
Yu2-087	0.35058 ± 0.00705	5.9553 ± 0.1568	1937.3 ± 38.9	1969.3 ± 51.9	0.19	M1-032	0.01451 ± 0.00071	0.1010 ± 0.0150	92.9 ± 4.6	97.7 ± 14.5	0.58
Yu2-088	0.32438 ± 0.00669	5.0844 ± 0.1573	1811.1 ± 37.3	1833.4 ± 56.7	0.38	M1-033	0.01597 ± 0.00086	0.1084 ± 0.0198	102.1 ± 5.5	104.5 ± 19.1	0.38
Yu2-089	0.31279 ± 0.00644	4.8634 ± 0.1492	1754.4 ± 36.1	1795.9 ± 55.1	0.28	M1-034	0.01503 ± 0.00065	0.0998 ± 0.0103	96.1 ± 4.1	96.6 ± 10.0	0.42
Yu2-090	0.37777 ± 0.00752	7.0900 ± 0.1734	2065.9 ± 41.1	2122.7 ± 51.9	0.41	M1-035	0.01432 ± 0.00071	0.0956 ± 0.0153	91.6 ± 4.6	92.7 ± 14.9	1.02
Yu2-091	0.03764 ± 0.00095	0.2643 ± 0.0298	238.2 ± 6.0	238.1 ± 26.9	0.64	M1-036	0.01681 ± 0.00073	0.1188 ± 0.0123	107.5 ± 4.6	114.0 ± 11.8	0.39
Yu2-092	0.02742 ± 0.00050	0.1945 ± 0.0115	174.4 ± 3.2	180.5 ± 10.7	0.41	M1-037	0.01671 ± 0.00079	0.1084 ± 0.0151	106.9 ± 5.0	104.5 ± 14.6	0.64
Yu2-093	0.28256 ± 0.00480	4.4734 ± 0.1485	1604.3 ± 27.3	1726.0 ± 57.3	0.32	M1-038	0.01441 ± 0.00052	0.1021 ± 0.0130	92.2 ± 3.3	98.7 ± 12.6	0.83
Yu2-094	0.03651 ± 0.00079	0.2808 ± 0.0269	231.2 ± 5.0	251.3 ± 24.1	0.77	M1-039	0.01794 ± 0.00068	0.1090 ± 0.0158	114.6 ± 4.3	105.0 ± 15.2	0.34
Yu2-095	0.47224 ± 0.00750	11.3142 ± 0.2912	2493.4 ± 39.6	2549.3 ± 65.6	0.55	M1-040	0.01518 ± 0.00064	0.1022 ± 0.0167	97.1 ± 4.1	98.8 ± 16.2	0.57
Yu2-096	0.03953 ± 0.00081	0.2800 ± 0.0246	249.9 ± 5.1	250.7 ± 22.0	0.82	M1-041	0.01594 ± 0.00074	0.1161 ± 0.0212	101.9 ± 4.7	111.5 ± 20.4	0.69
Yu2-097	0.02792 ± 0.00074	0.1864 ± 0.0133	177.5 ± 4.7	173.6 ± 12.4	0.38	M1-042	0.01482 ± 0.00058	0.1037 ± 0.0145	94.8 ± 3.7	100.1 ± 14.0	0.47
Yu2-098	0.04652 ± 0.00124	0.3448 ± 0.0240	293.1 ± 7.8	300.8 ± 20.9	0.33	M1-043	0.01729 ± 0.00037	0.1155 ± 0.0088	110.5 ± 2.4	111.0 ± 8.4	0.28
Yu2-099	0.03378 ± 0.00091	0.2571 ± 0.0181	214.2 ± 5.8	232.3 ± 16.4	0.67	M1-044	0.01640 ± 0.00048	0.1156 ± 0.0138	104.9 ± 3.1	111.1 ± 13.3	0.75
Yu2-100	0.33284 ± 0.00778	5.1492 ± 0.1856	1852.1 ± 43.3	1844.2 ± 66.5	0.40	M1-045	0.02058 ± 0.00051	0.1400 ± 0.0133	131.3 ± 3.2	133.1 ± 12.6	0.24
Yu2-101	0.26696 ± 0.00627	4.1236 ± 0.1520	1525.3 ± 35.9	1658.9 ± 61.2	0.35	M1-046	0.01339 ± 0.00031	0.0928 ± 0.0079	85.8 ± 2.0	90.1 ± 7.7	0.52

Grain No.	$^{206}Pb/^{238}U$	$^{207}Pb/^{235}U$	$^{238}U–^{206}Pb$ age (Ma)	$^{235}U–^{207}Pb$ age (Ma)	Th/U	Grain No.	$^{206}Pb/^{238}U$	$^{207}Pb/^{235}U$	$^{238}U–^{206}Pb$ age (Ma)	$^{235}U–^{207}Pb$ age (Ma)	Th/U
Yu2-102	0.35245 ± 0.00847	5.4469 ± 0.2191	1946.3 ± 46.8	1892.2 ± 76.1	0.31	M1-047	0.04340 ± 0.00133	0.2916 ± 0.0267	273.9 ± 8.4	259.8 ± 23.8	0.44
Yu2-103	0.02911 ± 0.00058	0.2069 ± 0.0143	185.0 ± 3.7	191.0 ± 13.2	0.17	M1-048	0.01450 ± 0.00057	0.0956 ± 0.0145	92.8 ± 3.7	92.7 ± 14.1	0.68
Yu2-104	0.04057 ± 0.00090	0.2868 ± 0.0237	256.4 ± 5.7	256.0 ± 21.2	0.74	M1-049	0.01918 ± 0.00071	0.1406 ± 0.0184	122.5 ± 4.6	133.6 ± 17.5	0.39
Yu2-105	0.02726 ± 0.00048	0.1807 ± 0.0102	173.4 ± 3.1	168.7 ± 9.5	0.49	M1-050	0.01590 ± 0.00063	0.1132 ± 0.0167	101.7 ± 4.0	108.9 ± 16.1	0.53
Yu2-106	0.04062 ± 0.00113	0.2842 ± 0.0327	256.7 ± 7.2	253.9 ± 29.2	0.32	M1-051	0.01541 ± 0.00050	0.1096 ± 0.0112	98.6 ± 3.2	105.6 ± 10.8	0.74
Yu2-107	0.39297 ± 0.00559	8.6158 ± 0.2025	2136.6 ± 30.4	2298.1 ± 54.0	0.16	M1-052	0.01589 ± 0.00047	0.1152 ± 0.0094	101.6 ± 3.0	110.7 ± 9.1	0.39
Yu2-108	0.03068 ± 0.00063	0.2249 ± 0.0162	194.8 ± 4.0	206.0 ± 14.8	0.28	M1-053	0.01580 ± 0.00056	0.1078 ± 0.0156	101.1 ± 3.6	103.9 ± 15.0	0.49
Yu2-109	0.03265 ± 0.00087	0.2174 ± 0.0239	207.1 ± 5.5	199.7 ± 22.0	0.20	M1-054	0.03423 ± 0.00101	0.2325 ± 0.0256	217.0 ± 6.4	212.2 ± 23.3	0.40
Yu2-110	0.32934 ± 0.00663	5.1880 ± 0.1623	1835.2 ± 36.9	1850.6 ± 57.9	0.12	M1-055	0.01704 ± 0.00057	0.1243 ± 0.0161	108.9 ± 3.7	119.0 ± 15.4	0.28
Yu2-111	0.37595 ± 0.00777	5.8510 ± 0.2030	2057.3 ± 42.5	1953.9 ± 67.8	0.37	M1-056	0.01728 ± 0.00076	0.1130 ± 0.0216	110.4 ± 4.8	108.7 ± 20.8	0.48
Yu2-112	0.02800 ± 0.00063	0.1950 ± 0.0111	178.0 ± 4.0	180.9 ± 10.3	0.40	M1-057	0.01556 ± 0.00083	0.1014 ± 0.0244	99.5 ± 5.3	98.0 ± 23.6	0.59
Yu2-113	0.04555 ± 0.00134	0.3024 ± 0.0326	287.1 ± 8.4	268.2 ± 28.9	0.69	M1-058	0.01577 ± 0.00043	0.0992 ± 0.0110	100.9 ± 2.7	96.0 ± 10.7	0.29
Yu2-114	0.03858 ± 0.00097	0.2926 ± 0.0219	244.0 ± 6.1	260.6 ± 19.5	0.12	M1-059	0.01564 ± 0.00051	0.0991 ± 0.0139	100.0 ± 3.2	95.9 ± 13.5	0.70

Appendix 3.

Grain No.	$^{206}Pb/^{238}U$	$^{207}Pb/^{235}U$	$^{238}U–^{206}Pb$ age (Ma)	$^{235}U–^{207}Pb$ age (Ma)	Th/U	Grain No.	$^{206}Pb/^{238}U$	$^{207}Pb/^{235}U$	$^{238}U–^{206}Pb$ age (Ma)	$^{235}U–^{207}Pb$ age (Ma)	Th/U
M1-060	0.01614 ± 0.00062	0.1054 ± 0.0181	103.2 ± 4.0	101.7 ± 17.4	0.74	M2-059	0.01816 ± 0.00078	0.1139 ± 0.0158	116.0 ± 5.0	109.5 ± 15.2	0.56
M1-061	0.01477 ± 0.00051	0.0957 ± 0.0144	94.5 ± 3.3	92.8 ± 13.9	0.84	M2-060	0.01481 ± 0.00062	0.1093 ± 0.0133	94.8 ± 3.9	105.3 ± 12.8	0.36
M1-062	0.01528 ± 0.00053	0.1009 ± 0.0142	97.8 ± 3.4	97.6 ± 13.7	0.70	M2-061	0.04857 ± 0.00221	0.3552 ± 0.0521	305.7 ± 13.9	308.6 ± 45.3	0.31
M1-063	0.01783 ± 0.00061	0.1186 ± 0.0162	113.9 ± 3.9	113.8 ± 15.5	0.32	M2-062	0.01883 ± 0.00066	0.1221 ± 0.0161	120.3 ± 4.2	116.9 ± 15.4	0.41
M1-064	0.01550 ± 0.00065	0.1107 ± 0.0194	99.1 ± 4.2	106.6 ± 18.7	0.50	M2-063	0.01665 ± 0.00049	0.1124 ± 0.0105	106.5 ± 3.1	108.1 ± 10.1	0.38
M1-065	0.01562 ± 0.00051	0.1152 ± 0.0143	99.9 ± 3.3	110.7 ± 13.8	0.41	M2-064	0.01611 ± 0.00064	0.0987 ± 0.0163	103.0 ± 4.1	95.6 ± 15.8	0.52
M1-066	0.01399 ± 0.00035	0.0970 ± 0.0076	89.6 ± 2.2	94.0 ± 7.4	0.46	M2-065	0.03960 ± 0.00147	0.2868 ± 0.0392	250.3 ± 9.3	256.0 ± 35.0	0.72
M1-067	0.01580 ± 0.00042	0.1120 ± 0.0103	101.1 ± 2.7	107.8 ± 9.9	0.28	M2-066	0.01632 ± 0.00054	0.1093 ± 0.0129	104.4 ± 3.5	105.3 ± 12.5	0.53
M1-068	0.01511 ± 0.00038	0.0971 ± 0.0080	96.7 ± 2.4	94.1 ± 7.8	0.88	M2-067	0.01412 ± 0.00059	0.0873 ± 0.0110	90.4 ± 3.8	85.0 ± 10.7	0.55
M1-069	0.01458 ± 0.00051	0.1032 ± 0.0136	93.3 ± 3.3	99.7 ± 13.2	0.47	M2-068	0.01603 ± 0.00067	0.1142 ± 0.0141	102.5 ± 4.3	109.8 ± 13.5	0.27
M1-070	0.01556 ± 0.00062	0.0950 ± 0.0163	99.6 ± 4.0	92.2 ± 15.8	0.46	M2-069	0.01575 ± 0.00070	0.1070 ± 0.0124	100.7 ± 4.5	103.2 ± 12.0	0.63
M1-071	0.01631 ± 0.00064	0.1193 ± 0.0185	104.3 ± 4.1	114.5 ± 17.8	0.48	M2-070	0.04096 ± 0.00211	0.2765 ± 0.0472	258.8 ± 13.3	247.8 ± 42.3	0.73
M1-072	0.01700 ± 0.00070	0.1138 ± 0.0196	108.7 ± 4.5	109.4 ± 18.8	0.67	M2-071	0.01293 ± 0.00056	0.0850 ± 0.0094	82.8 ± 3.6	82.8 ± 9.1	0.49
M1-073	0.01532 ± 0.00049	0.1060 ± 0.0124	98.0 ± 3.2	102.3 ± 11.9	0.39	M2-072	0.01539 ± 0.00079	0.1034 ± 0.0179	98.5 ± 5.1	99.9 ± 17.3	0.38
M1-074	0.01751 ± 0.00051	0.1202 ± 0.0119	111.9 ± 3.3	115.3 ± 11.4	0.23	M2-073	0.01686 ± 0.00075	0.1158 ± 0.0134	107.8 ± 4.8	111.3 ± 12.9	0.60
M1-075	0.04256 ± 0.00187	0.3201 ± 0.0614	268.7 ± 11.8	282.0 ± 54.1	0.50	M2-074	0.01598 ± 0.00084	0.1102 ± 0.0196	102.2 ± 5.4	106.1 ± 18.9	0.46
M1-076	0.01655 ± 0.00044	0.1190 ± 0.0136	105.8 ± 2.8	114.2 ± 13.0	0.75	M2-075	0.01455 ± 0.00051	0.1057 ± 0.0138	93.1 ± 3.2	102.0 ± 13.3	0.65
M1-077	0.01563 ± 0.00047	0.1055 ± 0.0142	100.0 ± 3.0	101.8 ± 13.7	0.66	M2-076	0.01504 ± 0.00065	0.1070 ± 0.0191	96.2 ± 4.2	103.2 ± 18.4	0.79
M1-078	0.01784 ± 0.00052	0.1138 ± 0.0152	114.0 ± 3.3	109.4 ± 14.7	0.47	M2-077	0.01365 ± 0.00052	0.0873 ± 0.0136	87.4 ± 3.3	84.9 ± 13.3	1.00
M1-079	0.01652 ± 0.00041	0.1187 ± 0.0124	105.6 ± 2.6	113.9 ± 11.9	0.35	M2-078	0.01698 ± 0.00060	0.1112 ± 0.0154	108.5 ± 3.8	107.1 ± 14.8	0.33
M1-080	0.01382 ± 0.00043	0.1011 ± 0.0135	88.5 ± 2.7	97.8 ± 13.1	0.56	Site 5 (Sample 160705-01) in Hattomaki Unit (M3) of Miyama Complex					
M1-081	0.01453 ± 0.00045	0.0897 ± 0.0093	93.0 ± 2.9	87.2 ± 9.0	0.39	M3-001	0.05850 ± 0.00729	0.4754 ± 0.1341	366.5 ± 45.7	394.9 ± 111.4	0.92
M1-082	0.01468 ± 0.00061	0.0966 ± 0.0160	93.9 ± 3.9	93.7 ± 15.5	0.76	M3-002	0.24019 ± 0.00998	3.5519 ± 0.2226	1387.7 ± 57.6	1538.8 ± 96.4	0.47

http://dx.doi.org/10.5772/67982

Grain No.	$^{206}Pb/^{238}U$	$^{207}Pb/^{235}U$	^{238}U–^{206}Pb age (Ma)	^{235}U–^{207}Pb age (Ma)	Th/U	Grain No.	$^{206}Pb/^{238}U$	$^{207}Pb/^{235}U$	^{238}U–^{206}Pb age (Ma)	^{235}U–^{207}Pb age (Ma)	Th/U
M1-083	0.01424 ± 0.00046	0.0996 ± 0.0107	91.2 ± 3.0	96.4 ± 10.3	0.74	M3-003	0.33403 ± 0.01341	5.8980 ± 0.3195	1857.9 ± 74.6	1960.9 ± 106.2	0.35
M1-084	0.03231 ± 0.00089	0.2323 ± 0.0161	205.0 ± 5.6	212.1 ± 14.7	0.54	M3-004	0.43957 ± 0.01787	9.5363 ± 0.5276	2348.7 ± 95.5	2391.0 ± 132.3	0.89
M1-085	0.01482 ± 0.00055	0.0912 ± 0.0130	94.9 ± 3.5	88.6 ± 12.7	0.57	M3-005	0.48244 ± 0.01937	11.2839 ± 0.4946	2537.9 ± 101.9	2546.8 ± 111.6	0.45
M1-086	0.01370 ± 0.00039	0.0953 ± 0.0073	87.7 ± 2.5	92.5 ± 7.1	0.61	M3-006	0.03281 ± 0.00184	0.2115 ± 0.0305	208.1 ± 11.7	194.8 ± 28.1	0.84
Site 4 (Sample 160705-02) in Gomadanzan Unit (M2) of Miyama Complex						M3-007	0.30011 ± 0.00706	4.6679 ± 0.1602	1691.9 ± 39.8	1761.4 ± 60.5	0.22
M2-001	0.01512 ± 0.00055	0.0922 ± 0.0142	96.7 ± 3.5	89.6 ± 13.8	0.56	M3-008	0.02922 ± 0.00082	0.1880 ± 0.0159	185.7 ± 5.2	174.9 ± 14.8	0.39
M2-002	0.04319 ± 0.00196	0.2801 ± 0.0544	272.6 ± 12.4	250.8 ± 48.7	0.42	M3-009	0.41899 ± 0.01069	8.5583 ± 0.3539	2255.9 ± 57.5	2292.1 ± 94.8	0.24
M2-003	0.01287 ± 0.00033	0.0869 ± 0.0068	82.4 ± 2.1	84.6 ± 6.7	0.39	M3-010	0.44011 ± 0.01046	8.1774 ± 0.2802	2351.1 ± 55.9	2250.8 ± 77.1	0.26
M2-004	0.04466 ± 0.00173	0.2934 ± 0.0436	281.6 ± 10.9	261.3 ± 38.8	0.51	M3-011	0.33916 ± 0.00888	5.1873 ± 0.2493	1882.6 ± 49.3	1850.5 ± 88.9	0.81
M2-005	0.01645 ± 0.00061	0.1174 ± 0.0161	105.2 ± 3.9	112.7 ± 15.4	0.35	M3-012	0.56269 ± 0.02040	13.7939 ± 1.1389	2877.7 ± 104.3	2735.6 ± 225.9	0.58
M2-006	0.01394 ± 0.00065	0.0932 ± 0.0181	89.3 ± 4.2	90.5 ± 17.5	0.59	M3-013	0.43205 ± 0.01588	9.3971 ± 0.7960	2314.9 ± 85.1	2377.5 ± 201.4	0.67
M2-007	0.01506 ± 0.00086	0.0910 ± 0.0168	96.3 ± 5.5	88.4 ± 16.3	0.51	M3-014	0.34000 ± 0.01231	5.5575 ± 0.4671	1886.7 ± 68.3	1909.5 ± 160.5	0.30
M2-008	0.01654 ± 0.00082	0.1066 ± 0.0238	105.7 ± 5.2	102.9 ± 23.0	0.79	M3-015	0.34359 ± 0.01245	5.4157 ± 0.4563	1903.9 ± 69.0	1887.3 ± 159.0	0.14
M2-009	0.01523 ± 0.00039	0.0963 ± 0.0080	97.4 ± 2.5	93.4 ± 7.8	0.33	M3-016	0.32813 ± 0.01185	5.6375 ± 0.4706	1829.3 ± 66.1	1921.8 ± 160.4	0.51
M2-010	0.01836 ± 0.00047	0.1168 ± 0.0097	117.3 ± 3.0	112.2 ± 9.3	0.29	M3-017	0.01212 ± 0.00059	0.0839 ± 0.0147	77.6 ± 3.8	81.8 ± 14.3	0.42
M2-011	0.01342 ± 0.00036	0.0858 ± 0.0078	85.9 ± 2.3	83.6 ± 7.6	0.32	M3-018	0.45878 ± 0.01704	10.9825 ± 0.9386	2434.2 ± 90.4	2521.6 ± 215.5	0.56
M2-012	0.01419 ± 0.00047	0.0948 ± 0.0123	90.8 ± 3.0	92.0 ± 12.0	0.53	M3-019	0.34373 ± 0.01266	6.1176 ± 0.5269	1904.6 ± 70.2	1992.7 ± 171.6	0.15
M2-013	0.01515 ± 0.00043	0.1041 ± 0.0103	97.0 ± 2.7	100.6 ± 10.0	0.31	M3-020	0.44305 ± 0.01588	8.8143 ± 0.7212	2364.3 ± 84.7	2318.9 ± 189.7	0.40
M2-014	0.01669 ± 0.00053	0.1140 ± 0.0135	106.7 ± 3.4	109.6 ± 13.0	0.60	M3-021	0.38735 ± 0.00861	7.1727 ± 0.2522	2110.5 ± 46.9	2133.0 ± 75.0	0.14
M2-015	0.01443 ± 0.00066	0.0874 ± 0.0174	92.3 ± 4.2	85.1 ± 16.9	0.47	M3-022	0.02723 ± 0.00083	0.2026 ± 0.0212	173.2 ± 5.3	187.4 ± 19.6	0.53
M2-016	0.01537 ± 0.00046	0.1075 ± 0.0097	98.3 ± 2.9	103.7 ± 9.3	0.35	M3-023	0.44212 ± 0.01106	10.0244 ± 0.4656	2360.1 ± 59.0	2436.9 ± 113.2	0.45
M2-017	0.01666 ± 0.00050	0.1137 ± 0.0104	106.5 ± 3.2	109.3 ± 10.0	0.29	M3-024	0.31680 ± 0.00691	5.0013 ± 0.1666	1774.1 ± 38.7	1819.5 ± 60.6	0.20
M2-018	0.01608 ± 0.00055	0.1074 ± 0.0132	102.9 ± 3.5	103.6 ± 12.7	0.30	M3-025	0.32397 ± 0.00708	5.6488 ± 0.1872	1809.1 ± 39.5	1923.5 ± 63.8	0.12
M2-019	0.01440 ± 0.00045	0.1040 ± 0.0089	92.2 ± 2.9	100.4 ± 8.6	0.32	M3-026	0.33157 ± 0.00721	5.6222 ± 0.1835	1846.0 ± 40.1	1919.4 ± 62.6	0.30
M2-020	0.01676 ± 0.00071	0.1231 ± 0.0179	107.1 ± 4.5	117.9 ± 17.2	0.57	M3-027	0.31506 ± 0.00997	5.0267 ± 0.2449	1765.6 ± 55.9	1823.8 ± 88.9	0.38

Grain No.	$^{206}Pb/^{238}U$	$^{207}Pb/^{235}U$	$^{238}U–^{206}Pb$ age (Ma)	$^{235}U–^{207}Pb$ age (Ma)	Th/U	Grain No.	$^{206}Pb/^{238}U$	$^{207}Pb/^{235}U$	$^{238}U–^{206}Pb$ age (Ma)	$^{235}U–^{207}Pb$ age (Ma)	Th/U
M2-021	0.01749 ± 0.00069	0.1130 ± 0.0146	111.8 ± 4.4	108.7 ± 14.0	0.38	M3-028	0.35280 ± 0.01140	5.6616 ± 0.2991	1948.0 ± 63.0	1925.5 ± 101.7	0.31
M2-022	0.01507 ± 0.00057	0.1074 ± 0.0117	96.4 ± 3.6	103.6 ± 11.3	0.33	M3-029	0.30026 ± 0.00966	4.7000 ± 0.2452	1692.6 ± 54.5	1767.2 ± 92.2	0.60
M2-023	0.01289 ± 0.00043	0.0881 ± 0.0062	82.6 ± 2.8	85.7 ± 6.0	0.28	M3-030	0.39751 ± 0.01278	8.6305 ± 0.4336	2157.6 ± 69.4	2299.7 ± 115.5	0.51
M2-024	0.01580 ± 0.00057	0.1023 ± 0.0100	101.0 ± 3.6	98.9 ± 9.7	0.36	M3-031	0.29785 ± 0.00968	4.8453 ± 0.2609	1680.6 ± 54.6	1792.7 ± 96.5	0.31
M2-025	0.01943 ± 0.00052	0.1320 ± 0.0148	124.0 ± 3.3	125.9 ± 14.1	0.45	M3-032	0.31148 ± 0.00991	4.8101 ± 0.2401	1748.0 ± 55.6	1786.6 ± 89.2	0.24
M2-026	0.45382 ± 0.00845	9.6216 ± 0.5603	2412.2 ± 44.9	2399.2 ± 139.7	0.28	M3-033	0.27492 ± 0.00864	4.4461 ± 0.2092	1565.7 ± 49.2	1720.9 ± 81.0	0.10
M2-027	0.01682 ± 0.00043	0.1182 ± 0.0124	107.5 ± 2.8	113.4 ± 11.9	0.37	M3-034	0.31298 ± 0.00415	4.8704 ± 0.1568	1755.3 ± 23.3	1797.1 ± 57.9	0.18
M2-028	0.01303 ± 0.00042	0.0847 ± 0.0119	83.4 ± 2.7	82.5 ± 11.6	0.44	M3-035	0.29957 ± 0.00424	4.8954 ± 0.1722	1689.2 ± 23.9	1801.4 ± 63.3	0.13
M2-029	0.01429 ± 0.00050	0.0912 ± 0.0144	91.5 ± 3.2	88.6 ± 14.0	0.92	M3-036	0.31724 ± 0.00543	4.9510 ± 0.2228	1776.2 ± 30.4	1810.9 ± 81.5	0.35
M2-030	0.01565 ± 0.00050	0.1055 ± 0.0145	100.1 ± 3.2	101.8 ± 14.0	0.47	M3-037	0.32048 ± 0.00462	5.0713 ± 0.1827	1792.1 ± 25.8	1831.2 ± 66.0	0.13
M2-031	0.01623 ± 0.00049	0.1079 ± 0.0140	103.8 ± 3.1	104.1 ± 13.5	0.51	M3-038	0.30604 ± 0.00495	4.8184 ± 0.2018	1721.2 ± 27.8	1788.1 ± 74.9	0.12
M2-032	0.01468 ± 0.00072	0.0885 ± 0.0160	93.9 ± 4.6	86.1 ± 15.6	0.56	M3-039	0.34634 ± 0.00556	5.7967 ± 0.1821	1917.1 ± 30.8	1945.8 ± 61.1	0.41
M2-033	0.01541 ± 0.00066	0.1054 ± 0.0137	98.6 ± 4.2	101.7 ± 13.2	0.74	M3-040	0.03477 ± 0.00068	0.2394 ± 0.0172	220.3 ± 4.3	217.9 ± 15.7	0.58
M2-034	0.01249 ± 0.00067	0.0753 ± 0.0158	80.0 ± 4.3	73.7 ± 15.5	1.50	M3-041	0.42758 ± 0.00795	8.7703 ± 0.3758	2294.8 ± 42.7	2314.3 ± 99.2	0.64
M2-035	0.01171 ± 0.00046	0.0850 ± 0.0085	75.0 ± 2.9	82.8 ± 8.3	1.17	M3-042	0.04292 ± 0.00084	0.3018 ± 0.0175	270.9 ± 5.3	267.8 ± 15.5	0.31
M2-036	0.01409 ± 0.00050	0.0924 ± 0.0130	90.2 ± 3.2	89.7 ± 12.6	0.77	M3-043	0.01543 ± 0.00056	0.1129 ± 0.0172	98.7 ± 3.6	108.6 ± 16.5	0.32
M2-037	0.01651 ± 0.00045	0.1223 ± 0.0104	105.6 ± 2.8	117.2 ± 10.0	0.39	M3-044	0.45494 ± 0.00826	9.8946 ± 0.4103	2417.2 ± 43.9	2424.9 ± 100.6	1.32
M2-038	0.01730 ± 0.00076	0.1203 ± 0.0222	110.6 ± 4.9	115.4 ± 21.3	0.48	M3-045	0.29874 ± 0.00526	4.8806 ± 0.1988	1685.1 ± 29.7	1798.9 ± 73.3	0.15
M2-039	0.01692 ± 0.00057	0.1053 ± 0.0142	108.1 ± 3.6	101.6 ± 13.7	0.40	M3-046	0.02860 ± 0.00137	0.2062 ± 0.0225	181.8 ± 8.7	190.3 ± 20.8	0.48
M2-040	0.04208 ± 0.00124	0.3021 ± 0.0308	265.7 ± 7.8	268.0 ± 27.4	0.44	M3-047	0.38733 ± 0.01758	7.8829 ± 0.5714	2110.5 ± 95.8	2217.7 ± 160.7	0.71
M2-041	0.01446 ± 0.00066	0.0903 ± 0.0182	92.6 ± 4.2	87.8 ± 17.7	0.55	M3-048	0.36623 ± 0.01640	7.1120 ± 0.4951	2011.6 ± 90.1	2125.5 ± 148.0	0.34
M2-042	0.01532 ± 0.00046	0.1080 ± 0.0115	98.0 ± 2.9	104.2 ± 11.1	0.27	M3-049	0.31207 ± 0.01382	4.8847 ± 0.3307	1750.9 ± 77.6	1799.6 ± 121.8	0.32
M2-043	0.01441 ± 0.00079	0.0959 ± 0.0222	92.2 ± 5.0	93.0 ± 21.5	0.52	M3-050	0.32301 ± 0.00981	5.1919 ± 0.2393	1804.4 ± 54.8	1851.2 ± 85.3	0.12
M2-044	0.01567 ± 0.00064	0.1094 ± 0.0167	100.3 ± 4.1	105.4 ± 16.1	0.39	M3-051	0.30477 ± 0.00990	4.6026 ± 0.2744	1714.9 ± 55.7	1749.7 ± 104.3	0.31
M2-045	0.01434 ± 0.00052	0.0871 ± 0.0113	91.8 ± 3.3	84.8 ± 11.0	0.47	M3-052	0.03053 ± 0.00125	0.2168 ± 0.0303	193.9 ± 7.9	199.2 ± 27.8	0.59

http://dx.doi.org/10.5772/67982

Grain No.	$^{206}Pb/^{238}U$	$^{207}Pb/^{235}U$	^{238}U–^{206}Pb age (Ma)	^{235}U–^{207}Pb age (Ma)	Th/U	Grain No.	$^{206}Pb/^{238}U$	$^{207}Pb/^{235}U$	^{238}U–^{206}Pb age (Ma)	^{235}U–^{207}Pb age (Ma)	Th/U
M2-046	0.01546 ± 0.00047	0.1050 ± 0.0089	98.9 ± 3.0	101.4 ± 8.6	0.31	M3-053	0.01582 ± 0.00066	0.1058 ± 0.0156	101.2 ± 4.2	102.1 ± 15.1	0.37
M2-047	0.01498 ± 0.00054	0.0960 ± 0.0122	95.9 ± 3.4	93.0 ± 11.8	0.50	M3-054	0.29973 ± 0.00894	4.6933 ± 0.1990	1690.0 ± 50.4	1766.0 ± 74.9	0.13
M2-048	0.01279 ± 0.00045	0.0914 ± 0.0099	81.9 ± 2.9	88.8 ± 9.7	2.02	M3-055	0.29581 ± 0.00954	4.6026 ± 0.2665	1670.5 ± 53.9	1749.7 ± 101.3	0.76
M2-049	0.01367 ± 0.00050	0.0976 ± 0.0127	87.5 ± 3.2	94.6 ± 12.3	0.59	M3-056	0.40681 ± 0.01124	7.5780 ± 0.3069	2200.3 ± 60.8	2182.2 ± 88.4	0.50
M2-050	0.01415 ± 0.00044	0.1004 ± 0.0092	90.6 ± 2.8	97.2 ± 8.9	0.60	M3-057	0.37011 ± 0.00991	7.6708 ± 0.2655	2029.9 ± 54.3	2193.1 ± 75.9	0.36
M2-051	0.03984 ± 0.00121	0.2941 ± 0.0255	251.9 ± 7.7	261.8 ± 22.7	0.53	M3-058	0.40937 ± 0.01147	8.6387 ± 0.3599	2212.0 ± 62.0	2300.6 ± 95.8	0.36
M2-052	0.01488 ± 0.00063	0.0896 ± 0.0159	95.2 ± 4.1	87.1 ± 15.5	0.43	M3-059	0.02851 ± 0.00091	0.2096 ± 0.0189	181.2 ± 5.8	193.2 ± 17.4	0.23
M2-053	0.39156 ± 0.01032	7.4980 ± 0.3398	2130.1 ± 56.2	2172.7 ± 98.5	0.29	M3-060	0.30442 ± 0.00827	4.9979 ± 0.1931	1713.2 ± 46.5	1818.9 ± 70.3	0.10
M2-054	0.01326 ± 0.00077	0.0896 ± 0.0250	84.9 ± 4.9	87.1 ± 24.4	0.31	M3-061	0.02895 ± 0.00155	0.2070 ± 0.0452	183.9 ± 9.8	191.0 ± 41.7	1.07
M2-055	0.01356 ± 0.00057	0.0867 ± 0.0075	86.8 ± 3.7	84.4 ± 7.3	0.55	M3-062	0.32053 ± 0.00915	4.9799 ± 0.2392	1792.3 ± 51.2	1815.8 ± 87.2	0.47
M2-056	0.01234 ± 0.00065	0.0844 ± 0.0146	79.0 ± 4.2	82.2 ± 14.3	1.18	M3-063	0.02883 ± 0.00087	0.2162 ± 0.0165	183.2 ± 5.5	198.8 ± 15.2	0.36
M2-057	0.01554 ± 0.00058	0.1080 ± 0.0099	99.4 ± 3.7	104.1 ± 9.5	0.27	M3-064	0.43490 ± 0.01188	9.1731 ± 0.5270	2327.8 ± 63.6	2355.3 ± 135.3	0.95
M2-058	0.01422 ± 0.00053	0.0996 ± 0.0089	91.1 ± 3.4	96.4 ± 8.6	0.34	M3-065	0.36458 ± 0.00819	6.5059 ± 0.2484	2003.8 ± 45.0	2046.6 ± 78.1	0.55

Appendix 4.

Grain No.	$^{206}Pb/^{238}U$	$^{207}Pb/^{235}U$	$^{238}U–^{206}Pb$ age (Ma)	$^{235}U–^{207}Pb$ age (Ma)	Th/U	Grain No.	$^{206}Pb/^{238}U$	$^{207}Pb/^{235}U$	$^{238}U–^{206}Pb$ age (Ma)	$^{235}U–^{207}Pb$ age (Ma)	Th/U
M3-066	0.44954 ± 0.00976	9.7085 ± 0.3250	2393.2 ± 52.0	2407.4 ± 80.6	0.22	Ry1-032	0.01637 ± 0.00043	0.1131 ± 0.0086	104.7 ± 2.7	108.8 ± 8.3	0.38
M3-067	0.02680 ± 0.00070	0.1844 ± 0.0147	170.5 ± 4.5	171.8 ± 13.7	0.32	Ry1-033	0.01476 ± 0.00044	0.1080 ± 0.0107	94.5 ± 2.8	104.1 ± 10.3	0.67
M3-068	0.27836 ± 0.00739	4.3495 ± 0.2610	1583.1 ± 42.0	1702.7 ± 102.2	0.29	Ry1-034	0.01354 ± 0.00059	0.0956 ± 0.0163	86.7 ± 3.8	92.7 ± 15.8	1.82
M3-069	0.30093 ± 0.00711	4.6220 ± 0.2114	1695.9 ± 40.1	1753.2 ± 80.2	0.11	Ry1-035	0.01480 ± 0.00043	0.1039 ± 0.0100	94.7 ± 2.8	100.4 ± 9.7	0.39
M3-070	0.37189 ± 0.00831	5.9032 ± 0.2256	2038.3 ± 45.5	1961.6 ± 75.0	0.44	Ry1-036	0.01440 ± 0.00052	0.0929 ± 0.0134	92.2 ± 3.3	90.2 ± 13.0	0.61
M3-071	0.27481 ± 0.00748	4.2010 ± 0.1892	1565.2 ± 42.6	1674.2 ± 75.4	0.44	Ry1-037	0.01287 ± 0.00039	0.0874 ± 0.0090	82.5 ± 2.5	85.0 ± 8.8	0.55
M3-072	0.01279 ± 0.00058	0.0859 ± 0.0160	81.9 ± 3.7	83.7 ± 15.6	0.65	Ry1-038	0.01293 ± 0.00120	0.0828 ± 0.0140	82.8 ± 7.7	80.8 ± 13.7	0.54
M3-073	0.03551 ± 0.00160	0.2404 ± 0.0436	225.0 ± 10.1	218.8 ± 39.7	0.50	Ry1-039	0.00981 ± 0.00089	0.0648 ± 0.0091	62.9 ± 5.7	63.8 ± 9.0	0.54
M3-074	0.34292 ± 0.00940	5.5050 ± 0.2515	1900.7 ± 52.1	1901.3 ± 86.9	0.13	Ry1-040	0.01336 ± 0.00124	0.0953 ± 0.0163	85.5 ± 8.0	92.5 ± 15.8	1.12
M3-075	0.24853 ± 0.00697	3.7852 ± 0.1902	1430.9 ± 40.1	1589.6 ± 79.9	0.49	Ry1-041	0.01319 ± 0.00039	0.0948 ± 0.0116	84.4 ± 2.5	92.0 ± 11.3	0.61
M3-076	0.31811 ± 0.00879	4.9171 ± 0.2330	1780.5 ± 49.2	1805.1 ± 85.5	0.39	Ry1-042	0.01312 ± 0.00063	0.0898 ± 0.0201	84.0 ± 4.0	87.3 ± 19.6	0.75
M3-077	0.02889 ± 0.00135	0.2111 ± 0.0193	183.6 ± 8.6	194.5 ± 17.8	0.12	Site 7 (Sample 160704-04) in Komatagawa Unit (Ry2) of Ryujin Complex					
M3-078	0.41717 ± 0.01858	8.9603 ± 0.5355	2247.6 ± 100.1	2333.9 ± 139.5	0.57	Ry2-001	0.01056 ± 0.00058	0.0658 ± 0.0174	67.7 ± 3.7	64.7 ± 17.1	0.52
M3-079	0.39475 ± 0.01757	7.3516 ± 0.4406	2144.8 ± 95.5	2155.0 ± 129.2	0.49	Ry2-002	0.01024 ± 0.00064	0.0703 ± 0.0204	65.7 ± 4.1	69.0 ± 20.0	0.74
M3-080	0.27689 ± 0.01243	4.4336 ± 0.2805	1575.7 ± 70.8	1718.6 ± 108.7	0.61	Ry2-003	0.01430 ± 0.00032	0.0973 ± 0.0072	91.6 ± 2.0	94.3 ± 7.0	0.81
M3-081	0.40293 ± 0.01797	8.0274 ± 0.4846	2182.5 ± 97.4	2234.0 ± 134.9	0.35	Ry2-004	0.01205 ± 0.00051	0.0820 ± 0.0154	77.2 ± 3.3	80.0 ± 15.0	0.59
M3-082	0.34039 ± 0.00796	5.1557 ± 0.2514	1888.5 ± 44.2	1845.3 ± 90.0	0.34	Ry2-005	0.01312 ± 0.00048	0.0961 ± 0.0145	84.0 ± 3.1	93.2 ± 14.0	0.69
M3-083	0.02869 ± 0.00091	0.2033 ± 0.0238	182.3 ± 5.8	187.9 ± 22.0	1.04	Ry2-006	0.35260 ± 0.00759	5.4637 ± 0.2551	1947.0 ± 41.9	1894.8 ± 88.5	0.18
M3-084	0.30129 ± 0.00635	4.7291 ± 0.1693	1697.7 ± 35.8	1772.3 ± 63.5	0.37	Ry2-007	0.01029 ± 0.00052	0.0676 ± 0.0143	66.0 ± 3.3	66.4 ± 14.1	0.62
M3-085	0.39364 ± 0.01455	7.5405 ± 0.4093	2139.7 ± 79.1	2177.7 ± 118.2	0.32	Ry2-008	0.01731 ± 0.00055	0.1248 ± 0.0110	110.6 ± 3.5	119.4 ± 10.5	0.44
M3-086	0.37331 ± 0.01382	6.1409 ± 0.3377	2045.0 ± 75.7	1996.0 ± 109.8	0.18	Ry2-009	0.01217 ± 0.00048	0.0823 ± 0.0106	78.0 ± 3.1	80.3 ± 10.3	0.51
M3-087	0.31720 ± 0.01210	5.1176 ± 0.3196	1776.1 ± 67.8	1839.0 ± 114.8	0.74	Ry2-010	0.01024 ± 0.00054	0.0712 ± 0.0151	65.7 ± 3.5	69.9 ± 14.8	0.48
M3-088	0.29584 ± 0.01113	4.6848 ± 0.2774	1670.6 ± 62.8	1764.5 ± 104.5	0.72	Ry2-011	0.01551 ± 0.00058	0.0988 ± 0.0111	99.2 ± 3.7	95.6 ± 10.8	0.35

http://dx.doi.org/10.5772/67982

Grain No.	$^{206}Pb/^{238}U$	$^{207}Pb/^{235}U$	^{238}U–^{206}Pb age (Ma)	^{235}U–^{207}Pb age (Ma)	Th/U	Grain No.	$^{206}Pb/^{238}U$	$^{207}Pb/^{235}U$	^{238}U–^{206}Pb age (Ma)	^{235}U–^{207}Pb age (Ma)	Th/U
M3-089	0.30734 ± 0.01186	4.8209 ± 0.3154	1727.6 ± 66.7	1788.5 ± 117.0	0.93	Ry2-012	0.01100 ± 0.00058	0.0656 ± 0.0152	70.5 ± 3.7	64.5 ± 14.9	0.82
M3-090	0.28392 ± 0.01097	4.5500 ± 0.2978	1611.1 ± 62.2	1740.1 ± 113.9	0.45	Ry2-013	0.03942 ± 0.00440	0.2833 ± 0.0457	249.2 ± 27.8	253.3 ± 40.9	0.44
M3-091	0.01459 ± 0.00072	0.0889 ± 0.0154	93.4 ± 4.6	86.4 ± 15.0	0.47	Ry2-014	0.01025 ± 0.00120	0.0627 ± 0.0147	65.8 ± 7.7	61.7 ± 14.5	0.61
M3-092	0.02902 ± 0.00057	0.2068 ± 0.0135	184.4 ± 3.6	190.8 ± 12.4	0.47	Ry2-015	0.01190 ± 0.00142	0.0749 ± 0.0197	76.2 ± 9.1	73.3 ± 19.3	0.76
M3-093	0.31623 ± 0.00499	4.9435 ± 0.1415	1771.3 ± 28.0	1809.7 ± 51.8	0.34	Ry2-016	0.00967 ± 0.00131	0.0637 ± 0.0178	62.0 ± 8.4	62.7 ± 17.5	0.93
M3-094	0.35706 ± 0.00549	5.8265 ± 0.1532	1968.2 ± 30.3	1950.3 ± 51.3	0.45	Ry2-017	0.01902 ± 0.00047	0.1293 ± 0.0092	121.5 ± 3.0	123.5 ± 8.8	0.44
M3-095	0.27993 ± 0.00433	4.2977 ± 0.1007	1591.0 ± 24.6	1692.9 ± 39.7	0.27	Ry2-018	0.01036 ± 0.00114	0.0742 ± 0.0172	66.4 ± 7.3	72.6 ± 16.8	0.64
M3-096	0.40651 ± 0.01377	8.3635 ± 0.4154	2198.9 ± 74.5	2271.2 ± 112.8	0.37	Ry2-019	0.01307 ± 0.00115	0.0945 ± 0.0184	83.7 ± 7.3	91.7 ± 17.9	0.54
M3-097	0.25072 ± 0.00854	3.7948 ± 0.1979	1442.2 ± 49.2	1591.6 ± 83.0	0.25	Ry2-020	0.01441 ± 0.00113	0.0973 ± 0.0179	92.2 ± 7.2	94.3 ± 17.3	0.38
M3-098	0.31836 ± 0.01143	5.3902 ± 0.3254	1781.7 ± 64.0	1883.2 ± 113.7	0.22	Ry2-021	0.01159 ± 0.00046	0.0799 ± 0.0113	74.3 ± 2.9	78.1 ± 11.1	0.62
M3-099	0.34825 ± 0.01189	5.6094 ± 0.2915	1926.2 ± 65.8	1917.5 ± 99.6	0.27	Ry2-022	0.01050 ± 0.00049	0.0626 ± 0.0123	67.3 ± 3.1	61.6 ± 12.1	0.74
M3-100	0.02659 ± 0.00115	0.1756 ± 0.0231	169.2 ± 7.3	164.3 ± 21.6	0.91	Ry2-023	0.01454 ± 0.00058	0.1005 ± 0.0145	93.1 ± 3.7	97.3 ± 14.0	0.66
M3-101	0.31664 ± 0.01077	5.0037 ± 0.2574	1773.3 ± 60.3	1819.9 ± 93.6	0.31	Ry2-024	0.33110 ± 0.00930	5.4844 ± 0.2161	1843.7 ± 51.8	1898.1 ± 74.8	0.37
M3-102	0.01405 ± 0.00055	0.0982 ± 0.0102	90.0 ± 3.5	95.1 ± 9.9	0.41	Ry2-025	0.01285 ± 0.00067	0.0939 ± 0.0186	82.3 ± 4.3	91.2 ± 18.1	0.56
M3-103	0.29647 ± 0.00521	4.7116 ± 0.1299	1673.8 ± 29.4	1769.2 ± 48.8	0.17	Ry2-026	0.01718 ± 0.00119	0.1056 ± 0.0312	109.8 ± 7.6	101.9 ± 30.1	0.39
M3-104	0.48278 ± 0.00896	11.4133 ± 0.3620	2539.4 ± 47.1	2557.4 ± 81.1	1.16	Ry2-027	0.03886 ± 0.00151	0.2803 ± 0.0361	245.8 ± 9.5	250.9 ± 32.3	0.79
M3-105	0.45489 ± 0.00849	10.3393 ± 0.3377	2417.0 ± 45.1	2465.5 ± 80.5	1.04	Ry2-028	0.01669 ± 0.00053	0.1073 ± 0.0091	106.7 ± 3.4	103.5 ± 8.8	0.44
M3-106	0.34227 ± 0.00614	6.8732 ± 0.1994	1897.6 ± 34.0	2095.1 ± 60.8	0.12	Ry2-029	0.01353 ± 0.00053	0.0895 ± 0.0123	86.6 ± 3.4	87.0 ± 11.9	0.44
M3-107	0.37280 ± 0.00674	7.1919 ± 0.2165	2042.6 ± 36.9	2135.4 ± 64.3	0.14	Ry2-030	0.01623 ± 0.00051	0.1066 ± 0.0089	103.8 ± 3.3	102.9 ± 8.5	0.58
M3-108	0.39657 ± 0.00721	8.7743 ± 0.2626	2153.2 ± 39.1	2314.7 ± 69.3	0.57	Ry2-031	0.01217 ± 0.00045	0.0759 ± 0.0098	78.0 ± 2.9	74.2 ± 9.6	0.61
M3-109	0.46478 ± 0.00776	10.1766 ± 0.2958	2460.6 ± 41.1	2450.9 ± 71.2	0.79	Ry2-032	0.01210 ± 0.00048	0.0753 ± 0.0112	77.6 ± 3.1	73.7 ± 10.9	1.10
M3-110	0.39449 ± 0.00645	7.0356 ± 0.2012	2143.6 ± 35.0	2115.9 ± 60.5	0.46	Ry2-033	0.01035 ± 0.00041	0.0700 ± 0.0102	66.4 ± 2.6	68.7 ± 10.0	0.96
M3-111	0.02597 ± 0.00052	0.1965 ± 0.0123	165.2 ± 3.3	182.2 ± 11.4	0.51	Ry2-034	0.01453 ± 0.00067	0.0958 ± 0.0192	93.0 ± 4.3	92.9 ± 18.6	0.43
M3-112	0.30986 ± 0.00486	4.9552 ± 0.1265	1740.0 ± 27.3	1811.6 ± 46.3	0.28	Ry2-035	0.01150 ± 0.00047	0.0805 ± 0.0136	73.7 ± 3.0	78.6 ± 13.3	0.76
M3-113	0.03669 ± 0.00119	0.2772 ± 0.0371	232.3 ± 7.5	248.4 ± 33.3	0.51	Ry2-036	0.41908 ± 0.00963	8.0401 ± 0.3381	2256.3 ± 51.8	2235.5 ± 94.0	0.26

Grain No.	$^{206}Pb/^{238}U$	$^{207}Pb/^{235}U$	^{238}U–^{206}Pb age (Ma)	^{235}U–^{207}Pb age (Ma)	Th/U	Grain No.	$^{206}Pb/^{238}U$	$^{207}Pb/^{235}U$	^{238}U–^{206}Pb age (Ma)	^{235}U–^{207}Pb age (Ma)	Th/U
M3-114	0.02673 ± 0.00061	0.2000 ± 0.0161	170.0 ± 3.9	185.1 ± 14.9	0.57	Ry2-037	0.01049 ± 0.00048	0.0667 ± 0.0137	67.3 ± 3.1	65.6 ± 13.5	1.20
M3-115	0.01207 ± 0.00034	0.0874 ± 0.0097	77.3 ± 2.1	85.1 ± 9.5	0.98	Ry2-038	0.01497 ± 0.00046	0.0921 ± 0.0104	95.8 ± 2.9	89.5 ± 10.1	0.77
M3-116	0.44019 ± 0.00887	9.4339 ± 0.3910	2351.5 ± 47.4	2381.1 ± 98.7	0.72	Ry2-039	0.01335 ± 0.00032	0.0909 ± 0.0082	85.5 ± 2.0	88.4 ± 8.0	0.26
M3-117	0.02934 ± 0.00071	0.2106 ± 0.0172	186.4 ± 4.5	194.1 ± 15.9	0.76	Ry2-040	0.01286 ± 0.00045	0.0913 ± 0.0215	82.3 ± 2.9	88.7 ± 20.9	0.82
M3-118	0.01595 ± 0.00039	0.1125 ± 0.0096	102.0 ± 2.5	108.2 ± 9.2	0.51	Ry2-041	0.01185 ± 0.00035	0.0728 ± 0.0123	75.9 ± 2.2	71.3 ± 12.1	0.53
M3-119	0.02722 ± 0.00058	0.2023 ± 0.0129	173.1 ± 3.7	187.0 ± 11.9	0.43	Ry2-042	0.35483 ± 0.00746	6.1989 ± 0.3038	1957.6 ± 41.2	2004.2 ± 98.2	0.29
Site 6 (Sample 160704-05) in Sohgawa Unit (Ry1) of Ryujin Complex						Ry2-043	0.01398 ± 0.00114	0.0879 ± 0.0210	89.5 ± 7.3	85.5 ± 20.4	0.81
Ry1-001	0.01688 ± 0.00068	0.1121 ± 0.0225	107.9 ± 4.3	107.9 ± 21.6	0.56	Ry2-044	0.01530 ± 0.00064	0.0959 ± 0.0152	97.9 ± 4.1	92.9 ± 14.7	0.80
Ry1-002	0.01499 ± 0.00077	0.0939 ± 0.0226	95.9 ± 4.9	91.1 ± 21.9	0.68	Ry2-045	0.01051 ± 0.00044	0.0686 ± 0.0107	67.4 ± 2.8	67.4 ± 10.5	0.87
Ry1-003	0.01331 ± 0.00047	0.0964 ± 0.0134	85.2 ± 3.0	93.4 ± 13.0	0.59	Ry2-046	0.39638 ± 0.01101	7.8925 ± 0.2777	2152.4 ± 59.8	2218.7 ± 78.1	0.11
Ry1-004	0.04290 ± 0.00147	0.3194 ± 0.0417	270.8 ± 9.3	281.4 ± 36.8	0.59	Ry2-047	0.01545 ± 0.00044	0.0990 ± 0.0132	98.9 ± 2.8	95.9 ± 12.8	0.25
Ry1-005	0.06143 ± 0.00155	0.4831 ± 0.0372	384.3 ± 9.7	400.2 ± 30.8	0.82	Ry2-048	0.40526 ± 0.00568	7.9106 ± 0.2542	2193.2 ± 30.7	2220.8 ± 71.4	0.65
Ry1-006	0.01350 ± 0.00045	0.0974 ± 0.0158	86.5 ± 2.9	94.4 ± 15.3	0.61	Ry2-049	0.02567 ± 0.00043	0.1825 ± 0.0113	163.4 ± 2.8	170.2 ± 10.6	0.41
Ry1-007	0.01817 ± 0.00059	0.1213 ± 0.0157	116.1 ± 3.8	116.3 ± 15.0	0.47	Ry2-050	0.01587 ± 0.00048	0.1110 ± 0.0150	101.5 ± 3.1	106.9 ± 14.5	0.40
Ry1-008	0.01728 ± 0.00050	0.1235 ± 0.0102	110.4 ± 3.2	118.2 ± 9.8	0.38	Ry2-051	0.01343 ± 0.00028	0.0885 ± 0.0076	86.0 ± 1.8	86.1 ± 7.4	0.54
Ry1-009	0.01557 ± 0.00050	0.1133 ± 0.0152	99.6 ± 3.2	109.0 ± 14.6	0.49	Ry2-052	0.04000 ± 0.00102	0.2721 ± 0.0254	252.8 ± 6.5	244.4 ± 22.8	0.41
Ry1-010	0.01404 ± 0.00060	0.1014 ± 0.0191	89.9 ± 3.9	98.1 ± 18.4	0.69	Ry2-053	0.01286 ± 0.00038	0.0909 ± 0.0110	82.4 ± 2.4	88.4 ± 10.7	0.91
Ry1-011	0.01258 ± 0.00055	0.0772 ± 0.0161	80.6 ± 3.6	75.5 ± 15.7	0.54	Ry2-054	0.35297 ± 0.00690	5.7242 ± 0.2042	1948.8 ± 38.1	1935.0 ± 69.0	0.56
Ry1-012	0.35684 ± 0.00588	5.5728 ± 0.2525	1967.2 ± 32.4	1911.8 ± 86.6	0.22	Ry2-055	0.01380 ± 0.00050	0.0944 ± 0.0148	88.3 ± 3.2	91.6 ± 14.3	0.55
Ry1-013	0.01472 ± 0.00028	0.0999 ± 0.0069	94.2 ± 1.8	96.7 ± 6.7	0.97	Ry2-056	0.01180 ± 0.00026	0.0774 ± 0.0060	75.6 ± 1.7	75.7 ± 5.9	0.41
Ry1-014	0.01769 ± 0.00041	0.1214 ± 0.0110	113.1 ± 2.6	116.3 ± 10.6	0.58	Ry2-057	0.01594 ± 0.00038	0.1019 ± 0.0091	102.0 ± 2.4	98.5 ± 8.8	0.29
Ry1-015	0.01539 ± 0.00043	0.0997 ± 0.0132	98.5 ± 2.8	96.5 ± 12.8	0.44	Ry2-058	0.28300 ± 0.00600	4.4133 ± 0.2232	1606.4 ± 34.1	1714.8 ± 86.7	0.52
Ry1-016	0.01704 ± 0.00057	0.1116 ± 0.0076	108.9 ± 3.7	107.4 ± 7.4	0.46	Site 8 (Sample 160704-03) in Yunohara Unit (Ry3) of Ryujin Complex					
Ry1-017	0.39488 ± 0.01271	7.1334 ± 0.2958	2145.4 ± 69.0	2128.1 ± 88.3	0.35	Ry3-001	0.32395 ± 0.01490	5.1223 ± 0.3477	1809.0 ± 83.2	1839.7 ± 124.9	0.19
Ry1-018	0.01345 ± 0.00061	0.0806 ± 0.0159	86.2 ± 3.9	78.7 ± 15.6	0.57	Ry3-002	0.35419 ± 0.01676	5.9175 ± 0.4121	1954.6 ± 92.5	1963.7 ± 136.7	0.14

Grain No.	$^{206}Pb/^{238}U$	$^{207}Pb/^{235}U$	$^{238}U–^{206}Pb$ age (Ma)	$^{235}U–^{207}Pb$ age (Ma)	Th/U	Grain No.	$^{206}Pb/^{238}U$	$^{207}Pb/^{235}U$	$^{238}U–^{206}Pb$ age (Ma)	$^{235}U–^{207}Pb$ age (Ma)	Th/U
Ry1-019	0.40363 ± 0.00797	6.8531 ± 0.3226	2185.7 ± 43.1	2092.5 ± 98.5	0.18	Ry3-003	0.01150 ± 0.00033	0.0752 ± 0.0070	73.7 ± 2.1	73.7 ± 6.8	0.42
Ry1-020	0.01017 ± 0.00042	0.0610 ± 0.0112	65.2 ± 2.7	60.1 ± 11.1	0.99	Ry3-004	0.01349 ± 0.00052	0.0894 ± 0.0141	86.4 ± 3.3	86.9 ± 13.7	0.69
Ry1-021	0.01240 ± 0.00036	0.0786 ± 0.0089	79.5 ± 2.3	76.8 ± 8.7	0.83	Ry3-005	0.32864 ± 0.00851	5.2744 ± 0.2519	1831.8 ± 47.4	1864.7 ± 89.1	0.35
Ry1-022	0.01124 ± 0.00033	0.0777 ± 0.0088	72.1 ± 2.1	76.0 ± 8.6	1.28	Ry3-006	0.01104 ± 0.00040	0.0736 ± 0.0097	70.8 ± 2.6	72.1 ± 9.5	0.58
Ry1-023	0.01557 ± 0.00034	0.1065 ± 0.0074	99.6 ± 2.2	102.8 ± 7.2	0.28	Ry3-007	0.01700 ± 0.00046	0.1243 ± 0.0079	108.6 ± 2.9	119.0 ± 7.5	0.37
Ry1-024	0.03569 ± 0.00090	0.2687 ± 0.0192	226.1 ± 5.7	241.7 ± 17.3	0.30	Ry3-008	0.01077 ± 0.00033	0.0678 ± 0.0065	69.1 ± 2.1	66.6 ± 6.4	0.55
Ry1-025	0.34481 ± 0.00849	5.5609 ± 0.3004	1909.8 ± 47.0	1910.0 ± 103.2	0.87	Ry3-009	0.01718 ± 0.00060	0.1235 ± 0.0149	109.8 ± 3.8	118.2 ± 14.2	0.37
Ry1-026	0.01163 ± 0.00052	0.0795 ± 0.0151	74.5 ± 3.3	77.7 ± 14.7	0.55	Ry3-010	0.01251 ± 0.00043	0.0812 ± 0.0099	80.1 ± 2.7	79.3 ± 9.7	0.32
Ry1-027	0.01514 ± 0.00062	0.1113 ± 0.0182	96.9 ± 4.0	107.2 ± 17.5	0.69	Ry3-011	0.01318 ± 0.00049	0.0821 ± 0.0105	84.4 ± 3.2	80.1 ± 10.2	0.63
Ry1-028	0.04386 ± 0.00136	0.2904 ± 0.0324	276.7 ± 8.6	258.9 ± 28.9	0.46	Ry3-012	0.01276 ± 0.00051	0.0864 ± 0.0123	81.7 ± 3.3	84.1 ± 12.0	0.28
Ry1-029	0.01323 ± 0.00060	0.0908 ± 0.0173	84.7 ± 3.8	88.2 ± 16.9	0.48	Ry3-013	0.35532 ± 0.01016	5.8872 ± 0.2474	1959.9 ± 56.0	1959.3 ± 82.3	0.32
Ry1-030	0.34661 ± 0.00789	5.6171 ± 0.2204	1918.4 ± 43.7	1918.7 ± 75.3	0.25	Ry3-014	0.01230 ± 0.00040	0.0813 ± 0.0068	78.8 ± 2.5	79.4 ± 6.7	0.36
Ry1-031	0.01358 ± 0.00036	0.0987 ± 0.0101	86.9 ± 2.3	95.6 ± 9.8	0.63	Ry3-015	0.01626 ± 0.00057	0.1144 ± 0.0114	104.0 ± 3.7	109.9 ± 10.9	0.36

Appendix 5.

Grain No.	$^{206}Pb/^{238}U$	$^{207}Pb/^{235}U$	^{238}U–^{206}Pb age (Ma)	^{235}U–^{207}Pb age (Ma)	Th/U	Grain No.	$^{206}Pb/^{238}U$	$^{207}Pb/^{235}U$	^{238}U–^{206}Pb age (Ma)	^{235}U–^{207}Pb age (Ma)	Th/U
Ry3-016	0.01320 ± 0.00057	0.0801 ± 0.0105	84.5 ± 3.6	78.2 ± 10.2	0.43	Ry3-030	0.01264 ± 0.00030	0.0856 ± 0.0096	81.0 ± 1.9	83.4 ± 9.4	0.62
Ry3-017	0.03781 ± 0.00113	0.2906 ± 0.0200	239.2 ± 7.1	259.0 ± 17.8	0.60	Ry3-031	0.15669 ± 0.00467	1.7004 ± 0.1571	938.4 ± 28.0	1008.7 ± 93.2	1.35
Ry3-018	0.01577 ± 0.00083	0.0995 ± 0.0228	100.9 ± 5.3	96.3 ± 22.1	0.74	Ry3-032	0.01299 ± 0.00149	0.0850 ± 0.0196	83.2 ± 9.6	82.8 ± 19.1	0.49
Ry3-019	0.00998 ± 0.00023	0.0671 ± 0.0053	64.0 ± 1.5	65.9 ± 5.3	0.43	Ry3-033	0.26867 ± 0.00523	4.2498 ± 0.1477	1534.0 ± 29.9	1683.6 ± 58.5	0.26
Ry3-020	0.01294 ± 0.00044	0.0941 ± 0.0126	82.9 ± 2.8	91.3 ± 12.3	0.57	Ry3-034	0.01264 ± 0.00073	0.0816 ± 0.0112	81.0 ± 4.7	79.6 ± 11.0	0.56
Ry3-021	0.01070 ± 0.00046	0.0715 ± 0.0112	68.6 ± 3.0	70.1 ± 11.0	1.04	Ry3-035	0.01129 ± 0.00049	0.0765 ± 0.0083	72.4 ± 3.1	74.8 ± 8.1	0.34
Ry3-022	0.01778 ± 0.00084	0.1218 ± 0.0138	113.6 ± 5.4	116.7 ± 13.2	0.31	Ry3-036	0.32362 ± 0.00570	5.1530 ± 0.1371	1807.4 ± 31.8	1844.8 ± 49.1	0.10
Ry3-023	0.31903 ± 0.00911	5.2675 ± 0.2267	1785.0 ± 51.0	1863.5 ± 80.2	0.19	Ry3-037	0.01240 ± 0.00039	0.0790 ± 0.0086	79.5 ± 2.5	77.2 ± 8.4	0.51
Ry3-024	0.29031 ± 0.00838	4.7035 ± 0.2101	1643.1 ± 47.4	1767.8 ± 79.0	0.41	Ry3-038	0.02713 ± 0.00085	0.2053 ± 0.0207	172.6 ± 5.4	189.6 ± 19.1	0.18
Ry3-025	0.01553 ± 0.00078	0.1050 ± 0.0101	99.4 ± 5.0	101.4 ± 9.8	0.29	Ry3-039	0.00978 ± 0.00033	0.0673 ± 0.0084	62.7 ± 2.1	66.1 ± 8.3	1.51
Ry3-026	0.01314 ± 0.00036	0.0896 ± 0.0054	84.1 ± 2.3	87.1 ± 5.3	0.36	Ry3-040	0.01301 ± 0.00042	0.0828 ± 0.0098	83.3 ± 2.7	80.8 ± 9.6	0.84
Ry3-027	0.01554 ± 0.00033	0.0998 ± 0.0101	99.4 ± 2.1	96.6 ± 9.8	0.56	Ry3-041	0.01274 ± 0.00042	0.0882 ± 0.0101	81.6 ± 2.7	85.9 ± 9.8	0.48
Ry3-028	0.35079 ± 0.00385	5.8962 ± 0.2697	1938.4 ± 21.2	1960.6 ± 89.7	0.25	Ry3-042	0.01322 ± 0.00058	0.0892 ± 0.0165	84.7 ± 3.7	86.8 ± 16.1	0.66
Ry3-029	0.01510 ± 0.00036	0.0995 ± 0.0114	96.6 ± 2.3	96.3 ± 11.0	0.75	Ry3-043	0.01090 ± 0.00041	0.0784 ± 0.0110	69.9 ± 2.6	76.7 ± 10.8	1.25

http://dx.doi.org/10.5772/67982

Author details

Tetsuya Tokiwa[1*], Makoto Takeuchi[2], Yusuke Shimura[1], Kazuho Shobu[1], Akari Ota[2], Koshi Yamamoto[2] and Hiroshi Mori[1]

*Address all correspondence to: tokiwa@shinshu-u.ac.jp

1 Faculty of Science, Shinshu University, Asahi, Japan

2 Graduate School of Environmental Studies, Nagoya University, Furocho, Chikusa, Japan

References

[1] Taira A, Katto J, Tashiro M, Okamura M, Kodama K. The Shimanto Belt in Shikoku, Japan: Evolution of Cretaceous to Miocene Accretionary Prism. Modern Geology. 1988; **12**: 5-46.

[2] Matsuoka A, Yamakita S, Sakakibara M, Hisada K. Unit Division for the Chichibu Composite Belt from a View Point of Accretionary Tectonics and Geology of Western Shikoku, Japan. Journal of the Geological Society of Japan. 1998; **104**: 634-653.

[3] Hirata T, Nesbitt RW. U–Pb Isotope Geochronology of Zircon: Evaluation of the Laser Probe-Inductively Coupled Plasma Mass Spectrometry Technique. Geochimica et Cosmochimica Acta. 1995; **59**: 2491-2500. DOI: 10.1016/0016-7037(95)00144-1.

[4] Košler J, Sylvester PJ. Present Trends and the Future of Zircon in Geochronology: Laser Ablation ICPMS. Reviews in Mineralogy and Geochemistry. 2003; **53**: 243-275. DOI: 10.2113/0530243.

[5] Sato D, Matsuura H, Yamamoto T. Timing of the Late Cretaceous ignimbrite Flare-up at the Eastern Margin of the Eurasian Plate: New Zircon U–Pb Ages from the Aioi-Arima-Koto Region of SW Japan. Journal of Volcanology and Geothermal Research. 2016; **310**: 89-97. DOI: 10.1016/j.jvolgeores.2015.11.014.

[6] Hara H, Kon Y, Usuki T, Lan CY, Kamata Y, Hisada K, Ueno K, Charoentitirat T, Charusiri P. U–Pb Ages of detrital zircons within the Inthanon Zone of the Paleo-Tethyan subduction zone, northern Thailand: New constraints on accretionary age and arc activity. Journal of Asian Earth Sciences. 2013; **74**: 50-61. DOI: 10.1016/j.jseaes.2013.06.006.

[7] Wang Z, Xu D, Hu G, Yu L, Wu C, Zhang Z, Cai J, Shan Q, Hou M, Chen H. Detrital zircon U–Pb ages of the Proterozoic Metaclastic-Sedimentary Rocks in Hainan Province of South China: New Constraints on the Depositional Time, Source Area, and Tectonic Setting of the Shilu Fe-Co-Cu Ore District. Journal of Asian Earth Sciences. 2015; **113**: 1143-1161. DOI: 10.1016/j.jseaes.2015.04.014.

[8] Kishu Shimanto Research Group. New Perspective on the Study of the Cretaceous to Neogene Shimanto Accretionary Prism in the Kii Peninsula, Southwest Japan. The Association for the Geological Collaboration in Japan, Monograph. 2012; **59**: 295.

[9] Kishu Shimanto Research Group. Yukawa and Miyama Formations of the Hidakagawa Group in the Eastern-Middle Part of Wakayama Prefecture: The Study of the Shimanto Terrain in the Kii Peninsula, Southwest Japan (Part 12). Earth Science. 1991; **45**: 19-38.

[10] Kishu Shimanto Research Group. Geology of the Hidakagawa Belt in the Central-Western Part of Wakayama Prefecture, Southwest Japan: Study of the Shimanto Superbelt, Kii Peninsula (Part 13). Earth Science. 2006; **60**: 355-374.

[11] Kishu Shimanto Research Group. Proposal of the Yukawa Accretionary Complex: Albian to Cenomanian Accretionary Prism. The Association for the Geological Collaboration in Japan, Monograph. 2012; **59**: 25-34.

[12] Kishu Shimanto Research Group. Tectonostratigraphic Division of the Ryujin Accretionary Complex: Upper Campanian to Middle Maastrichtian Accretionary Prism. The Association for the Geological Collaboration in Japan, Monograph. 2012; **59**: 43-50.

[13] Suzuki H. On the Geologic Age of the Ryujin Formation in the Hidakagawa Belt, Shimanto Terrane of Southwest Japan. The Science and Engineering Review of Doshisha University. 1992; **32**: 350-361.

[14] Tokiwa T, Mori Y, Suzuki H. Cretaceous Radiolarian Fossils from the Ryujin Formation of the Shimanto Belt in the Kawabe area, Wakayama Prefecture, Southwest Japan. Journal of Earth and Planetary Sciences, Nagoya University. 2005; **52**: 11-23.

[15] Tokiwa T, Mori Y, Suzuki H, Niwa K. Late Campanian Radiolarians Obtained from the Ryujin Formation in the Hidakagawa Group, Northern Shimanto Belt, Kii Peninsula, Japan. Journal of the Geological Society of Japan. 2007; **113**: 270-273. DOI: 10.5575/geosoc.113.270.

[16] Sakamoto T, Bessho T, Yamamoto T. Sedimentary Process of Felsic Tuff in the Ryujin Accretionary Complex, Shimanto Belt, Kii Peninsula, Southwest Japan. The Association for the Geological Collaboration in Japan, Monograph. 2012; **59**: 175-183.

[17] Yamamoto T, Suzuki H. Hanazono Accretionary Complex in the northern margin of the Shimanto Belt in the Kii Peninsula, Southwest Japan. The Association for the Geological Collaboration in Japan, Monograph. 2012; **59**: 1-14.

[18] Kumon F, Bessho T, Barry PR. Modal and Chemical Characteristics of the Coarse Clastic Rocks from the Shimanto terrane in Kii Peninsula, Southwest Japan. The Association for the Geological Collaboration in Japan, Monograph. 2012; **59**: 193-216.

[19] Wiedenbeck M, Alle P, Corfu F, Griffin WL, Meier M, Oberli F, Von Quadt A, Roddick JC, Spiegel W. Three Natural Zircon Standards for U-Th-Pb, Lu-Hf, Trace Element and REE Analyses. Geostandards Newsletter. 1995; **19**: 1-23. DOI: 10.1111/j.1751-908X.1995.tb00147.x.

[20] Orihashi Y, Nakai S, Hirata TU–Pb Age Determination for Seven Standard Zircons using Inductively Coupled Plasma-Mass Spectrometry Coupled with Frequency Quintupled Nd-YAG (Lambda=213 nm) Laser Ablation System: Comparison with LA-ICP-MS Zircon Analyses with a NIST Glass Reference Material. Resource Geology. 2008; **58**: 101-123. DOI: 10.1111/j.1751-3928.2008.00052.x.

[21] Kouchi Y, Orihashi Y, Obara H, Fujimoto T, Haruta Y, Yamamoto K. Zircon U–Pb Dating by 213 nm Nd: YAG Laser Ablation Inductively Coupled Plasma Mass Spectrometry: Optimization of the Analytical Condition to use NIST SRM 610 for Pb/U Fractionation Correction. Geochemistry. 2015; **49**: 1-17. DOI: 10.14934/chikyukagaku.49.1.

[22] Cao L, Jiang T, Wang Z, Zhang Y, Sun H. Provenance of Upper Miocene Sediments in the Yinggehai and Qiongdongnan Basins, Northwestern South China Sea: Evidence from REE, Heavy Minerals and Zircon U–Pb Ages. Marine Geology. 2015; **361**: 136-146. DOI: 10.1016/j.margeo.2015.01.007.

[23] Wang W, Liu X, Zhao Y, Zheng G, Chen L. U–Pb Zircon Ages and Hf Isotopic Compositions of Metasedimentary Rocks from the Grove Subglacial Highlands, East Antarctica: Constraints on the Provenance of Protoliths and Timing of Sedimentation and Metamorphism. Precambrian Research. 2016; **275**: 135-150. DOI: 10.1016/j.precamres.2015.12.018.

[24] Zhang J, Zhang B, Zhao H. Timing of Amalgamation of the Alxa Block and the North China Block: Constraints Based on Detrital Zircon U–Pb Ages and Sedimentologic and Structural Evidence. Tectonophysics. 2016; **668-669**: 65-81. DOI: 10.1016/j.tecto.2015.12.006.

[25] Ludwig KR. Isoplot 3.75: A Geochronological Toolkit for Microsoft Excel. Berkeley Geochronology Center Special Publication. 2012; **5**: 75.

[26] Dickinson WR, Gehrels GE. Use of U–Pb ages of Detrital Zircons to Infer Maximum Depositional Ages of Strata: A Test Against a Colorado Plateau Mesozoic Database. Earth and Planetary Science Letters. 2009; **288**: 115-125. DOI: 10.1016/j.epsl.2009.09.013.

[27] Sambridge MS, Compston W. Mixture Modeling of Multi-Component Data Sets with Application to Ion-Probe Zircon Ages. Earth and Planetary Science Letters. 1994; **128**: 373-390. DOI: 10.1016/0012-821X(94)90157-0.

[28] Stelten ME, Cooper KM, Vazquez JA, Calvert AT, Glessner JJG. Mechanisms and Timescales of Generating Eruptible Rhyolitic Magmas at Yellowstone Caldera from Zircon and Sanidine Geochronology and Geochemistry. Journal of Petrology. 2015; **56**: 1607-1642. DOI: 10.1093/petrology/egv047.

[29] Henning J, Hal R, Armstrong RA. U–Pb Zircon Geochronology of Rocks from West Central Sulawesi, Indonesia: Extension-Related Metamorphism and Magmatism During the Early Stages of Mountain Building. Gondwana Research. 2016; **32**: 41-63. DOI: 10.1016/j.gr.2014.12.012.

[30] Pankhurst RJ, Hervé F, Fanning CM, Calderón M, Niemeyer H, Griem-Klee S, Soto F. The pre-Mesozoic Rocks of Northern Chile: U–Pb Ages, and Hf and O Isotopes. Earth-Science Reviews. 2016; **152**: 88-105. DOI: 10.1016/j.earscirev.2015.11.009.

[31] Suzuki K, Adachi M. Denudation History of the High T/P Ryoke Metamorphic Belt, Southwest Japan: Constraints from CHIME Monazite Ages of Gneisses and Granitoids. Journal of Metamorphic Geology. 1998; **16**: 23-37. DOI: 10.1111/j.1525-1314.1998.00057.x.

[32] Iida K, Iwamori H, Orihashi Y, Park T, Jwa YJ, Kwon ST, Danhara T, Iwano H. Tectonic Reconstruction of Batholith Formation Based on the Spatiotemporal Distribution of Cretaceous—Paleogene Granitic Rocks in Southwestern Japan. Island Arc. 2015; **24**: 205-220. DOI: 10.1111/iar.12103.

[33] Sato K, Santosh M, Tsunogae T, Kon Y, Yamamoto S, Hirata T. Laser Ablation ICP Mass Spectrometry for Zircon U–Pb Geochronology of Ultrahigh-Temperature Gneisses and A-Type Granites from the Achankovil Suture Zone, Southern India. Journal of Geodynamics. 2010; **50**: 286-299. DOI: 10.1016/j.jog.2010.04.001.

[34] Kumon F. Coarse Clastic Rocks of the Shimanto Supergroup in Eastern Shikoku and Kii Peninsula, Southwest Japan. Memoirs of the Faculty of Science, Kyoto University. Series of Geology and Mineralogy. 1983; **49**: 63-109.

[35] Kiminami K, Oyaizu A, Ishihama S, Miura K. Chemical Composition of Sandstones from the Cretaceous Shimanto Belt, Western Shikoku, Japan, and Correlation of Petrofacies Units in the Northern Shimanto Belt. Memoir of the Geological Society of Japan. 2000; **57**: 107-117.

[36] Corfu F, Hanchar JM, Hoskin PWO, Kinny P. Atlas of Zircon Textures. In: Hanchar JM, Hoskin PWO, editors. Zircon. Mineralogical Society of America Reviews in Mineralogy and Geochemistry. 2003; **53**: 469-500. DOI: 10.2113/0530469.